Robot Mechanisms and Mechanical Devices Illustrated

Paul E. Sandin

McGraw-Hill
New York | Chicago | San Francisco | Lisbon | London | Madrid
Mexico City | Milan | New Delhi | San Juan | Seoul | Singapore | Sydney | Toronto

The McGraw-Hill Companies

Cataloging-in-Publication Data is on file with the Library of Congress

1 2 3 4 5 6 7 8 9 0 DOC/DOC 0 9 8 7 6 5 4 3

ISBN 0-07-141200-X

The sponsoring editor for this book was *Judy Bass* and the production supervisor was *Sherri Souffrance*. It was set in Times Roman by *TopDesk Publishers' Group*.

Printed and bound by RR Donnelley.

This book is printed on recycled, acid-free paper containing a minimum of 50% recycled, de-inked fiber.

For Vicky, Conor, and Alex

Contents

Introduction

This book is meant to be interesting, helpful, and educational to hobbyists, students, educators, and midlevel engineers studying or designing mobile robots that do real work. It is primarily focused on mechanisms and devices that relate to vehicles that move around by themselves and actually do things autonomously, i.e. a robot. Making a vehicle that can autonomously drive around, both indoors and out, seems, at first, like a simple thing. Build a chassis, add drive wheels, steering wheels, a power source (usually batteries), some control code that includes some navigation and obstacle avoidance routines or some other way to control it, throw some bump sensors on it, and presto! a robot.

Unfortunately, soon after these first attempts, the designer will find the robot getting stuck on what seem to be innocuous objects or bumps, held captive under a chair or fallen tree trunk, incapable of doing anything useful, or with a manipulator that crushes every beer can it tries to pick up. Knowledge of the mechanics of sensors, manipulators, and the concept of mobility will help reduce these problems. This book provides that knowledge with the aid of hundreds of sketches showing drive layouts and manipulator geometries and their work envelope. It discusses what mobility really is and how to increase it without increasing the size of the robot, and how the shape of the robot can have a dramatic effect on its performance. Interspersed throughout the book are unusual mechanisms and devices, included to entice the reader to think outside the box. It is my sincere hope that this book will decrease the time it takes to produce a working robot, reduce the struggles and effort required to achieve that goal, and, therefore, increase the likelihood that your project will be a success.

Building, designing, and working with practical mobile robots requires knowledge in three major engineering fields: mechanical, electrical, and software. Many books have been written on robots, some focusing on the complete robot system, others giving a cookbook approach allowing a novice to take segments of chapters and put together a unique robot. While there are books describing the electric circuits used in robots, and books that teach the software and control code for robots, there are few that are focused entirely on the mechanisms and mechanical devices used in mobile robots.

This book intends to fill the gap in the literature of mobile robots by containing, in a single reference, complete graphically presented information on the mechanics of a mobile robot. It is written in laymen's language and filled with sketches so novices and those not trained in mechanical engineering can acquire some understanding of this interesting field. It also includes clever schemes and mechanisms that mid-level mechanical engineers should find new and useful. Since mobile robots are being called on to perform more and more complex and practical tasks, and many are now carrying one or even two manipulators, this book has a section on manipulators and grippers for mobile robots. It shows why a manipulator used on a robot is different in several ways from a manipulator used in industry.

Autonomous robots place special demands on their mobility system because of the unstructured and highly varied environment the robot might drive through, and the fact that even the best sensors are poor in comparison to a human's ability to see, feel, and balance. This means the mobility system of a robot that relies on those sensors will have much less information about the environment and will encounter obstacles that it must deal with on its own. In many cases, the microprocessor controlling the robot will only be telling the mobility system "go over there" without regard to what lays directly in that path. This forces the mobility system to be able to handle anything that comes along.

In contrast, a human driver has very acute sensors: eyes for seeing things and ranging distances, force sensors to sense acceleration, and balance to sense levelness. A human expects certain things of an automobile's (car, truck, jeep, HumVee, etc.) mobility system (wheels, suspension, and steering) and uses those many and powerful sensors to guide that mobility system's efforts to traverse difficult terrain. The robot's mobility system must be passively very capable, the car's mobility system must feel right to a human.

For these reasons, mobility systems on mobile robots can be both simpler and more complex than those found in automobiles. For example, the Ackerman steering system in automobiles is not actually suited for high mobility. It feels right to a human, and it is well suited to higher speed travel, but a robot doesn't care about feeling right, not yet, at least! The best mobility system for a robot to have is one that effectively accomplishes the required task, without regard to how well a human could use it.

There are a few terms specific to mobile robots that must be defined to avoid confusion. First, the term robot itself has unfortunately come to have at least three different meanings. In this book, the word robot means an autonomous or semi-autonomous mobile land vehicle that may or may not have a manipulator or other device for affecting its environment. Colin Angle, CEO of iRobot Corp. defines a robot as a mobile thing with sensors that looks at those sensors and decides on its own what actions to take.

In the manufacturing industry, however, the word robot means a reprogrammable stationary manipulator with few, if any sensors, commonly found in large industrial manufacturing plants. The third common meaning of robot is a teleoperated vehicle similar to but more sophisticated than a radio controlled toy car or truck. This form of robot usually has a microprocessor on it to aid in controlling the vehicle itself, perform some autonomous or automatic tasks, and aid in controlling the manipulator if one is onboard.

This book mainly uses the first meaning of robot and focuses on things useful to making robots, but it also includes several references to mechanisms useful to both of the other types of robots. Robot and mobile robot are used interchangeably throughout the book. Autonomous, in this book, means acting completely independent of any human input. Therefore, autonomous robot means a self-controlled, self-powered, mobile vehicle that makes its own decisions based on inputs from sensors. There are very few truly autonomous robots, and no known autonomous robots with manipulators on them whose manipulators are also autonomous. The more common form of mobile robot today is semiautonomous, where the robot has some sensors and acts partially on its own, but there is always a human in the control loop through a radio link or tether. Another name for this type of control structure is telerobotic, as opposed to a teleoperated robot, where there are no, or very few, sensors on the vehicle that it uses to make decisions. Specific vehicles in this book that do not use sensors to make decisions are labeled telerobotic or teleoperated to differentiate them from autonomous robots. It is important to note that the mechanisms and mechanical devices that are shown in this book can be applied, in their appropriate category, to almost any vehicle or manipulator whether autonomous or not.

Another word, which gets a lot of use in the robot world, is mobility. Mobility is defined in this book as a drive system's ability to deal with the effects of heat and ice, ground cover, slopes or staircases, and to negotiate obstacles. Chapter Nine focuses entirely on comparing drive systems' mobility based on a wide range of common obstacles found in outdoor and indoor environments, some of which can be any size (like rocks), others that cannot (like stair cases).

I intentionally left out the whole world of hydraulics. While hydraulic power can be the answer when very compact actuators or high power density motors are required, it is potentially more dangerous, and definitely messier, to work with than electrically powered devices. It is also much less efficient—a real problem for battery powered robots. There are many texts on hydraulic power and its uses. If hydraulics is being considered in a design, the reader is referred to Industrial Fluid Power (4 volumes) 3rd ed., published by Womack Education Publications.

The designer has powerful tools to aid in the design process beyond the many tricks, mechanical devices, and techniques shown in this book. These tools include 3D design tools like SolidWorks and Pro-Engineer and also new ways to produce prototypes of the mechanisms themselves. This is commonly called Rapid Prototyping (RP).

NEW PROCESSES EXPAND CHOICES FOR RAPID PROTOTYPING

New concepts in rapid prototyping (RP) have made it possible to build many different kinds of 3D prototype models faster and cheaper than by traditional methods. The 3D models are fashioned automatically from such materials as plastic or paper, and they can be full size or scaled-down versions of larger objects. Rapid-prototyping techniques make use of computer programs derived from computer-aided design (CAD) drawings of the object. The completed models, like those made by machines and manual wood carving, make it easier for people to visualize a new or redesigned product. They can be passed around a conference table and will be especially valuable during discussions among product design team members, manufacturing managers, prospective suppliers, and customers.

At least nine different RP techniques are now available commercially, and others are still in the development stage. Rapid prototyping models can be made by the owners of proprietary equipment, or the work can be contracted out to various RP centers, some of which are owned by the RP equipment manufacturers. The selection of the most appropriate RP method for any given modeling application usually depends on the urgency of the design project, the relative costs of each RP process, and the anticipated time and cost savings RP will offer over conventional model-making practice. New and improved RP methods are being introduced regularly, so the RP field is in a state of change, expanding the range of designer choices.

Three-dimensional models can be made accurately enough by RP methods to evaluate the design process and eliminate interference fits or dimensioning errors before production tooling is ordered. If design flaws or omissions are discovered, changes can be made in the source CAD program and a replacement model can be produced quickly to verify that the corrections or improvements have been made. Finished models are useful in evaluations of the form, fit, and function of the product design and for organizing the necessary tooling, manufacturing, or even casting processes.

Most of the RP technologies are additive; that is, the model is made automatically by building up contoured laminations sequentially from materials such as photopolymers, extruded or beaded plastic, and even paper until they reach the desired height. These processes can be used to form internal cavities, overhangs, and complex convoluted geometries as well as simple planar or curved shapes. By contrast, a subtractive RP process involves milling the model from a block of soft material, typically plastic or aluminum, on a computer-controlled milling machine with commands from a CAD-derived program.

In the additive RP processes, photopolymer systems are based on successively depositing thin layers of a liquid resin, which are then solidified by exposure to a specific wavelengths of light. Thermoplastic systems are based on procedures for successively melting and fusing solid filaments or beads of wax or plastic in layers, which harden in the air to form the finished object. Some systems form layers by applying adhesives or binders to materials such as paper, plastic powder, or coated ceramic beads to bond them.

The first commercial RP process introduced was *stereolithography* in 1987, followed by a succession of others. Most of the commercial RP processes are now available in Europe and Japan as well as the United States. They have become multinational businesses through branch offices, affiliates, and franchises.

Each of the RP processes focuses on specific market segments, taking into account their requirements for model size, durability, fabrication speed, and finish in the light of anticipated economic benefits and cost. Some processes are not effective in making large models, and each process results in a model with a different finish. This introduces an economic tradeoff of higher price for smoother surfaces versus additional cost and labor of manual or machine finishing by sanding or polishing.

Rapid prototyping is now also seen as an integral part of the even larger but not well defined rapid tooling (RT) market. Concept modeling addresses the early stages of the design process, whereas RT concentrates on production tooling or mold making.

Some concept modeling equipment, also called 3D or office printers, are self-contained desktop or benchtop manufacturing units small enough and inexpensive enough to permit prototype fabrication to be done in an office environment. These units include provision for the containment or venting of any smoke or noxious chemical vapors that will be released during the model's fabrication.

Computer-Aided Design Preparation

The RP process begins when the object is drawn on the screen of a CAD workstation or personal computer to provide the digital data base. Then, in a post-design data processing step, computer software slices the object mathematically into a finite number of horizontal layers in generating an STL (Solid Transfer Language) file. The thickness of the "slices" can range from 0.0025 to 0.5 in. (0.06 to 13 mm) depending on the RP process selected. The STL file is then converted to a file that is compatible with the specific 3D "printer" or processor that will construct the model.

The digitized data then guides a laser, X-Y table, optics, or other apparatus that actually builds the model in a process comparable to building a high-rise building one story at a time. Slice thickness might have to be modified in some RP processes during model building to compensate for material shrinkage.

Prototyping Choices

All of the commercial RP methods depend on computers, but four of them depend on laser beams to cut or fuse each lamination, or provide enough heat to sinter or melt certain kinds of materials. The four processes that make use of lasers are Directed-Light Fabrication (DLF), Laminated-Object Manufacturing (LOM), Selective Laser Sintering (SLS), and Stereolithography (SL); the five processes that do not require lasers are Ballistic Particle Manufacturing (BPM), Direct-Shell Production Casting (DSPC), Fused-Deposition Modeling (FDM), Solid-Ground Curing (SGC), and 3D Printing (3DP).

Stereolithography (SL)

The stereolithographic (SL) process is performed on the equipment shown in Figure 1. The movable platform on which the 3D model is formed is initially immersed in a vat of liquid photopolymer resin to a level just below its surface so that a thin layer of the resin covers it. The SL equipment is located in a sealed chamber to prevent the escape of fumes from the resin vat.

The resin changes from a liquid to a solid when exposed to the ultraviolet (UV) light from a low-power, highly focused laser. The UV laser beam is focused on an X-Y mirror in a computer-controlled beam-shaping and scanning system so that it draws the outline of the lowest cross-section layer of the object being built on the film of photopolymer resin.

After the first layer is completely traced, the laser is then directed to scan the traced areas of resin to solidify the model's first cross section. The laser beam can harden the layer down to a depth of 0.0025 to 0.0300 in. (0.06 to 0.8 mm). The laser beam scans at speeds up to 350 in./s (890 cm/s). The photopolymer not scanned by the laser beam remains a liquid. In general, the thinner the resin film (slice thickness), the higher the resolution or more refined the finish of the completed model. When model surface finish is important, layer thicknesses are set for 0.0050 in. (0.13 mm) or less.

The table is then submerged under computer control to the specified depth so that the next layer of liquid polymer flows over the first hardened layer. The tracing, hardening, and recoating steps are repeated, layer-by-layer, until the complete 3D model is built on the platform within the resin vat.

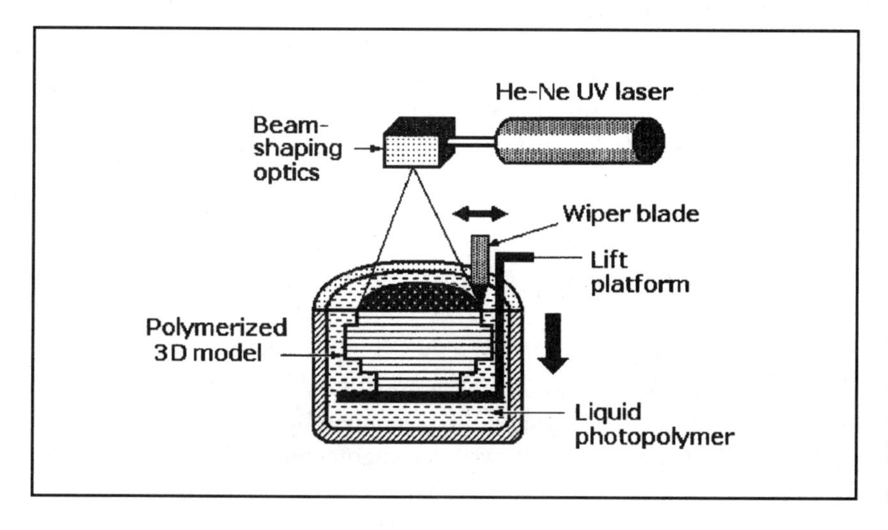

Figure 1 Stereolithography (SL): A computer-controlled neon–helium ultraviolet light (UV)–emitting laser outlines each layer of a 3D model in a thin liquid film of UV-curable photopolymer on a platform submerged a vat of the resin. The laser then scans the outlined area to solidify the layer, or "slice." The platform is then lowered into the liquid to a depth equal to layer thickness, and the process is repeated for each layer until the 3D model is complete. Photopolymer not exposed to UV remains liquid. The model is them removed for finishing.

Because the photopolymer used in the SL process tends to curl or sag as it cures, models with overhangs or unsupported horizontal sections must be reinforced with supporting structures: walls, gussets, or columns. Without support, parts of the model can sag or break off before the polymer has fully set. Provision for forming these supports is included in the digitized fabrication data. Each scan of the laser forms support layers where necessary while forming the layers of the model.

When model fabrication is complete, it is raised from the polymer vat and resin is allowed to drain off; any excess can be removed manually from the model's surfaces. The SL process leaves the model only partially polymerized, with only about half of its fully cured strength. The model is then finally cured by exposing it to intense UV light in the enclosed chamber of post-curing apparatus (PCA). The UV completes the hardening or curing of the liquid polymer by linking its molecules in chainlike formations. As a final step, any supports that were required are removed, and the model's surfaces are sanded or polished. Polymers such as urethane acrylate resins can be milled, drilled, bored, and tapped, and their outer surfaces can be polished, painted, or coated with sprayed-on metal.

The liquid SL photopolymers are similar to the photosensitive UV-curable polymers used to form masks on semiconductor wafers for etching and plating features on integrated circuits. Resins can be formulated to solidify under either UV or visible light.

The SL process was the first to gain commercial acceptance, and it still accounts for the largest base of installed RP systems. 3D Systems of Valencia, California, is a company that manufactures stereolithography equipment for its proprietary SLA process. It offers the *ThermoJet Solid Object Printer*. The SLA process can build a model within a volume measuring 10 × 7.5 × 8 in. (25 × 19 × 20 cm). It also offers the SLA 7000 system, which can form objects within a volume of 20 × 20 × 23.62 in. (51 × 51 × 60 cm). Aaroflex, Inc. of Fairfax, Virginia, manufactures the Aacura 22 solid-state SL system and operates AIM, an RP manufacturing service.

Solid Ground Curing (SGC)

Solid ground curing (SGC) (or the "solider process") is a multistep in-line process that is diagrammed in Figure 2. It begins when a photomask for the first layer of the 3D model is generated by the equipment shown at the far left. An electron gun writes a charge pattern of the photomask on a clear glass plate, and opaque toner is transferred electrostatically to the plate to form the photolithographic pattern in a xerographic process.

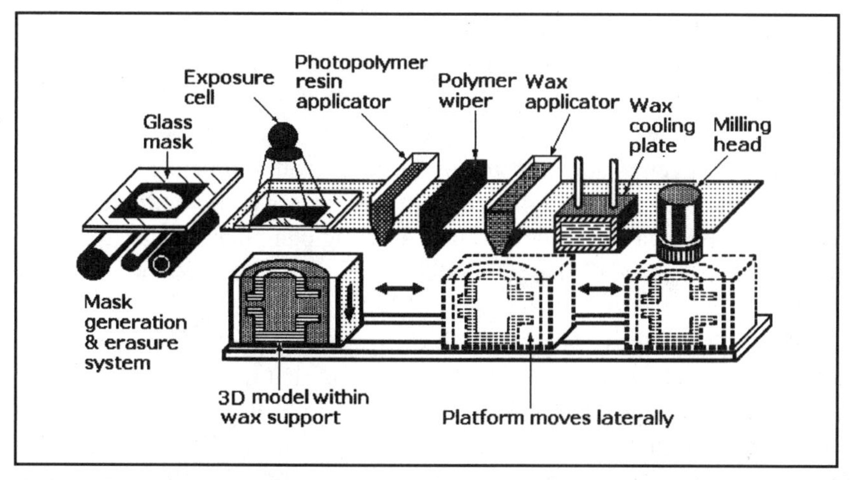

Figure 2 Solid Ground Curing (SGC): First, a photomask is generated on a glass plate by a xerographic process. Liquid photopolymer is applied to the work platform to form a layer, and the platform is moved under the photomask and a strong UV source that defines and hardens the layer. The platform then moves to a station for excess polymer removal before wax is applied over the hardened layer to fill in margins and spaces. After the wax is cooled, excess polymer and wax are milled off to form the first "slice." The first photomask is erased, and a second mask is formed on the same glass plate. Masking and layer formation are repeated with the platform being lowered and moved back and forth under the stations until the 3D model is complete. The wax is then removed by heating or immersion in a hot water bath to release the prototype.

The photomask is then moved to the exposure station, where it is aligned over a work platform and under a collimated UV lamp.

Model building begins when the work platform is moved to the right to a resin application station where a thin layer of photopolymer resin is applied to the top surface of the work platform and wiped to the desired thickness. The platform is then moved left to the exposure station, where the UV lamp is then turned on and a shutter is opened for a few seconds to expose the resin layer to the mask pattern. Because the UV light is so intense, the layer is fully cured and no secondary curing is needed.

The platform is then moved back to the right to the wiper station, where all of resin that was not exposed to UV is removed and discarded. The platform then moves right again to the wax application station, where melted wax is applied and spread into the cavities left by the removal of the uncured resin. The wax is hardened at the next station by pressing it against a cooling plate. After that, the platform is moved right again to the milling station, where the resin and wax layer are milled to a precise thickness. The platform piece is then returned to the resin application station, where it is lowered a depth equal to the thickness of the next layer and more resin is applied.

Meanwhile, the opaque toner has been removed from the glass mask and a new mask for the next layer is generated on the same plate. The complete cycle is repeated, and this will continue until the 3D model encased in the wax matrix is completed. This matrix supports any overhangs or undercuts, so extra support structures are not needed.

After the prototype is removed from the process equipment, the wax is either melted away or dissolved in a washing chamber similar to a dishwasher. The surface of the 3D model is then sanded or polished by other methods.

The SGC process is similar to *drop on demand inkjet plotting*, a method that relies on a dual inkjet subsystem that travels on a precision X-Y drive carriage and deposits both thermoplastic and wax materials onto the build platform under CAD program control. The drive carriage also energizes a flatbed milling subsystem for obtaining the precise vertical height of each layer and the overall object by milling off the excess material.

Cubital America Inc., Troy, Michigan, offers the *Solider 4600/5600* equipment for building prototypes with the SGC process.

Selective Laser Sintering (SLS)

Selective laser sintering (SLS) is another RP process similar to stereolithography (SL). It creates 3D models from plastic, metal, or ceramic powders with heat generated by a carbon dioxide infrared (IR)–emitting laser, as shown in Figure 3. The prototype is fabricated in a cylinder with a piston, which acts as a moving platform, and it is positioned next to a cylinder filled with preheated powder. A piston within the powder delivery system rises to eject powder, which is spread by a roller over the top of the build cylinder. Just before it is applied, the powder is heated further until its temperature is just below its melting point

When the laser beam scans the thin layer of powder under the control of the optical scanner system, it raises the temperature of the powder even further until it melts or sinters and flows together to form a solid layer in a pattern obtained from the CAD data.

As in other RP processes, the piston or supporting platform is lowered upon completion of each layer and the roller spreads the next layer of powder over the previously deposited layer. The process is repeated, with each layer being fused to the underlying layer, until the 3D prototype is completed.

The unsintered powder is brushed away and the part removed. No final curing is required, but because the objects are sintered they are porous. Wax, for example, can be applied to the inner and outer porous

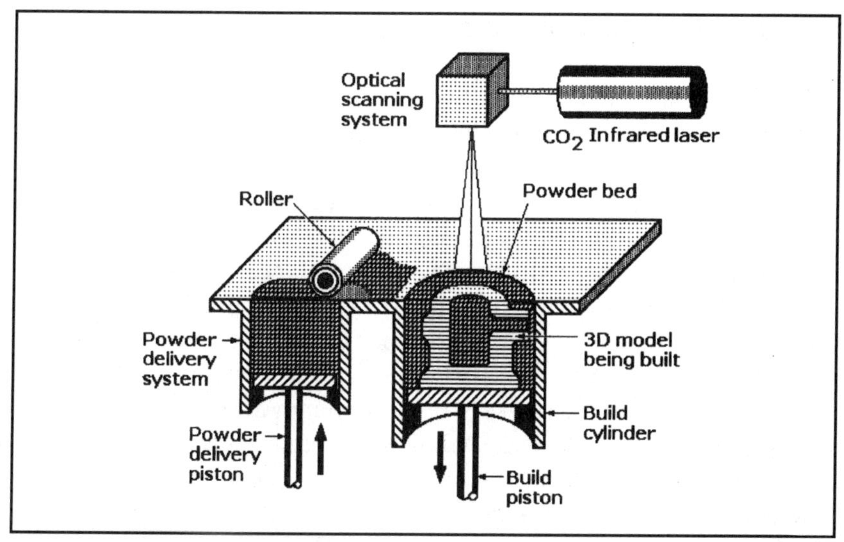

Figure 3 Selective Laser Sintering (SLS): Loose plastic powder from a reservoir is distributed by roller over the surface of piston in a build cylinder positioned at a depth below the table equal to the thickness of a single layer. The powder layer is then scanned by a computer-controlled carbon dioxide infrared laser that defines the layer and melts the powder to solidify it. The cylinder is again lowered, more powder is added, and the process is repeated so that each new layer bonds to the previous one until the 3D model is completed. It is then removed and finished. All unbonded plastic powder can be reused.

surfaces, and it can be smoothed by various manual or machine grinding or melting processes. No supports are required in SLS because overhangs and undercuts are supported by the compressed unfused powder within the build cylinder.

Many different powdered materials have been used in the SLS process, including polycarbonate, nylon, and investment casting wax. Polymer-coated metal powder is also being studied as an alternative. One advantage of the SLS process is that materials such as polycarbonate and nylon are strong and stable enough to permit the model to be used in limited functional and environmental testing. The prototypes can also serve as molds or patterns for casting parts.

SLS process equipment is enclosed in a nitrogen-filled chamber that is sealed and maintained at a temperature just below the melting point of the powder. The nitrogen prevents an explosion that could be caused by the rapid oxidation of the powder.

The SLS process was developed at the University of Texas at Austin, and it has been licensed by the DTM Corporation of Austin, Texas. The company makes a *Sinterstation 2500plus*. Another company participating in SLS is EOS GmbH of Germany.

Laminated-Object Manufacturing (LOM)

The Laminated-Object Manufacturing (LOM) process, diagrammed in Figure 4, forms 3D models by cutting, stacking, and bonding successive layers of paper coated with heat-activated adhesive. The carbon-dioxide laser beam, directed by an optical system under CAD data control, cuts cross-sectional outlines of the prototype in the layers of paper, which are bonded to previous layers to become the prototype.

The paper that forms the bottom layer is unwound from a supply roll and pulled across the movable platform. The laser beam cuts the outline of each lamination and cross-hatches the waste material within and around the lamination to make it easier to remove after the prototype is completed. The outer waste material web from each lamination is continuously removed by a take-up roll. Finally, a heated roller applies pressure to bond the adhesive coating on each layer cut from the paper to the previous layer.

A new layer of paper is then pulled from a roll into position over the previous layer, and the cutting, cross hatching, web removal, and bonding procedure is repeated until the model is completed. When all the layers have been cut and bonded, the excess cross-hatched material in the

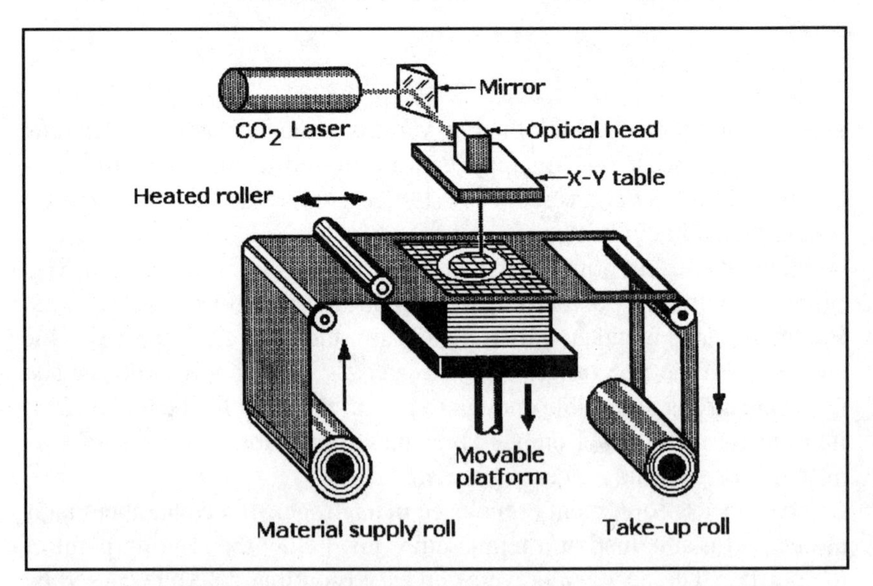

Figure 4 Laminated Object Manufacturing (LOM): Adhesive-backed paper is fed across an elevator platform and a computer-controlled carbon dioxide infrared-emitting laser cuts the outline of a layer of the 3D model and cross-hatches the unused paper. As more paper is fed across the first layer, the laser cuts the outline and a heated roller bonds the adhesive of the second layer to the first layer. When all the layers have been cut and bonded, the cross-hatched material is removed to expose the finished model. The complete model can then be sealed and finished.

form of stacked segments is removed to reveal the finished 3D model. The models made by the LOM have woodlike finishes that can be sanded or polished before being sealed and painted.

Using inexpensive, solid-sheet materials makes the 3D LOM models more resistant to deformity and less expensive to produce than models made by other processes, its developers say. These models can be used directly as patterns for investment and sand casting, and as forms for silicone molds. The objects made by LOM can be larger than those made by most other RP processes—up to 30 × 20 × 20 in. (75 × 50 × 50 cm).

The LOM process is limited by the ability of the laser to cut through the generally thicker lamination materials and the additional work that must be done to seal and finish the model's inner and outer surfaces. Moreover, the laser cutting process burns the paper, forming smoke that must be removed from the equipment and room where the LOM process is performed.

Helysys Corporation, Torrance, California, manufactures the LOM-2030H LOM equipment. Alternatives to paper including sheet plastic and ceramic and metal-powder-coated tapes have been developed.

Other companies offering equipment for building prototypes from paper laminations are the Schroff Development Corporation, Mission, Kansas, and CAM-LEM, Inc. Schroff manufactures the *JP System 5* to permit desktop rapid prototyping.

Fused Deposition Modeling (FDM)

The Fused Deposition Modeling (FDM) process, diagrammed in Figure 5, forms prototypes from melted thermoplastic filament. This filament, with a diameter of 0.070 in. (1.78 mm), is fed into a temperature-controlled FDM extrusion head where it is heated to a semi-liquid state. It is then extruded and deposited in ultrathin, precise layers on a fixtureless platform under X-Y computer control. Successive laminations ranging in thickness from 0.002 to 0.030 in. (0.05 to 0.76 mm) with wall thicknesses of 0.010 to 0.125 in. (0.25 to 3.1 mm) adhere to each by thermal fusion to form the 3D model.

Structures needed to support overhanging or fragile structures in FDM modeling must be designed into the CAD data file and fabricated as part of the model. These supports can easily be removed in a later secondary operation.

All components of FDM systems are contained within temperature-controlled enclosures. Four different kinds of inert, nontoxic filament materials are being used in FDM: ABS polymer (acrylonitrile butadiene styrene), high-impact-strength ABS (ABSi), investment casting wax, and

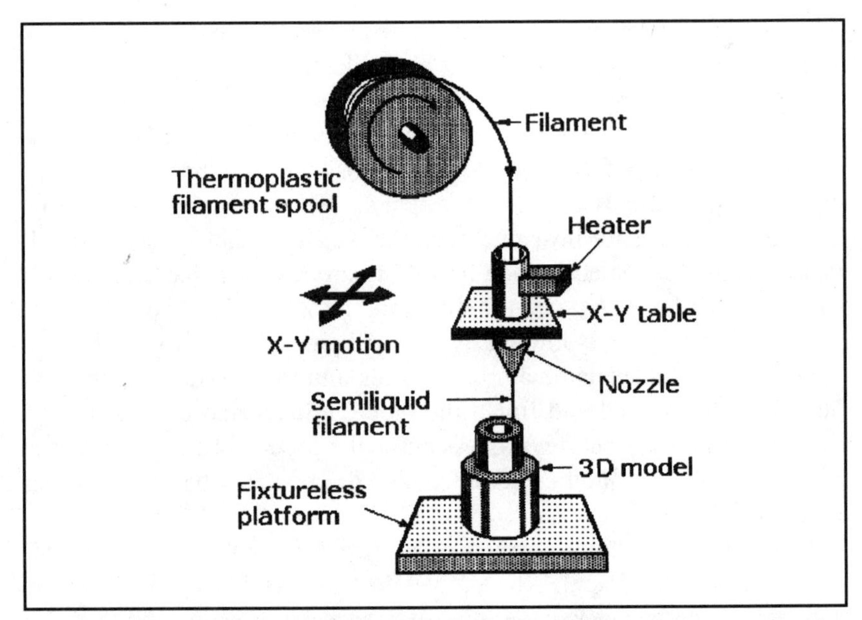

Figure 5 Fused Deposition Modeling (FDM): Filaments of thermoplastic are unwound from a spool, passed through a heated extrusion nozzle mounted on a computer-controlled X-Y table, and deposited on the fixtureless platform. The 3D model is formed as the nozzle extruding the heated filament is moved over the platform. The hot filament bonds to the layer below it and hardens. This laserless process can be used to form thin-walled, contoured objects for use as concept models or molds for investment casting. The completed object is removed and smoothed to improve its finish.

elastomer. These materials melt at temperatures between 180 and 220°F (82 and 104°C).

FDM is a proprietary process developed by Stratasys, Eden Prairie, Minnesota. The company offers four different systems. Its *Genisys* benchtop 3D printer has a build volume as large as 8 × 8 × 8 in. (20 × 20 × 20 cm), and it prints models from square polyester wafers that are stacked in cassettes. The material is heated and extruded through a 0.01-in. (0.25-mm)–diameter hole at a controlled rate. The models are built on a metallic substrate that rests on a table. Stratasys also offers four systems that use spooled material. The *FDM2000*, another benchtop system, builds parts up to 10 in^3 (164 cm^3) while the *FDM3000*, a floor-standing system, builds parts up to 10 × 10 × 16 in. (26 × 26 × 41 cm).

Two other floor-standing systems are the *FDM 8000*, which builds models up to 18 × 18 × 24 in. (46 × 46 × 61 cm), and the *FDM Quantum* system, which builds models up to 24 × 20 × 24 in. (61 × 51 × 61 cm). All of these systems can be used in an office environment.

Stratasys offers two options for forming and removing supports: a breakaway support system and a water-soluble support system. The water-soluble supports are formed by a separate extrusion head, and they can be washed away after the model is complete.

Three-Dimensional Printing (3DP)

The Three-Dimensional Printing (3DP) or inkjet printing process, diagrammed in Figure 6, is similar to Selective Laser Sintering (SLS) except that a multichannel inkjet head and liquid adhesive supply replaces the laser. The powder supply cylinder is filled with starch and cellulose powder, which is delivered to the work platform by elevating a delivery piston. A roller rolls a single layer of powder from the powder cylinder to the upper surface of a piston within a build cylinder. A multichannel inkjet head sprays a water-based liquid adhesive onto the surface of the powder to bond it in the shape of a horizontal layer of the model.

In successive steps, the build piston is lowered a distance equal to the thickness of one layer while the powder delivery piston pushes up fresh powder, which the roller spreads over the previous layer on the build pis-

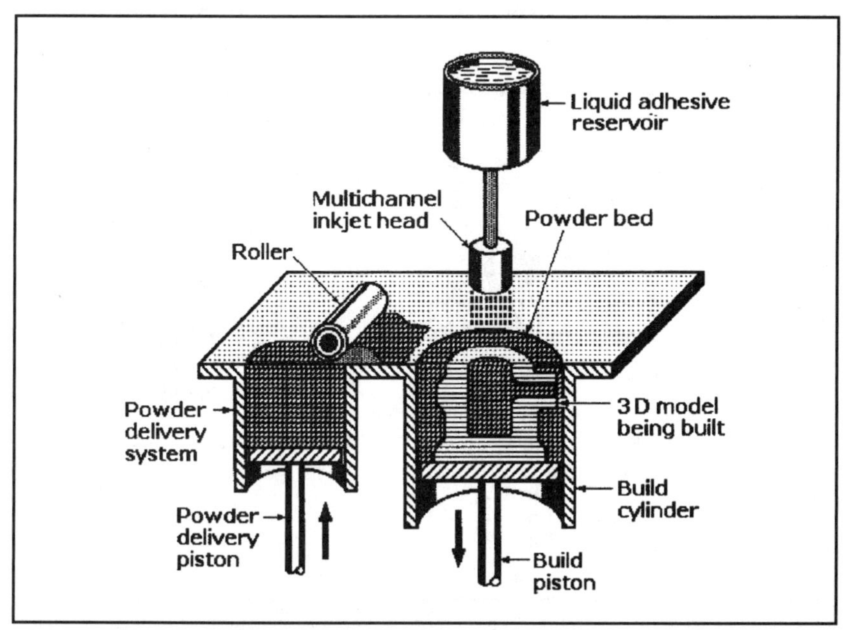

Figure 6 Three-Dimensional Printing (3DP): Plastic powder from a reservoir is spread across a work surface by roller onto a piston of the build cylinder recessed below a table to a depth equal to one layer thickness in the 3DP process. Liquid adhesive is then sprayed on the powder to form the contours of the layer. The piston is lowered again, another layer of powder is applied, and more adhesive is sprayed, bonding that layer to the previous one. This procedure is repeated until the 3D model is complete. It is then removed and finished.

ton. This process is repeated until the 3D model is complete. Any loose excess powder is brushed away, and wax is coated on the inner and outer surfaces of the model to improve its strength.

The 3DP process was developed at the Three-Dimensional Printing Laboratory at the Massachusetts Institute of Technology, and it has been licensed to several companies. One of those firms, the Z Corporation of Somerville, Massachusetts, uses the original MIT process to form 3D models. It also offers the Z402 3D modeler. Soligen Technologies has modified the 3DP process to make ceramic molds for investment casting. Other companies are using the process to manufacture implantable drugs, make metal tools, and manufacture ceramic filters.

Direct-Shell Production Casting (DSPC)

The Direct Shell Production Casting (DSPC) process, diagrammed in Figure 7, is similar to the 3DP process except that it is focused on forming molds or shells rather than 3D models. Consequently, the actual 3D model or prototype must be produced by a later casting process. As in the 3DP process, DSPC begins with a CAD file of the desired prototype.

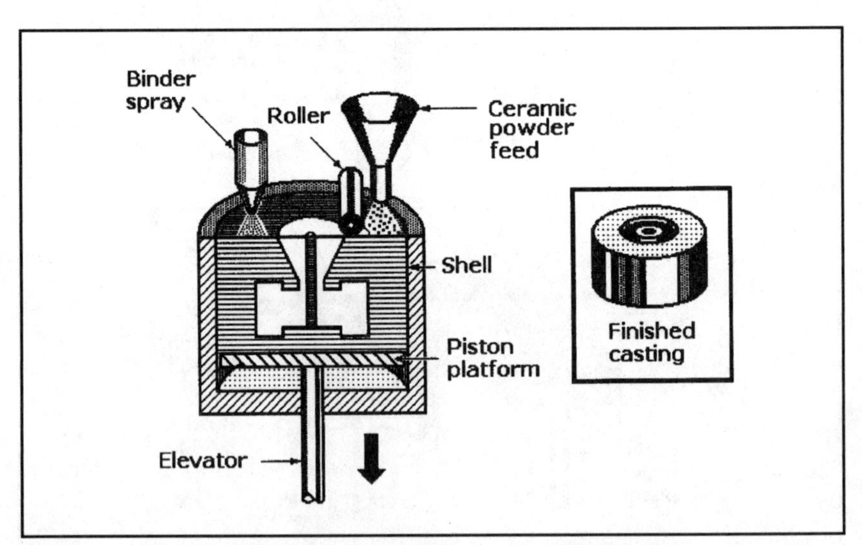

Figure 7 Direct Shell Production Casting (DSPC): Ceramic molds rather than 3D models are made by DSPC in a layering process similar to other RP methods. Ceramic powder is spread by roller over the surface of a movable piston that is recessed to the depth of a single layer. Then a binder is sprayed on the ceramic powder under computer control. The next layer is bonded to the first by the binder. When all of the layers are complete, the bonded ceramic shell is removed and fired to form a durable mold suitable for use in metal casting. The mold can be used to cast a prototype. The DSPC process is considered to be an RP method because it can make molds faster and cheaper than conventional methods.

Two specialized kinds of equipment are needed for DSPC: a dedicated computer called a shell-design unit (SDU) and a shell- or mold-processing unit (SPU). The CAD file is loaded into the SDU to generate the data needed to define the mold. SDU software also modifies the original design dimensions in the CAD file to compensate for ceramic shrinkage. This software can also add fillets and delete such features as holes or keyways that must be machined after the prototype is cast.

The movable platform in DSPC is the piston within the build cylinder. It is lowered to a depth below the rim of the build cylinder equal to the thickness of each layer. Then a thin layer of fine aluminum oxide (alumina) powder is spread by roller over the platform, and a fine jet of colloidal silica is sprayed precisely onto the powder surface to bond it in the shape of a single mold layer. The piston is then lowered for the next layer and the complete process is repeated until all layers have been formed, completing the entire 3D shell. The excess powder is then removed, and the mold is fired to convert the bonded powder to monolithic ceramic.

After the mold has cooled, it is strong enough to withstand molten metal and can function like a conventional investment-casting mold. After the molten metal has cooled, the ceramic shell and any cores or gating are broken away from the prototype. The casting can then be finished by any of the methods usually used on metal castings.

DSPC is a proprietary process of Soligen Technologies, Northridge, California. The company also offers a custom mold manufacturing service.

Ballistic Particle Manufacturing (BPM)

There are several different names for the Ballistic Particle Manufacturing (BPM) process, diagrammed in Figure 8. Variations of it are also called *inkjet methods*. The molten plastic used to form the model and the hot wax for supporting overhangs or indentations are kept in heated tanks above the build station and delivered to computer-controlled jet heads through thermally insulated tubing. The jet heads squirt tiny droplets of the materials on the work platform as it is moved by an X-Y table in the pattern needed to form each layer of the 3D object. The droplets are deposited only where directed, and they harden rapidly as they leave the jet heads. A milling cutter is passed over the layer to mill it to a uniform thickness. Particles that are removed by the cutter are vacuumed away and deposited in a collector.

Nozzle operation is monitored carefully by a separate fault-detection system. After each layer has been deposited, a stripe of each material is deposited on a narrow strip of paper for thickness measurement by opti-

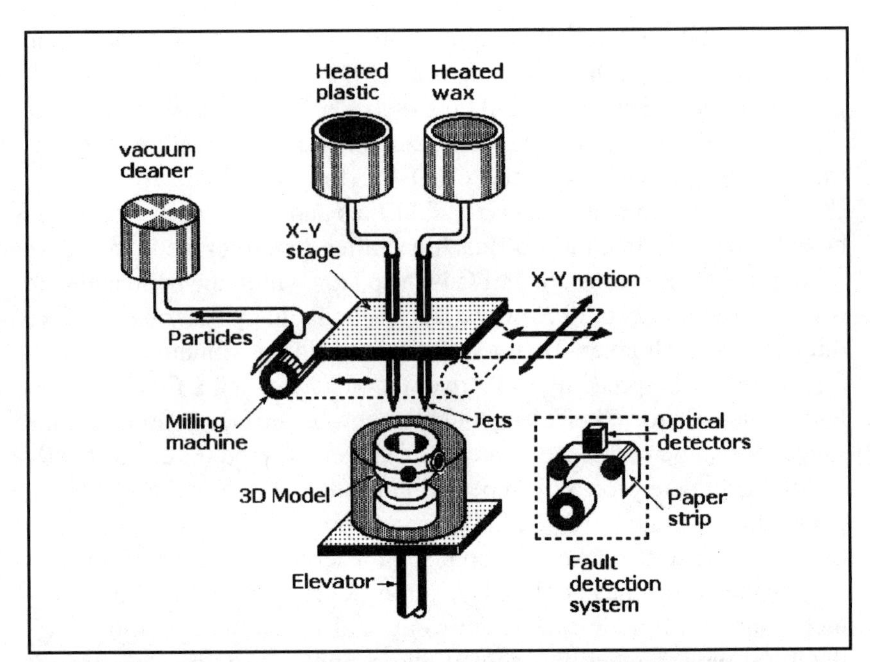

Figure 8 Ballistic Particle Manufacturing (BPM): Heated plastic and wax are deposited on a movable work platform by a computer-controlled X-Y table to form each layer. After each layer is deposited, it is milled to a precise thickness. The platform is lowered and the next layer is applied. This procedure is repeated until the 3D model is completed. A fault detection system determines the quality and thickness of the wax and plastic layers and directs rework if a fault is found. The supporting wax is removed from the 3D model by heating or immersion in a hot liquid bath.

cal detectors. If the layer meets specifications, the work platform is lowered a distance equal to the required layer thickness and the next layer is deposited. However, if a clot is detected in either nozzle, a jet cleaning cycle is initiated to clear it. Then the faulty layer is milled off and that layer is redeposited. After the 3D model is completed, the wax material is either melted from the object by radiant heat or dissolved away in a hot water wash.

The BPM system is capable of producing objects with fine finishes, but the process is slow. With this RP method, a slower process that yields a 3D model with a superior finish is traded off against faster processes that require later manual finishing.

The version of the BPM system shown in Figure 8 is called *Drop on Demand Inkjet Plotting* by Sanders Prototype Inc, Merrimac, New Hampshire. It offers the *ModelMaker II* processing equipment, which produces 3D models with this method. AeroMet Corporation builds titanium parts directly from CAD renderings by fusing titanium powder with an 18-kW carbon dioxide laser, and 3D Systems of Valencia,

California, produces a line of inkjet printers that feature multiple jets to speed up the modeling process.

Directed Light Fabrication (DLF)

The Directed Light Fabrication (DLF) process, diagrammed in Figure 9, uses a neodymium YAG (Nd:YAG) laser to fuse powdered metals to build 3D models that are more durable than models made from paper or plastics. The metal powders can be finely milled 300 and 400 series stainless steel, tungsten, nickel aluminides, molybdenum disilicide, copper, and aluminum. The technique is also called *Direct-Metal Fusing, Laser Sintering,* and *Laser Engineered Net Shaping (LENS).*

The laser beam under X-Y computer control fuses the metal powder fed from a nozzle to form dense 3D objects whose dimensions are said to be within a few thousandths of an inch of the desired design tolerance.

DLF is an outgrowth of nuclear weapons research at the Los Alamos National Laboratory (LANL), Los Alamos, New Mexico, and it is still in the development stage. The laboratory has been experimenting with the

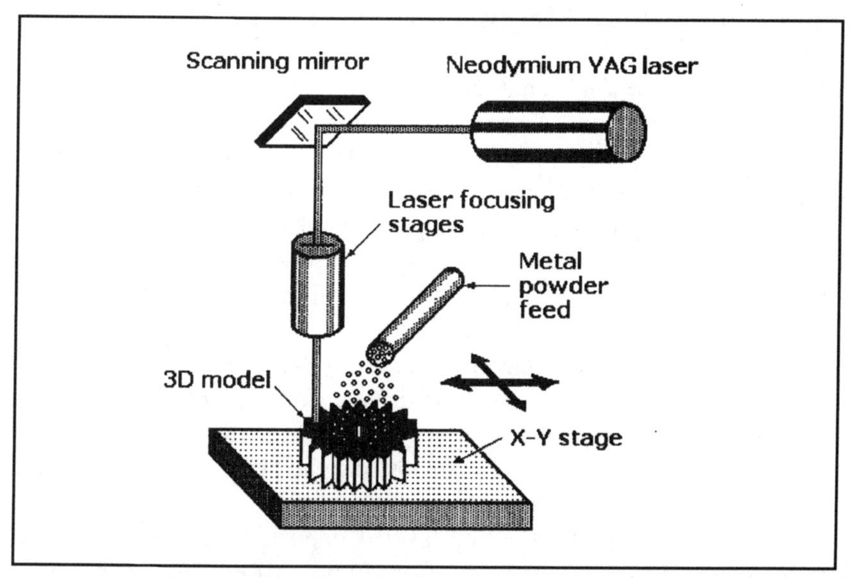

Figure 9 Directed Light Fabrication (DLF): Fine metal powder is distributed on an X-Y work platform that is rotated under computer control beneath the beam of a neodymium YAG laser. The heat from the laser beam melts the metal powder to form thin layers of a 3D model or prototype. By repeating this process, the layers are built up and bonded to the previous layers to form more durable 3D objects than can be made from plastic. Powdered aluminum, copper, stainless steel, and other metals have been fused to make prototypes as well as practical tools or parts that are furnace-fired to increase their bond strength.

laser fusing of ceramic powders to fabricate parts as an alternative to the use of metal powders. A system that would regulate and mix metal powder to modify the properties of the prototype is also being investigated.

Optomec Design Company, Albuquerque, New Mexico, has announced that direct fusing of metal powder by laser in its LENS process is being performed commercially. Protypes made by this method have proven to be durable and they have shown close dimensional tolerances.

Research and Development in RP

Many different RP techniques are still in the experimental stage and have not yet achieved commercial status. At the same time, practical commercial processes have been improved. Information about this research has been announced by the laboratories doing the work, and some of the research is described in patents. This discussion is limited to two techniques, SDM and Mold SDM, that have shown commercial promise.

Shape Deposition Manufacturing (SDM)

The Shape Deposition Manufacturing (SDM) process, developed at the SDM Laboratory of Carnegie Mellon University, Pittsburgh, Pennsylvania, produces functional metal prototypes directly from CAD data. This process, diagrammed in Figure 10, forms successive layers of metal on a platform without masking, and is also called *solid free- form* (SFF) fabrication. It uses hard metals to form more rugged prototypes that are then accurately machined under computer control during the process.

The first steps in manufacturing a part by SDM are to reorganize or destructure the CAD data into slices or layers of optimum thickness that will maintain the correct 3D contours of the outer surfaces of the part and then decide on the sequence for depositing the primary and supporting materials to build the object.

The primary metal for the first layer is deposited by a process called *microcasting* at the deposition station, Figure 10(a). The work is then moved to a machining station (b), where a computer-controlled milling machine or grinder removes deposited metal to shape the first layer of the part. Next, the work is moved to a stress-relief station (c), where it is shot- peened to relieve stresses that have built up in the layer. The work is then transferred back to the deposition station (a) for simultaneous deposition of primary metal for the next layer and sacrificial support

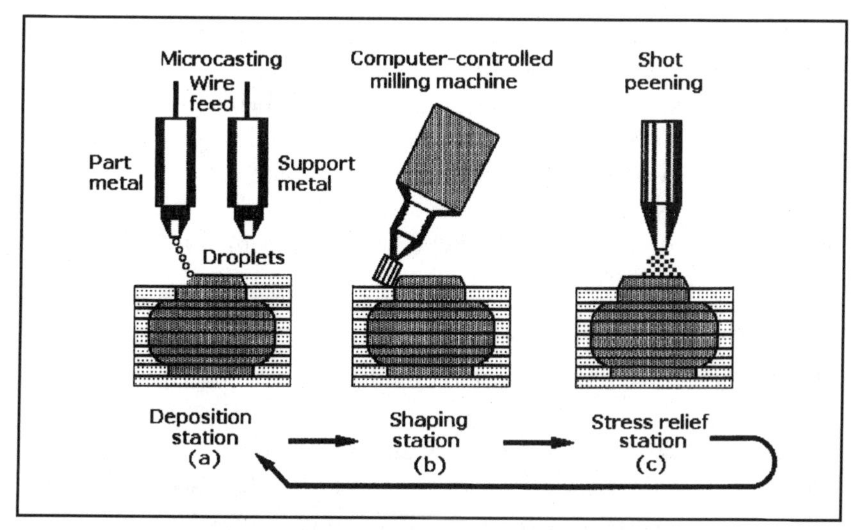

Figure 10 Shape Deposition Manufacturing (SDM): Functional metal parts or tools can be formed in layers by repeating three basic steps repetitively until the part is completed. Hot metal droplets of both primary and sacrificial support material form layers by a thermal metal spraying technique (a). They retain their heat long enough to remelt the underlying metal on impact to form strong metallurgical interlayer bonds. Each layer is machined under computer control (b) and shot-peened (c) to relieve stress buildup before the work is returned for deposition of the next layer. The sacrificial metal supports any undercut features. When deposition of all layers is complete, the sacrificial metal is removed by acid etching to release the completed part.

metal. The support material protects the part layers from the deposition steps that follow, stabilizes the layer for further machining operations, and provides a flat surface for milling the next layer. This SDM cycle is repeated until the part is finished, and then the sacrificial metal is etched away with acid. One combination of metals that has been successful in SDM is stainless steel for forming the prototype and copper for forming the support structure

The SDM Laboratory investigated many thermal techniques for depositing high-quality metals, including thermal spraying and plasma or laser welding, before it decided on microcasting, a compromise between these two techniques that provided better results than either technique by itself. The metal droplets in microcasting are large enough (1 to 3 mm in diameter) to retain their heat longer than the 50-mm droplets formed by conventional thermal spraying. The larger droplets remain molten and retain their heat long enough so that when they impact the metal surfaces they remelt them to form a strong metallurgical interlayer bond. This process overcame the low adhesion and low mechanical strength problems encountered with conventional thermal metal spraying. Weld-based deposition easily remelted the substrate material to form metallurgical bonds, but the larger amount of heat transferred tended to warp the substrate or delaminate it.

The SDM laboratory has produced custom-made functional mechanical parts and has embedded prefabricated mechanical parts, electronic components, electronic circuits, and sensors in the metal layers during the SDM process. It has also made custom tools such as injection molds with internal cooling pipes and metal heat sinks with embedded copper pipes for heat redistribution.

Mold SDM

The Rapid Prototyping Laboratory at Stanford University, Palo Alto, California, has developed its own version of SDM, called Mold SDM, for building layered molds for casting ceramics and polymers. Mold SDM, as diagrammed in Figure 11, uses wax to form the molds. The wax occupies the same position as the sacrificial support metal in SDM, and water-soluble photopolymer sacrificial support material occupies and supports the mold cavity. The photopolymer corresponds to the primary metal deposited to form the finished part in SDM. No machining is performed in this process.

The first step in the Mold SDM process begins with the decomposition of CAD mold data into layers of optimum thickness, which depends on the complexity and contours of the mold. The actual processing begins at Figure 11(a), which shows the results of repetitive cycles of the deposition of wax for the mold and sacrificial photopolymer in each layer to occupy the mold cavity and support it. The polymer is hardened by an ultraviolet (UV) source. After the mold and support structures are built up, the work is moved to a station (b) where the photopolymer is removed by dissolving it in water. This exposes the wax mold cavity into which the final part material is cast. It can be any compatible castable material. For example, ceramic parts can be formed by pouring a gel-casting ceramic slurry into the wax mold (c) and then curing the slurry. The wax mold is then removed (d) by melting it, releasing the "green" ceramic part for furnace firing. In step (e), after firing, the vents and sprues are removed as the final step.

Mold SDM has been expanded into making parts from a variety of polymer materials, and it has also been used to make preassembled mechanisms, both in polymer and ceramic materials.

For the designer just getting started in the wonderful world of mobile robots, it is suggested s/he follow the adage "prototype early, prototype often." This old design philosophy is far easier to use with the aid of RP tools. A simpler, cheaper, and more basic method, though, is to use

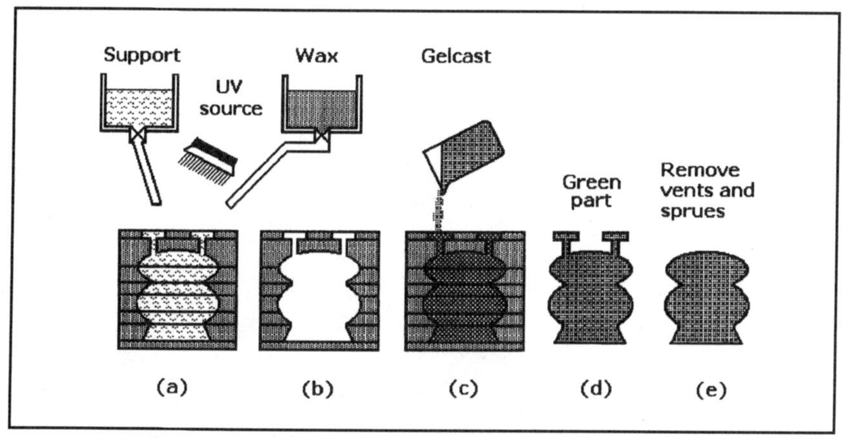

Figure 11 Mold Shape Deposition Manufacturing (MSDM): Casting molds can be formed in successive layers: Wax for the mold and water-soluble photopolymer to support the cavity are deposited in a repetitive cycle to build the mold in layers whose thickness and number depend on the mold's shape (a). UV energy solidifies the photopolymer. The photopolymer support material is removed by soaking it in hot water (b). Materials such as polymers and ceramics can be cast in the wax mold. For ceramic parts, a gelcasting ceramic slurry is poured into the mold to form green ceramic parts, which are then cured (c). The wax mold is then removed by heat or a hot liquid bath and the green ceramic part released (d). After furnace firing (e) any vents and sprues are removed.

Popsicle sticks, crazy glue, hot glue, shirt cardboard, packing tape, clay, or one of the many construction toy sets, etc. Fast, cheap, and surprisingly useful information on the effectiveness of whatever concept has been dreamed up can be achieved with very simple prototypes. There's nothing like holding the thing in your hand, even in a crude form, to see if it has any chance of working as originally conceived.

Robots can be very complicated in final form, especially those that do real work without aid of humans. Start simple and test ideas one at a time, then assemble those pieces into subassemblies and test those. Learn as much as possible about the actual obstacles that might be found in the environment for which the robot is destined. Design the mobility system to handle more difficult terrain because there will always be obstacles that will cause problems even in what appears to be a simple environment. Learn as much as possible about the required task, and design the manipulator and end effector to be only as complex as will accomplish that task.

Trial and error is the best method in many fields of design, and is especially so for robots. Prototype early, prototype often, and test everything. Mobile robots are inherently complex devices with many interactions within themselves and with their environment. The result of the effort, though, is exciting, fun, and rewarding. There is nothing like seeing an autonomous robot happily driving around, doing some useful task completely on its own.

Acknowledgments

This book would not even have been considered and would never have been completed without the encouragement and support of my loving wife, Victoria. Thank you so much.

In addition to the support of my wife, I would like to thank Joe Jones for his input, criticism, and support. Thank you for putting up with my many questions. Thanks also goes to Lee Sword, Chi Won, Tim Ohm, and Scott Miller for input on many of the ideas and layouts. The process of writing this book was made much easier by iRobot allowing me to use their office machines. And, lastly, thanks to my extended family, especially my Dad and Jenny for their encouragement and patience.

Chapter 1 Motor and Motion Control Systems

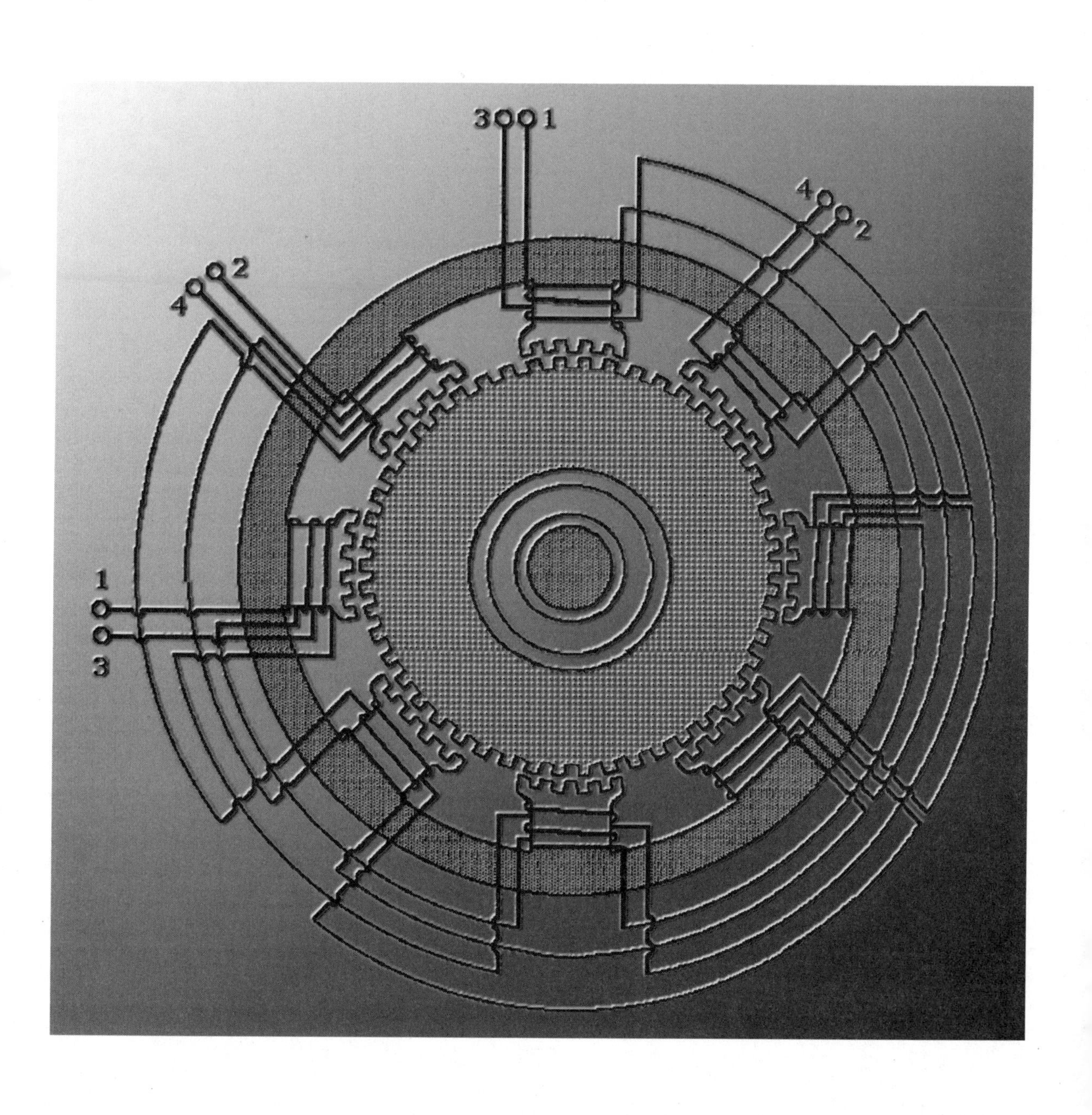

INTRODUCTION

A modern motion control system typically consists of a motion controller, a motor drive or amplifier, an electric motor, and feedback sensors. The system might also contain other components such as one or more belt-, ballscrew-, or leadscrew-driven linear guides or axis stages. A motion controller today can be a standalone programmable controller, a personal computer containing a motion control card, or a programmable logic controller (PLC).

All of the components of a motion control system must work together seamlessly to perform their assigned functions. Their selection must be based on both engineering and economic considerations. Figure 1-1 illustrates a typical multiaxis X-Y-Z motion platform that includes the three linear axes required to move a load, tool, or end effector precisely through three degrees of freedom. With additional mechanical or electro-

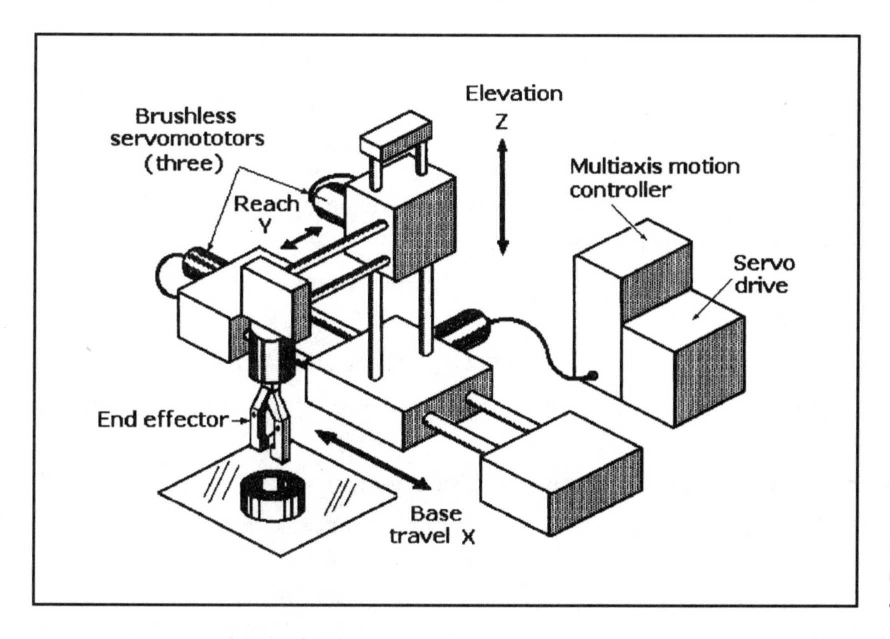

Figure 1-1 This multiaxis X-Y-Z motion platform is an example of a motion control system.

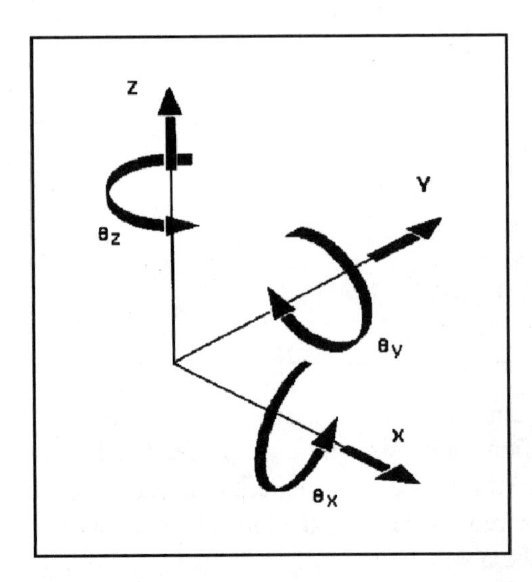

Figure 1-2 The right-handed coordinate system showing six degrees of freedom.

mechanical components on each axis, rotation about the three axes can provide up to six degrees of freedom, as shown in Figure 1-2.

Motion control systems today can be found in such diverse applications as materials handling equipment, machine tool centers, chemical and pharmaceutical process lines, inspection stations, robots, and injection molding machines.

Merits of Electric Systems

Most motion control systems today are powered by electric motors rather than hydraulic or pneumatic motors or actuators because of the many benefits they offer:

- More precise load or tool positioning, resulting in fewer product or process defects and lower material costs.
- Quicker changeovers for higher flexibility and easier product customizing.
- Increased throughput for higher efficiency and capacity.
- Simpler system design for easier installation, programming, and training.
- Lower downtime and maintenance costs.
- Cleaner, quieter operation without oil or air leakage.

Electric-powered motion control systems do not require pumps or air compressors, and they do not have hoses or piping that can leak hydraulic fluids or air. This discussion of motion control is limited to electric-powered systems.

Motion Control Classification

Motion control systems can be classified as *open-loop* or *closed-loop*. An open-loop system does not require that measurements of any output variables be made to produce error-correcting signals; by contrast, a closed-loop system requires one or more feedback sensors that measure and respond to errors in output variables.

Closed-Loop System

A *closed-loop motion control system*, as shown in block diagram Figure 1-3, has one or more feedback loops that continuously compare the system's response with input commands or settings to correct errors in motor and/or load speed, load position, or motor torque. Feedback sensors provide the electronic signals for correcting deviations from the desired input commands. Closed-loop systems are also called servosystems.

Each motor in a servosystem requires its own feedback sensors, typically encoders, resolvers, or tachometers that close loops around the motor and load. Variations in velocity, position, and torque are typically caused by variations in load conditions, but changes in ambient temperature and humidity can also affect load conditions.

A *velocity control loop,* as shown in block diagram Figure 1-4, typically contains a tachometer that is able to detect changes in motor speed. This sensor produces error signals that are proportional to the positive or negative deviations of motor speed from its preset value. These signals are sent

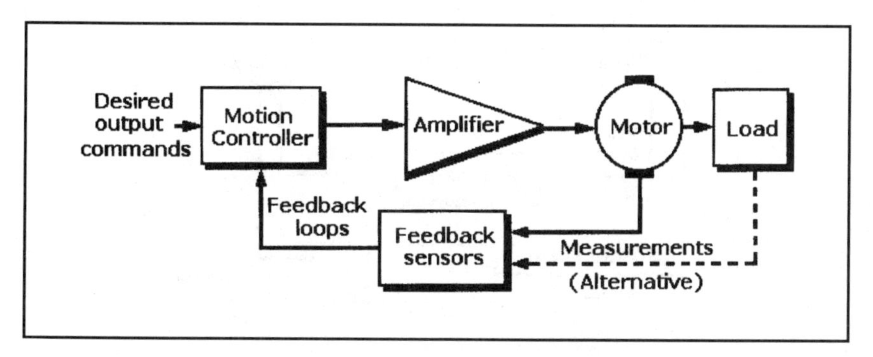

Figure 1-3 Block diagram of a basic closed-loop control system.

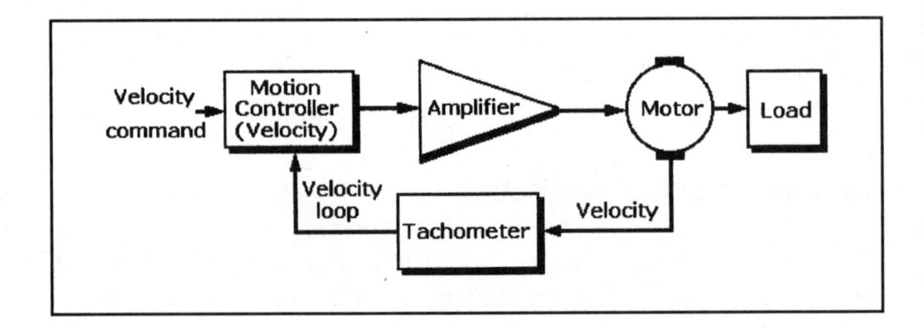

Figure 1-4 Block diagram of a velocity-control system.

to the motion controller so that it can compute a corrective signal for the amplifier to keep motor speed within those preset limits despite load changes.

A *position-control loop,* as shown in block diagram Figure 1-5, typically contains either an encoder or resolver capable of direct or indirect measurements of load position. These sensors generate error signals that are sent to the motion controller, which produces a corrective signal for amplifier. The output of the amplifier causes the motor to speed up or slow down to correct the position of the load. Most position control closed-loop systems also include a velocity-control loop.

The *ballscrew slide mechanism*, shown in Figure 1-6, is an example of a mechanical system that carries a load whose position must be controlled in a closed-loop servosystem because it is not equipped with position sensors. Three examples of feedback sensors mounted on the ballscrew mechanism that can provide position feedback are shown in Figure 1-7: (a) is a rotary optical encoder mounted on the motor housing with its shaft coupled to the motor shaft; (b) is an optical linear encoder with its gradu-

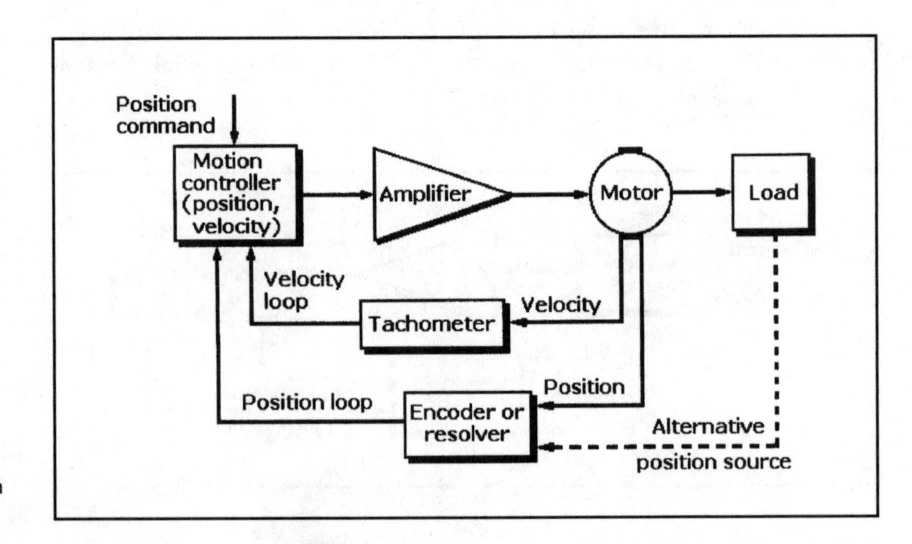

Figure 1-5 Block diagram of a position-control system.

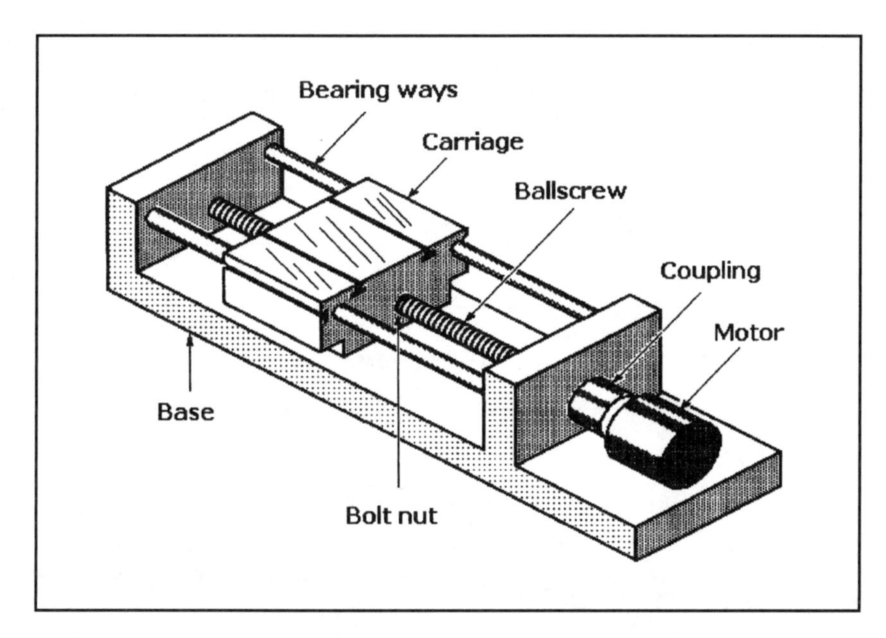

Figure 1-6 Ballscrew-driven single-axis slide mechanism without position feedback sensors.

ated scale mounted on the base of the mechanism; and (c) is the less commonly used but more accurate and expensive laser interferometer.

A *torque-control loop* contains electronic circuitry that measures the input current applied to the motor and compares it with a value proportional to the torque required to perform the desired task. An error signal from the circuit is sent to the motion controller, which computes a corrective signal for the motor amplifier to keep motor current, and hence torque, constant. Torque- control loops are widely used in machine tools where the load can change due to variations in the density of the material being machined or the sharpness of the cutting tools.

Trapezoidal Velocity Profile

If a motion control system is to achieve smooth, high-speed motion without overstressing the ser-

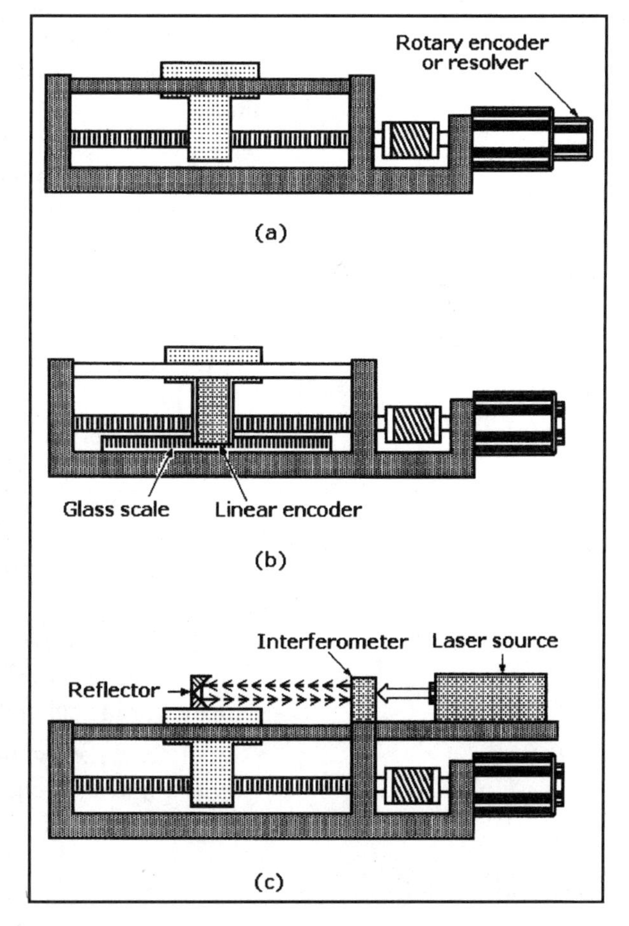

Figure 1-7 Examples of position feedback sensors installed on a ballscrew-driven slide mechanism: (a) rotary encoder, (b) linear encoder, and (c) laser interferometer.

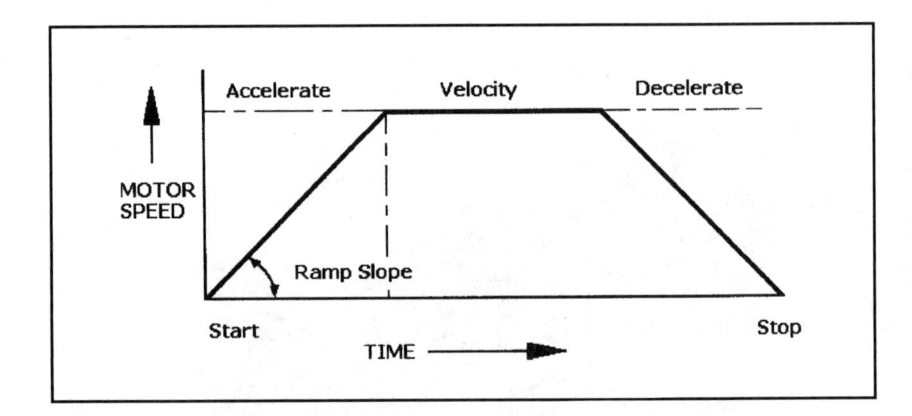

Figure 1-8 Servomotors are accelerated to constant velocity and decelerated along a trapezoidal profile to assure efficient operation.

vomotor, the motion controller must command the motor amplifier to ramp up motor velocity gradually until it reaches the desired speed and then ramp it down gradually until it stops after the task is complete. This keeps motor acceleration and deceleration within limits.

The trapezoidal profile, shown in Figure 1-8, is widely used because it accelerates motor velocity along a positive linear "up-ramp" until the desired constant velocity is reached. When the motor is shut down from the constant velocity setting, the profile decelerates velocity along a negative "down ramp" until the motor stops. Amplifier current and output voltage reach maximum values during acceleration, then step down to lower values during constant velocity and switch to negative values during deceleration.

Closed-Loop Control Techniques

The simplest form of feedback is *proportional control*, but there are also *derivative* and *integral control* techniques, which compensate for certain steady-state errors that cannot be eliminated from proportional control. All three of these techniques can be combined to form *proportional-integral-derivative (PID) control.*

- In *proportional control* the signal that drives the motor or actuator is directly proportional to the linear difference between the input command for the desired output and the measured actual output.
- In *integral control* the signal driving the motor equals the *time integral* of the difference between the input command and the measured actual output.

- In *derivative control* the signal that drives the motor is proportional to the *time derivative* of the difference between the input command and the measured actual output.
- In *proportional-integral-derivative (PID) control* the signal that drives the motor equals the weighted sum of the difference, the time integral of the difference, and the time derivative of the difference between the input command and the measured actual output.

Open-Loop Motion Control Systems

A typical *open-loop motion control system* includes a stepper motor with a programmable indexer or pulse generator and motor driver, as shown in Figure 1-9. This system does not need feedback sensors because load

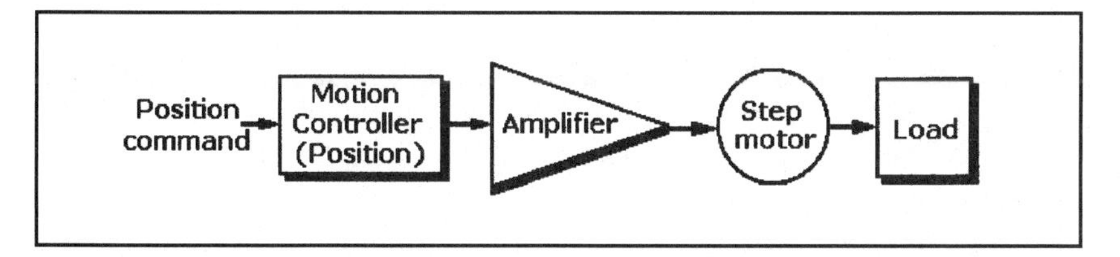

Figure 1-9 Block diagram of an open-loop motion control system.

position and velocity are controlled by the predetermined number and direction of input digital pulses sent to the motor driver from the controller. Because load position is not continuously sampled by a feedback sensor (as in a closed-loop servosystem), load positioning accuracy is lower and position errors (commonly called step errors) accumulate over time. For these reasons open-loop systems are most often specified in applications where the load remains constant, load motion is simple, and low positioning speed is acceptable.

Kinds of Controlled Motion

There are five different kinds of motion control: *point-to-point, sequencing, speed, torque,* and *incremental.*

- In *point-to-point motion control* the load is moved between a sequence of numerically defined positions where it is stopped before it is moved to the next position. This is done at a constant speed, with both velocity and distance monitored by the motion controller. Point-to-point positioning can be performed in single-axis or multiaxis systems with servomotors in closed loops or stepping motors in open loops. X-Y tables and milling machines position their loads by multi-axis point-to-point control.

- *Sequencing control* is the control of such functions as opening and closing valves in a preset sequence or starting and stopping a conveyor belt at specified stations in a specific order.
- *Speed control* is the control of the velocity of the motor or actuator in a system.
- *Torque control* is the control of motor or actuator current so that torque remains constant despite load changes.
- *Incremental motion control* is the simultaneous control of two or more variables such as load location, motor speed, or torque.

Motion Interpolation

When a load under control must follow a specific path to get from its starting point to its stopping point, the movements of the axes must be coordinated or interpolated. There are three kinds of interpolation: *linear, circular, and contouring.*

Linear interpolation is the ability of a motion control system having two or more axes to move the load from one point to another in a straight line. The motion controller must determine the speed of each axis so that it can coordinate their movements. True linear interpolation requires that the motion controller modify axis acceleration, but some controllers approximate true linear interpolation with programmed acceleration profiles. The path can lie in one plane or be three dimensional.

Circular interpolation is the ability of a motion control system having two or more axes to move the load around a circular trajectory. It requires that the motion controller modify load acceleration while it is in transit. Again the circle can lie in one plane or be three dimensional.

Contouring is the path followed by the load, tool, or end- effector under the coordinated control of two or more axes. It requires that the motion controller change the speeds on different axes so that their trajectories pass through a set of predefined points. Load speed is determined along the trajectory, and it can be constant except during starting and stopping.

Computer-Aided Emulation

Several important types of programmed computer-aided motion control can emulate mechanical motion and eliminate the need for actual gears or cams. *Electronic gearing* is the control by software of one or more axes to impart motion to a load, tool, or end effector that simulates the speed changes that can be performed by actual gears. *Electronic camming* is the control by software of one or more axes to impart a motion to a load, tool, or end effector that simulates the motion changes that are typically performed by actual cams.

Mechanical Components

The mechanical components in a motion control system can be more influential in the design of the system than the electronic circuitry used to control it. Product flow and throughput, human operator requirements, and maintenance issues help to determine the mechanics, which in turn influence the motion controller and software requirements.

Mechanical actuators convert a motor's rotary motion into linear motion. Mechanical methods for accomplishing this include the use of leadscrews, shown in Figure 1-10, ballscrews, shown in Figure 1-11, worm-drive gearing, shown in Figure 1-12, and belt, cable, or chain drives. Method selection is based on the relative costs of the alternatives and consideration for the possible effects of backlash. All actuators have finite levels of torsional and axial stiffness that can affect the system's frequency response characteristics.

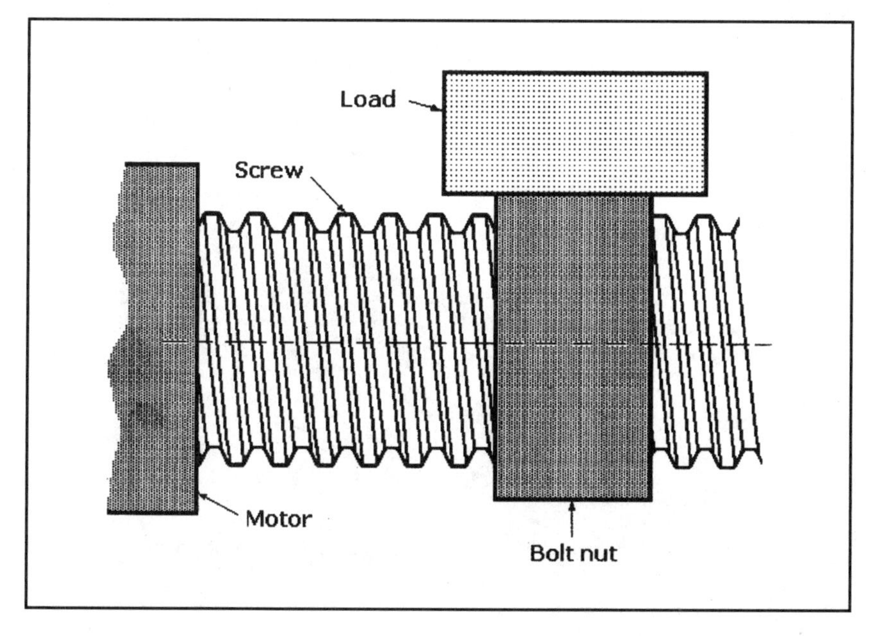

Figure 1-10 Leadscrew drive: As the leadscrew rotates, the load is translated in the axial direction of the screw.

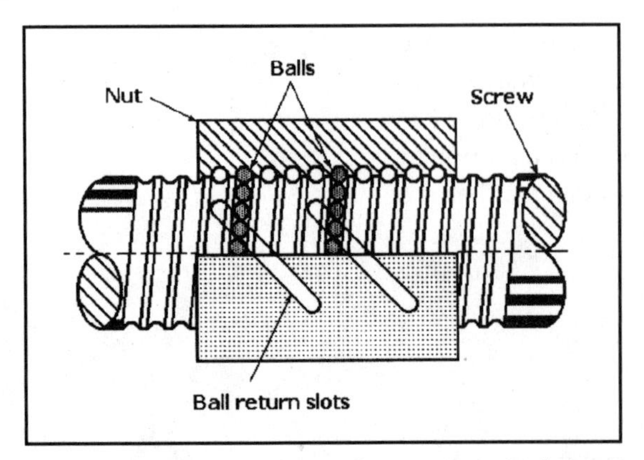

Figure 1-11 Ballscrew drive: Ballscrews use recirculating balls to reduce friction and gain higher efficiency than conventional leadscrews.

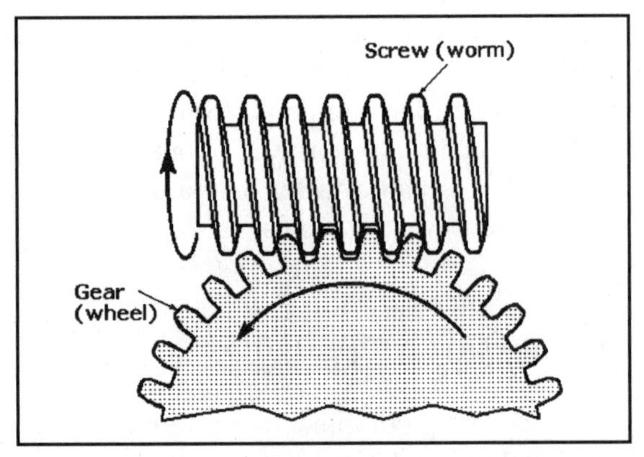

Figure 1-12 Worm-drive systems can provide high speed and high torque.

Linear guides or stages constrain a translating load to a single degree of freedom. The linear stage supports the mass of the load to be actuated and assures smooth, straight-line motion while minimizing friction. A common example of a linear stage is a ballscrew-driven single-axis stage, illustrated in Figure 1-13. The motor turns the ballscrew, and its rotary motion is translated into the linear motion that moves the carriage and load by the stage's bolt nut. The bearing ways act as linear guides. As shown in Figure 1-7, these stages can be equipped with sensors such as a rotary or linear encoder or a laser interferometer for feedback.

A ballscrew-driven single-axis stage with a rotary encoder coupled to the motor shaft provides an indirect measurement. This method ignores

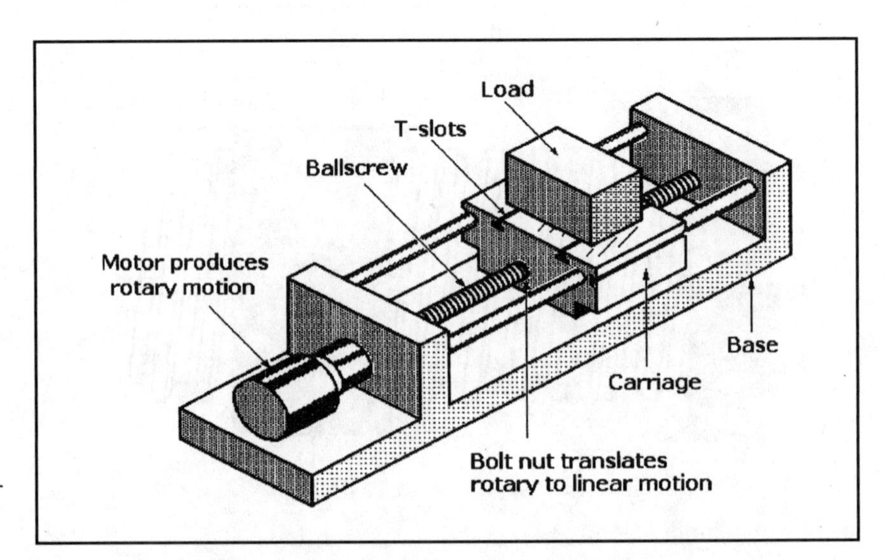

Figure 1-13 Ballscrew-driven single-axis slide mechanism translates rotary motion into linear motion.

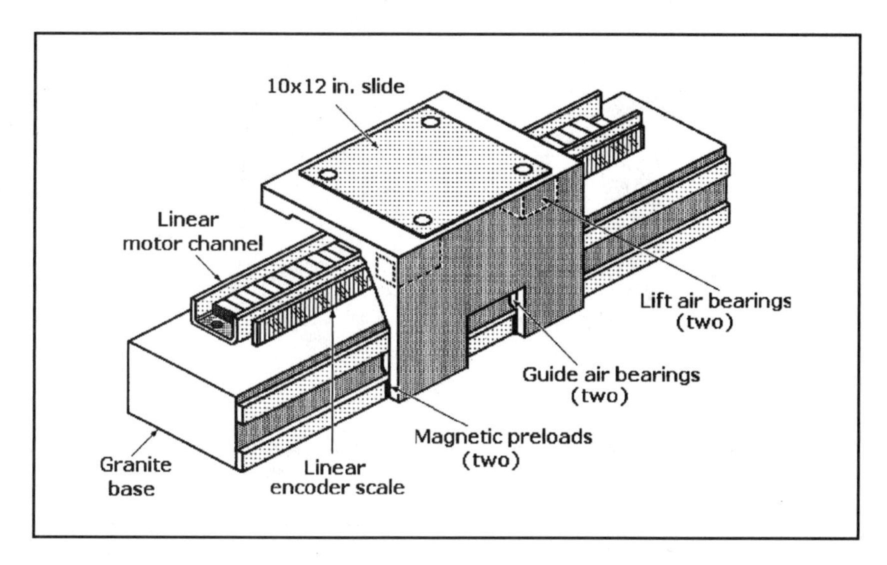

Figure 1-14 This single-axis linear guide for load positioning is supported by air bearings as it moves along a granite base.

the tolerance, wear, and compliance in the mechanical components between the carriage and the position encoder that can cause deviations between the desired and true positions. Consequently, this feedback method limits position accuracy to ballscrew accuracy, typically ±5 to 10 µm per 300 mm.

Other kinds of single-axis stages include those containing antifriction rolling elements such as recirculating and nonrecirculating balls or rollers, sliding (friction contact) units, air-bearing units, hydrostatic units, and magnetic levitation (Maglev) units.

A single-axis air-bearing guide or stage is shown in Figure 1-14. Some models being offered are 3.9 ft (1.2 m) long and include a carriage for mounting loads. When driven by a linear servomotors the loads can reach velocities of 9.8 ft/s (3 m/s). As shown in Figure 1-7, these stages can be equipped with feedback devices such as cost-effective linear encoders or ultra-high-resolution laser interferometers. The resolution of this type of stage with a noncontact linear encoder can be as fine as 20 nm and accuracy can be ±1 µm. However, these values can be increased to 0.3 nm resolution and submicron accuracy if a laser interferometer is installed.

The pitch, roll, and yaw of air-bearing stages can affect their resolution and accuracy. Some manufacturers claim ±1 arc-s per 100 mm as the limits for each of these characteristics. Large air-bearing surfaces provide excellent stiffness and permit large load-carrying capability.

The important attributes of all these stages are their dynamic and static friction, rigidity, stiffness, straightness, flatness, smoothness, and load capacity. Also considered is the amount of work needed to prepare the host machine's mounting surface for their installation.

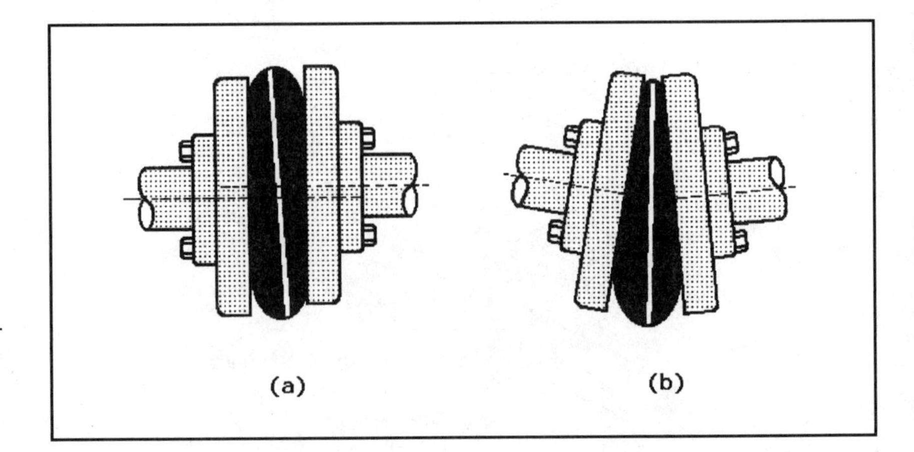

Figure 1-15 Flexible shaft couplings adjust for and accommodate parallel misalignment (a) and angular misalignment between rotating shafts (b).

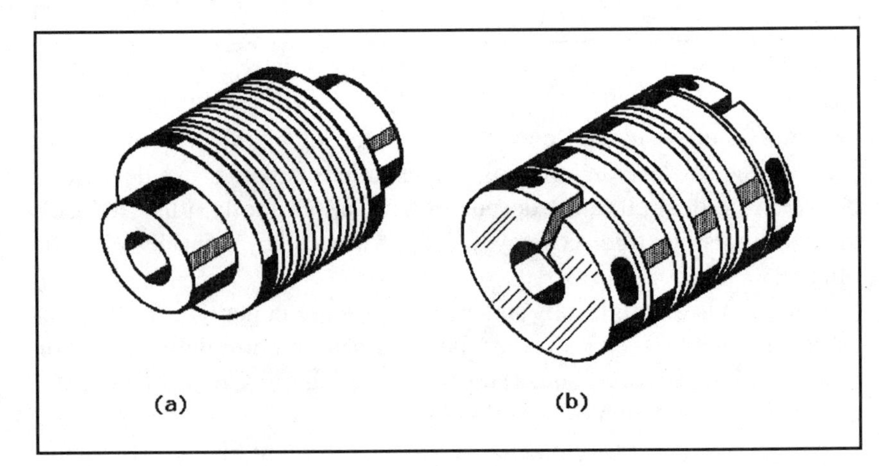

Figure 1-16 Bellows couplings (a) are acceptable for light-duty applications. Misalignments can be 9° angular or 1⁄4 in. parallel. Helical couplings (b) prevent backlash and can operate at constant velocity with misalignment and be run at high speed.

The structure on which the motion control system is mounted directly affects the system's performance. A properly designed base or host machine will be highly damped and act as a compliant barrier to isolate the motion system from its environment and minimize the impact of external disturbances. The structure must be stiff enough and sufficiently damped to avoid resonance problems. A high static mass to reciprocating mass ratio can also prevent the motion control system from exciting its host structure to harmful resonance.

Any components that move will affect a system's response by changing the amount of inertia, damping, friction, stiffness, or resonance. For example, a flexible shaft coupling, as shown in Figure 1-15, will compensate for minor parallel (a) and angular (b) misalignment between rotating shafts. Flexible couplings are available in other configurations such as bellows and helixes, as shown in Figure 1-16. The bellows configuration (a) is acceptable for light-duty applications where misalignments can be as great as 9° angular or ¼ in. parallel. By contrast, helical couplings (b) prevent backlash at constant velocity with some misalignment, and they can also be run at high speed.

Other moving mechanical components include cable carriers that retain moving cables, end stops that restrict travel, shock absorbers to dissipate energy during a collision, and way covers to keep out dust and dirt.

Electronic System Components

The motion controller is the "brain" of the motion control system and performs all of the required computations for motion path planning, servo-loop closure, and sequence execution. It is essentially a computer dedicated to motion control that has been programmed by the end user for the performance of assigned tasks. The motion controller produces a low-power motor command signal in either a digital or analog format for the motor driver or amplifier.

Significant technical developments have led to the increased acceptance of programmable motion controllers over the past five to ten years: These include the rapid decrease in the cost of microprocessors as well as dramatic increases in their computing power. Added to that are the decreasing cost of more advanced semiconductor and disk memories. During the past five to ten years, the capability of these systems to improve product quality, increase throughput, and provide just-in-time delivery has improved has improved significantly.

The motion controller is the most critical component in the system because of its dependence on software. By contrast, the selection of most motors, drivers, feedback sensors, and associated mechanisms is less critical because they can usually be changed during the design phase or even later in the field with less impact on the characteristics of the intended system. However, making field changes can be costly in terms of lost productivity.

The decision to install any of the three kinds of motion controllers should be based on their ability to control both the number and types of motors required for the application as well as the availability of the software that will provide the optimum performance for the specific application. Also to be considered are the system's multitasking capabilities, the number of input/output (I/O) ports required, and the need for such features as linear and circular interpolation and electronic gearing and camming.

In general, a motion controller receives a set of operator instructions from a host or operator interface and it responds with corresponding command signals for the motor driver or drivers that control the motor or motors driving the load.

Motor Selection

The most popular motors for motion control systems are stepping or stepper motors and permanent-magnet (PM) DC brush-type and brushless DC servomotors. Stepper motors are selected for systems because they can run open-loop without feedback sensors. These motors are indexed or partially rotated by digital pulses that turn their rotors a fixed fraction or a revolution where they will be clamped securely by their inherent holding torque. Stepper motors are cost-effective and reliable choices for many applications that do not require the rapid acceleration, high speed, and position accuracy of a servomotor.

However, a feedback loop can improve the positioning accuracy of a stepper motor without incurring the higher costs of a complete servosystem. Some stepper motor motion controllers can accommodate a closed loop.

Brush and brushless PM DC servomotors are usually selected for applications that require more precise positioning. Both of these motors can reach higher speeds and offer smoother low-speed operation with finer position resolution than stepper motors, but both require one or more feedback sensors in closed loops, adding to system cost and complexity.

Brush-type permanent-magnet (PM) DC servomotors have wound armatures or rotors that rotate within the magnetic field produced by a PM stator. As the rotor turns, current is applied sequentially to the appropriate armature windings by a mechanical commutator consisting of two or more brushes sliding on a ring of insulated copper segments. These motors are quite mature, and modern versions can provide very high performance for very low cost.

There are variations of the brush-type DC servomotor with its iron-core rotor that permit more rapid acceleration and deceleration because of their low-inertia, lightweight cup- or disk-type armatures. The disk-type armature of the pancake-frame motor, for example, has its mass concentrated close to the motor's faceplate permitting a short, flat cylindrical housing. This configuration makes the motor suitable for faceplate mounting in restricted space, a feature particularly useful in industrial robots or other applications where space does not permit the installation of brackets for mounting a motor with a longer length dimension.

The brush-type DC motor with a cup-type armature also offers lower weight and inertia than conventional DC servomotors. However, the trade-off in the use of these motors is the restriction on their duty cycles because the epoxy-encapsulated armatures are unable to dissipate heat buildup as easily as iron-core armatures and are therefore subject to damage or destruction if overheated.

However, any servomotor with brush commutation can be unsuitable for some applications due to the electromagnetic interference (EMI) caused by brush arcing or the possibility that the arcing can ignite nearby flammable fluids, airborne dust, or vapor, posing a fire or explosion hazard. The EMI generated can adversely affect nearby electronic circuitry. In addition, motor brushes wear down and leave a gritty residue that can contaminate nearby sensitive instruments or precisely ground surfaces. Thus brush-type motors must be cleaned constantly to prevent the spread of the residue from the motor. Also, brushes must be replaced periodically, causing unproductive downtime.

Brushless DC PM motors overcome these problems and offer the benefits of electronic rather than mechanical commutation. Built as inside-out DC motors, typical brushless motors have PM rotors and wound stator coils. Commutation is performed by internal noncontact Hall-effect devices (HEDs) positioned within the stator windings. The HEDs are wired to power transistor switching circuitry, which is mounted externally in separate modules for some motors but is mounted internally on circuit cards in other motors. Alternatively, commutation can be performed by a commutating encoder or by commutation software resident in the motion controller or motor drive.

Brushless DC motors exhibit low rotor inertia and lower winding thermal resistance than brush-type motors because their high-efficiency magnets permit the use of shorter rotors with smaller diameters. Moreover, because they are not burdened with sliding brush-type mechanical contacts, they can run at higher speeds (50,000 rpm or greater), provide higher continuous torque, and accelerate faster than brush-type motors. Nevertheless, brushless motors still cost more than comparably rated brush-type motors (although that price gap continues to narrow) and their installation adds to overall motion control system cost and complexity. Table 1-1 summarizes some of the outstanding characteristics of stepper, PM brush, and PM brushless DC motors.

The linear motor, another drive alternative, can move the load directly, eliminating the need for intermediate motion translation mechanism. These motors can accelerate rapidly and position loads accurately at high speed because they have no moving parts in contact with each other. Essentially rotary motors that have been sliced open and unrolled, they have many of the characteristics of conventional motors. They can replace conventional rotary motors driving leadscrew-, ballscrew-, or belt-driven single-axis stages, but they cannot be coupled to gears that could change their drive characteristics. If increased performance is

Table 1-1 Stepping and Permanent-Magnet DC Servomotors Compared.

	Stepping	PM Brush	PM Brushless
Cost	Low	Medium	High
Smoothness	Low to	Good to excellent	Good to excellent
Speed range	0-1500 rpm (typical)	0-6000 rpm	0-10,000 rpm
Torque	High- (falls off with speed)	Medium	High
Required feedback	None	Position or velocity	Commutation and position or velocity
Maintenance	None	Yes	None
Cleanliness	Excellent	Brush dust	Excellent

required from a linear motor, the existing motor must be replaced with a larger one.

Linear motors must operate in closed feedback loops, and they typically require more costly feedback sensors than rotary motors. In addition, space must be allowed for the free movement of the motor's power cable as it tracks back and forth along a linear path. Moreover, their applications are also limited because of their inability to dissipate heat as readily as rotary motors with metal frames and cooling fins, and the exposed magnetic fields of some models can attract loose ferrous objects, creating a safety hazard.

Motor Drivers (Amplifiers)

Motor drivers or amplifiers must be capable of driving their associated motors—stepper, brush, brushless, or linear. A drive circuit for a stepper motor can be fairly simple because it needs only several power transistors to sequentially energize the motor phases according to the number of digital step pulses received from the motion controller. However, more advanced stepping motor drivers can control phase current to permit "microstepping," a technique that allows the motor to position the load more precisely.

Servodrive amplifiers for brush and brushless motors typically receive analog voltages of ±10-VDC signals from the motion controller. These signals correspond to current or voltage commands. When amplified, the signals control both the direction and magnitude of the current in the motor windings. Two types of amplifiers are generally used in closed-loop servosystems: linear and pulse-width modulated (PWM).

Pulse-width modulated amplifiers predominate because they are more efficient than linear amplifiers and can provide up to 100 W. The transistors in PWM amplifiers (as in PWM power supplies) are optimized for switchmode operation, and they are capable of switching amplifier output voltage at frequencies up to 20 kHz. When the power transistors are switched on (on state), they saturate, but when they are off, no current is drawn. This operating mode reduces transistor power dissipation and boosts amplifier efficiency. Because of their higher operating frequencies, the magnetic components in PWM amplifiers can be smaller and lighter than those in linear amplifiers. Thus the entire drive module can be packaged in a smaller, lighter case.

By contrast, the power transistors in linear amplifiers are continuously in the on state although output power requirements can be varied. This operating mode wastes power, resulting in lower amplifier efficiency while subjecting the power transistors to thermal stress. However, linear amplifiers permit smoother motor operation, a requirement for some sensitive motion control systems. In addition linear amplifiers are better at driving low-inductance motors. Moreover, these amplifiers generate less EMI than PWM amplifiers, so they do not require the same degree of filtering. By contrast, linear amplifiers typically have lower maxi-mum power ratings than PWM amplifiers.

Feedback Sensors

Position feedback is the most common requirement in closed-loop motion control systems, and the most popular sensor for providing this information is the rotary optical encoder. The axial shafts of these encoders are mechanically coupled to the drive shafts of the motor. They generate either sine waves or pulses that can be counted by the motion controller to determine the motor or load position and direction of travel at any time to permit precise positioning. Analog encoders produce sine waves that must be conditioned by external circuitry for counting, but digital encoders include circuitry for translating sine waves into pulses.

Absolute rotary optical encoders produce binary words for the motion controller that provide precise position information. If they are stopped accidentally due to power failure, these encoders preserve the binary word because the last position of the encoder code wheel acts as a memory.

Linear optical encoders, by contrast, produce pulses that are proportional to the actual linear distance of load movement. They work on the same principles as the rotary encoders, but the graduations are engraved on a stationary glass or metal scale while the read head moves along the scale.

Tachometers are generators that provide analog signals that are directly proportional to motor shaft speed. They are mechanically coupled to the motor shaft and can be located within the motor frame. After tachometer output is converted to a digital format by the motion controller, a feedback signal is generated for the driver to keep motor speed within preset limits.

Other common feedback sensors include resolvers, linear variable differential transformers (LVDTs), Inductosyns, and potentiomcters. Less common are the more accurate laser interferometers. Feedback sensor selection is based on an evaluation of the sensor's accuracy, repeatability, ruggedness, temperature limits, size, weight, mounting requirements, and cost, with the relative importance of each determined by the application.

Installation and Operation of the System

The design and implementation of a cost-effective motion-control system require a high degree of expertise on the part of the person or persons responsible for system integration. It is rare that a diverse group of components can be removed from their boxes, installed, and interconnected to form an instantly effective system. Each servosystem (and many stepper systems) must be tuned (stabilized) to the load and environmental conditions. However, installation and development time can be minimized if the customer's requirements are accurately defined, optimum components are selected, and the tuning and debugging tools are applied correctly. Moreover, operators must be properly trained in formal classes or, at the very least, must have a clear understanding of the information in the manufacturers' technical manuals gained by careful reading.

SERVOMOTORS, STEPPER MOTORS, AND ACTUATORS FOR MOTION CONTROL

Many different kinds of electric motors have been adapted for use in motion control systems because of their linear characteristics. These include both conventional rotary and linear alternating current (AC) and direct current (DC) motors. These motors can be further classified into those that must be operated in closed-loop servosystems and those that can be operated open-loop.

The most popular servomotors are permanent magnet (PM) rotary DC servomotors that have been adapted from conventional PM DC motors. These servomotors are typically classified as brush-type and brushless. The brush-type PM DC servomotors include those with wound rotors and those with lighter weight, lower inertia cup- and disk coil-type armatures. Brushless servomotors have PM rotors and wound stators.

Some motion control systems are driven by two-part linear servomotors that move along tracks or ways. They are popular in applications where errors introduced by mechanical coupling between the rotary motors and the load can introduce unwanted errors in positioning. Linear motors require closed loops for their operation, and provision must be made to accommodate the back-and-forth movement of the attached data and power cable.

Stepper or stepping motors are generally used in less demanding motion control systems, where positioning the load by stepper motors is not critical for the application. Increased position accuracy can be obtained by enclosing the motors in control loops.

Permanent-Magnet DC Servomotors

Permanent-magnet (PM) field DC rotary motors have proven to be reliable drives for motion control applications where high efficiency, high starting torque, and linear speed–torque curves are desirable characteristics. While they share many of the characteristics of conventional rotary series, shunt, and compound-wound brush-type DC motors, PM DC servomotors increased in popularity with the introduction of stronger ceramic and rare-earth magnets made from such materials as neodymium–iron–boron and the fact that these motors can be driven easily by microprocessor-based controllers.

The replacement of a wound field with permanent magnets eliminates both the need for separate field excitation and the electrical losses that occur in those field windings. Because there are both brush-type and brushless DC servomotors, the term *DC motor* implies that it is brush-type or requires mechanical commutation unless it is modified by the term *brushless.* Permanent-magnet DC brush-type servomotors can also have armatures formed as laminated coils in disk or cup shapes. They are lightweight, low-inertia armatures that permit the motors to accelerate faster than the heavier conventional wound armatures.

The increased field strength of the ceramic and rare-earth magnets permitted the construction of DC motors that are both smaller and lighter

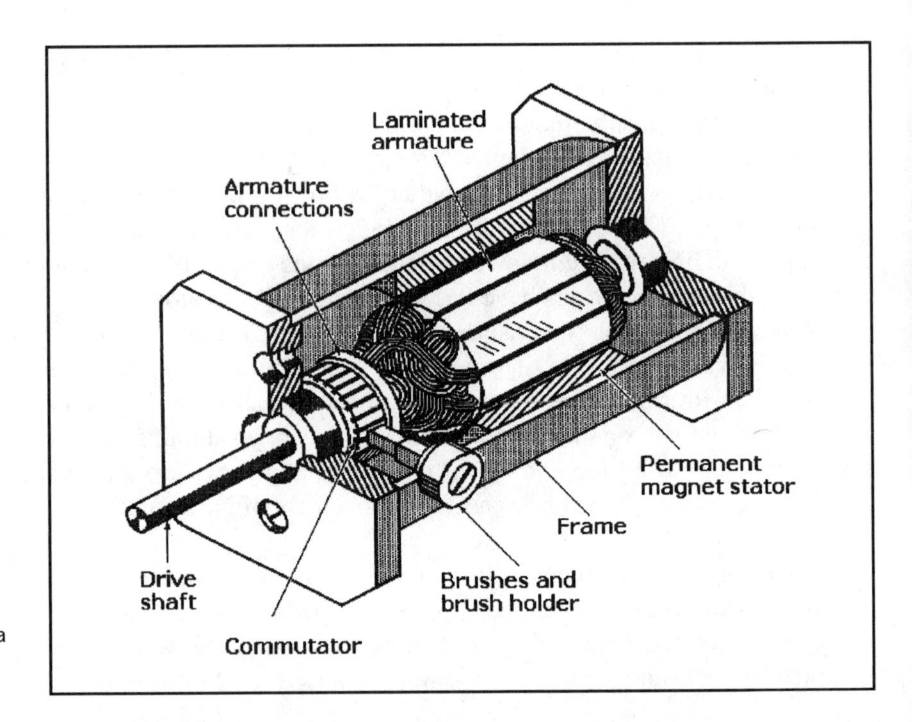

Figure 1-17 Cutaway view of a fractional horsepower permanent-magnet DC servomotor.

than earlier generation comparably rated DC motors with alnico (aluminum–nickel–cobalt or AlNiCo) magnets. Moreover, integrated circuitry and microprocessors have increased the reliability and cost-effectiveness of digital motion controllers and motor drivers or amplifiers while permitting them to be packaged in smaller and lighter cases, thus reducing the size and weight of complete, integrated motion-control systems.

Brush-Type PM DC Servomotors

The design feature that distinguishes the brush-type PM DC servomotor, as shown in Figure 1-17, from other brush-type DC motors is the use of a permanent-magnet field to replace the wound field. As previously stated, this eliminates both the need for separate field excitation and the electrical losses that typically occur in field windings.

Permanent-magnet DC motors, like all other mechanically commutated DC motors, are energized through brushes and a multisegment commutator. While all DC motors operate on the same principles, only PM DC motors have the linear speed–torque curves shown in Figure 1-18, making them ideal for closed-loop and variable-speed servomotor applications. These linear characteristics conveniently describe the full range of motor perform-

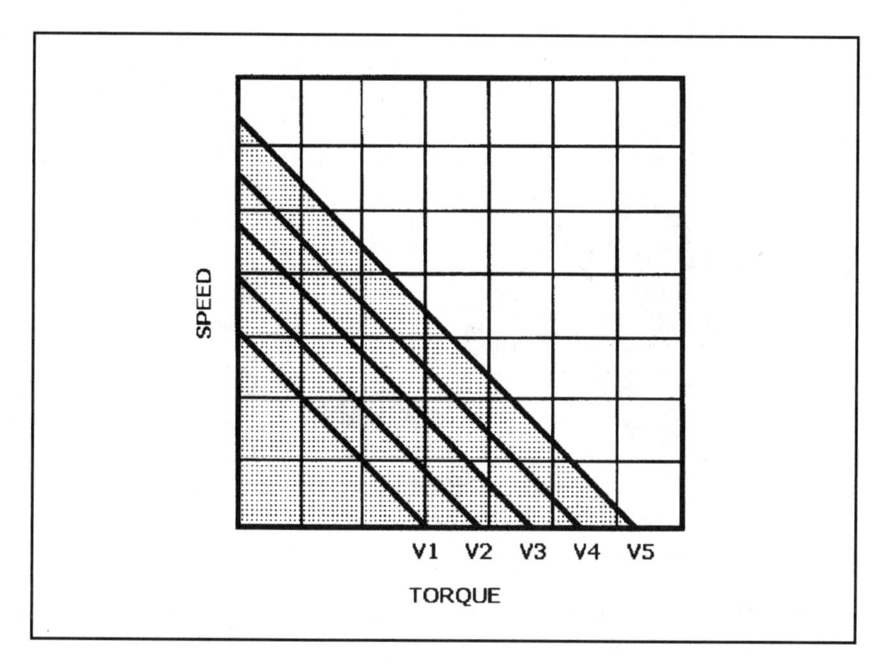

Figure 1-18 A typical family of speed/torque curves for a permanent-magnet DC servomotor at different voltage inputs, with voltage increasing from left to right (V1 to V5).

ance. It can be seen that both speed and torque increase linearly with applied voltage, indicated in the diagram as increasing from V1 to V5.

The stators of brush-type PM DC motors are magnetic pole pairs. When the motor is powered, the opposite polarities of the energized windings and the stator magnets attract, and the rotor rotates to align itself with the stator. Just as the rotor reaches alignment, the brushes move across the commutator segments and energize the next winding. This sequence continues as long as power is applied, keeping the rotor in continuous motion. The commutator is staggered from the rotor poles, and the number of its segments is directly proportional to the number of windings. If the connections of a PM DC motor are reversed, the motor will change direction, but it might not operate as efficiently in the reversed direction.

Disk-Type PM DC Motors

The disk-type motor shown exploded view in Figure 1-19 has a disk-shaped armature with stamped and laminated windings. This nonferrous laminated disk is made as a copper stamping bonded between epoxy–glass insulated layers and fastened to an axial shaft. The stator field can either be a ring of many individual ceramic magnet cylinders, as shown, or a ring-type ceramic magnet attached to the dish-shaped end

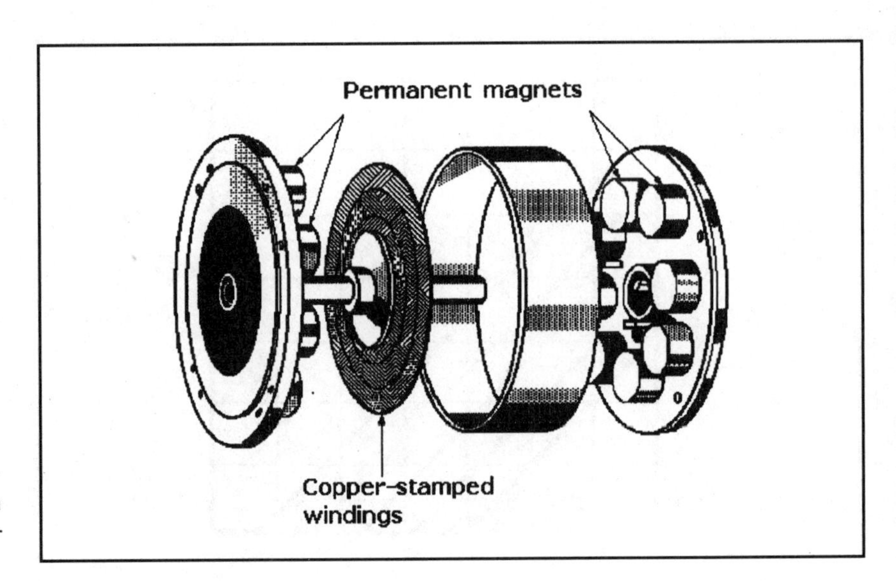

Figure 1-19 Exploded view of a permanent-magnet DC servomotor with a disk-type armature.

bell, which completes the magnetic circuit. The spring-loaded brushes ride directly on stamped commutator bars.

These motors are also called *pancake motors* because they are housed in cases with thin, flat form factors whose diameters exceed their lengths, suggesting pancakes. Earlier generations of these motors were called *printed-circuit motors* because the armature disks were made by a printed-circuit fabrication process that has been superseded. The flat motor case concentrates the motor's center of mass close to the mounting plate, permitting it to be easily surface mounted. This eliminates the awkward motor overhang and the need for supporting braces if a conventional motor frame is to be surface mounted. Their disk-type motor form factor has made these motors popular as axis drivers for industrial robots where space is limited.

The principal disadvantage of the disk-type motor is the relatively fragile construction of its armature and its inability to dissipate heat as rapidly as iron-core wound rotors. Consequently, these motors are usually limited to applications where the motor can be run under controlled conditions and a shorter duty cycle allows enough time for armature heat buildup to be dissipated.

Cup- or Shell-Type PM DC Motors

Cup- or shell-type PM DC motors offer low inertia and low inductance as well as high acceleration characteristics, making them useful in many

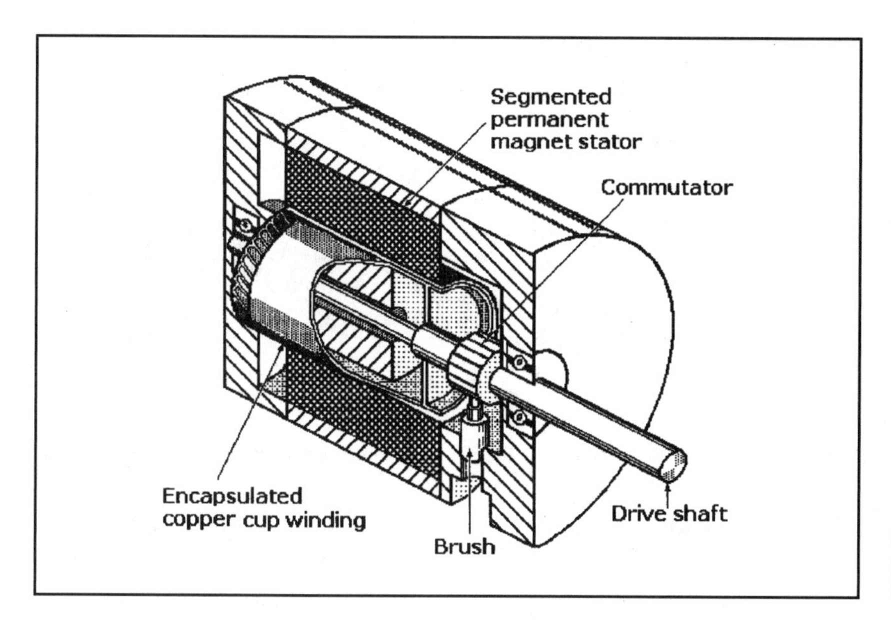

Figure 1-20 Cutaway view of a permanent-magnet DC servomotor with a cup-type armature.

servo applications. They have hollow cylindrical armatures made as aluminum or copper coils bonded by polymer resin and fiberglass to form a rigid "ironless cup," which is fastened to an axial shaft. A cutaway view of this class of servomotor is illustrated in Figure1-20.

Because the armature has no iron core, it, like the disk motor, has extremely low inertia and a very high torque-to-inertia ratio. This permits the motor to accelerate rapidly for the quick response required in many motion-control applications. The armature rotates in an air gap within very high magnetic flux density. The magnetic field from the stationary magnets is completed through the cup-type armature and a stationary ferrous cylindrical core connected to the motor frame. The shaft rotates within the core, which extends into the rotating cup. Spring-brushes commutate these motors.

Another version of a cup-type PM DC motor is shown in the exploded view in Figure 1-21. The cup type armature is rigidly fastened to the shaft by a disk at the right end of the winding, and the magnetic field is also returned through a ferrous metal housing. The brush assembly of this motor is built into its end cap or flange, shown at the far right.

The principal disadvantage of this motor is also the inability of its bonded armature to dissipate internal heat buildup rapidly because of its low thermal conductivity. Without proper cooling and sensitive control circuitry, the armature could be heated to destructive temperatures in seconds.

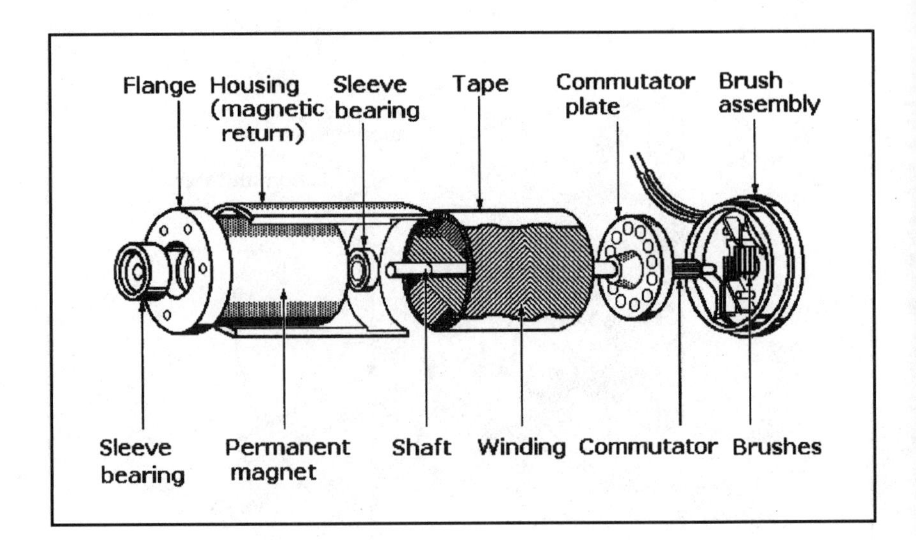

Figure 1-21 Exploded view of a fractional horsepower brush-type DC servomotor.

Brushless PM DC Motors

Brushless DC motors exhibit the same linear speed–torque characteristics as the brush-type PM DC motors, but they are electronically commutated. The construction of these motors, as shown in Figure 1-22, differs from that of a typical brush-type DC motor in that they are "inside-out." In other words, they have permanent magnet rotors instead of stators, and the stators rather than the rotors are wound. Although this geometry is required for brushless DC motors, some manufacturers have adapted this design for brush-type DC motors.

The mechanical brush and bar commutator of the brushless DC motor is replaced by electronic sensors, typically Hall-effect devices (HEDs). They are located within the stator windings and wired to solid-state transistor switching circuitry located either on circuit cards mounted within the motor housings or in external packages. Generally, only fractional horsepower brushless motors have switching circuitry within their housings.

The cylindrical magnet rotors of brushless DC motors are magnetized laterally to form opposing north and south poles across the rotor's diameter. These rotors are typically made from neodymium–iron–boron or samarium–cobalt rare-earth magnetic materials, which offer higher flux densities than alnico magnets. These materials permit motors offering higher performance to be packaged in the same frame sizes as earlier motor designs or those with the same ratings to be packaged in smaller frames than the earlier designs. Moreover, rare-earth or ceramic magnet

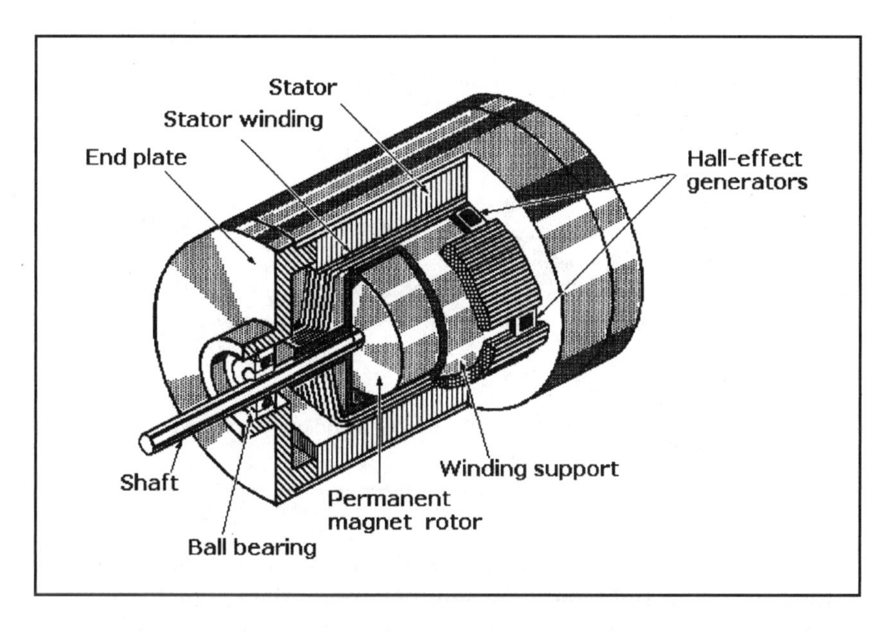

Figure 1-22 Cutaway view of a brushless DC motor.

rotors can be made with smaller diameters than those earlier models with alnico magnets, thus reducing their inertia.

A simplified diagram of a DC brushless motor control with one Hall-effect device (HED) for the electronic commutator is shown in Figure 1-23. The HED is a Hall-effect sensor integrated with an ampli-

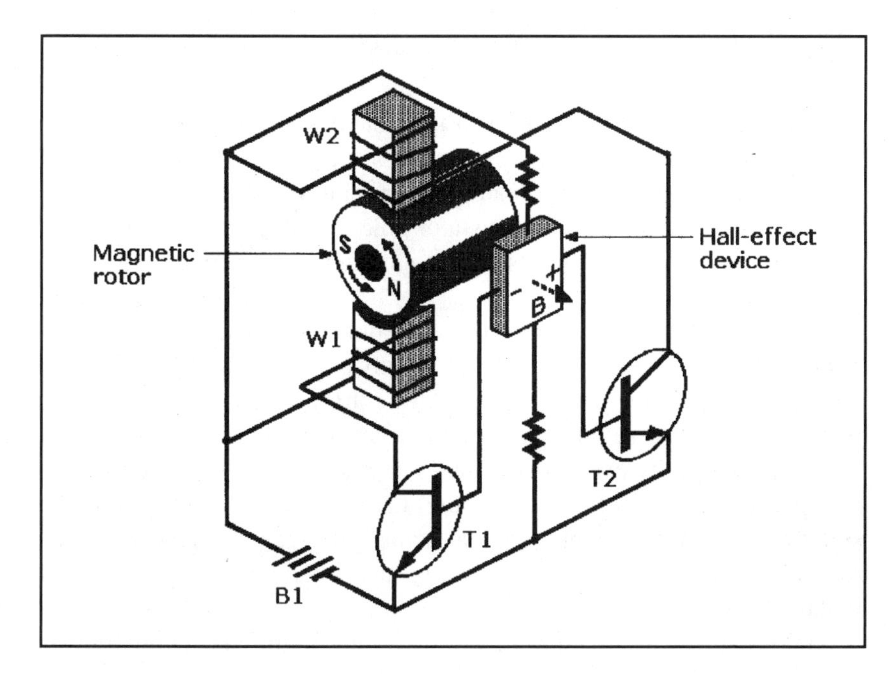

Figure 1-23 Simplified diagram of Hall-effect device (HED) commutation of a brushless DC motor.

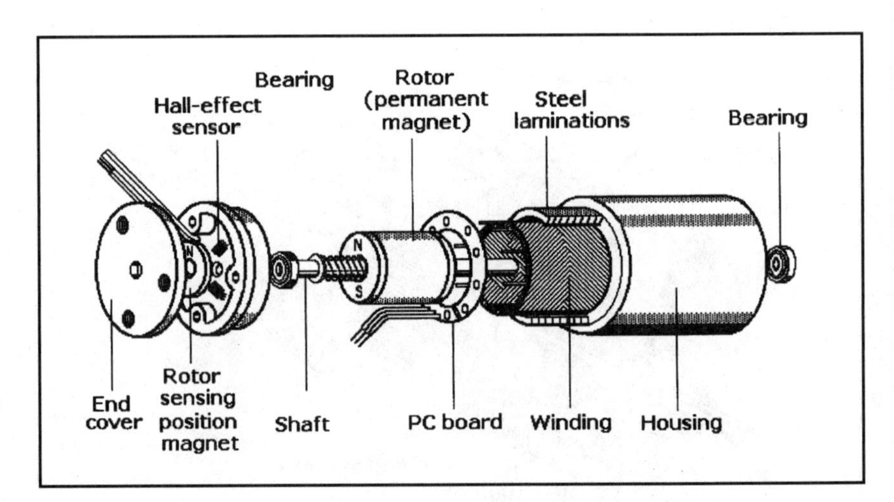

Figure 1-24 Exploded view of a brushless DC motor with Hall-effect device (HED) commutation.

fier in a silicon chip. This IC is capable of sensing the polarity of the rotor's magnetic field and then sending appropriate signals to power transistors T1 and T2 to cause the motor's rotor to rotate continuously. This is accomplished as follows:

1. With the rotor motionless, the HED detects the rotor's north magnetic pole, causing it to generate a signal that turns on transistor T2. This causes current to flow, energizing winding W2 to form a south-seeking electromagnetic rotor pole. This pole then attracts the rotor's north pole to drive the rotor in a counterclockwise (CCW) direction.
2. The inertia of the rotor causes it to rotate past its neutral position so that the HED can then sense the rotor's south magnetic pole. It then switches on transistor T1, causing current to flow in winding W1, thus forming a north-seeking stator pole that attracts the rotor's south pole, causing it to continue to rotate in the CCW direction.

The transistors conduct in the proper sequence to ensure that the excitation in the stator windings W2 and W1 always leads the PM rotor field to produce the torque necessary keep the rotor in constant rotation. The windings are energized in a pattern that rotates around the stator.

There are usually two or three HEDs in practical brushless motors that are spaced apart by 90 or 120° around the motor's rotor. They send the signals to the motion controller that actually triggers the power transistors, which drive the armature windings at a specified motor current and voltage level.

The brushless motor in the exploded view Figure 1-24 illustrates a design for a miniature brushless DC motor that includes Hall-effect commutation. The stator is formed as an ironless sleeve of copper coils bonded together in polymer resin and fiberglass to form a rigid structure similar to cup-type rotors. However, it is fastened inside the steel laminations within the motor housing.

This method of construction permits a range of values for starting current and specific speed (rpm/V) depending on wire gauge and the number of turns. Various terminal resistances can be obtained, permitting the user to select the optimum motor for a specific application. The Hall-effect sensors and a small magnet disk that is magnetized widthwise are mounted on a disk-shaped partition within the motor housing.

Position Sensing in Brushless Motors

Both magnetic sensors and resolvers can sense rotor position in brushless motors. The diagram in Figure 1-25 shows how three magnetic sensors can sense rotor position in a three-phase electronically commutated brushless DC motor. In this example the magnetic sensors are located inside the end-bell of the motor. This inexpensive version is adequate for simple controls.

In the alternate design shown in Figure 1-26, a resolver on the end cap of the motor is used to sense rotor position when greater positioning accuracy is required. The high-resolution signals from the resolver can

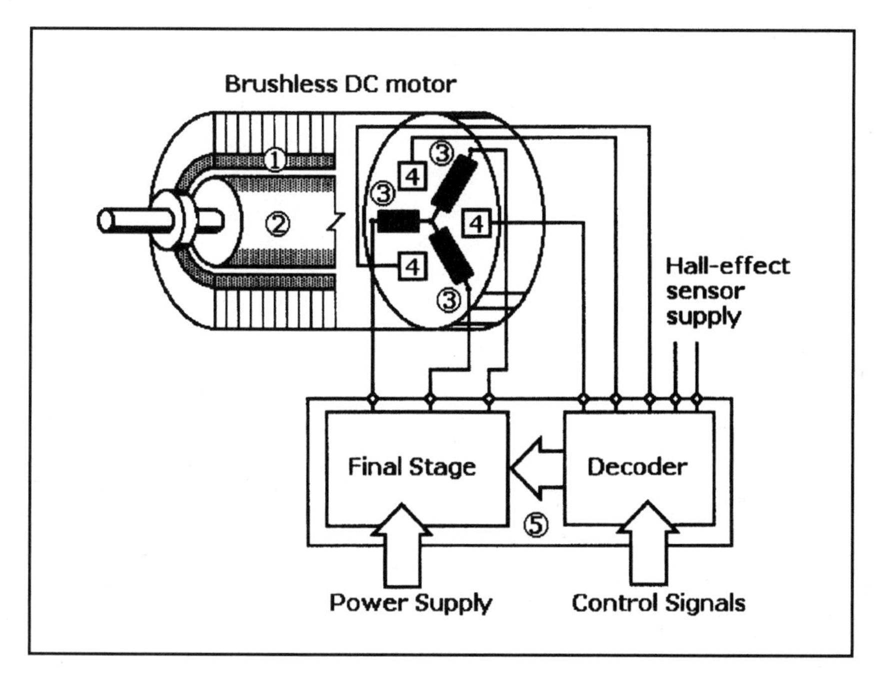

Figure 1-25 A magnetic sensor as a rotor position indicator: stationary brushless motor winding (1), permanent-magnet motor rotor (2), three-phase electronically commutated field (3), three magnetic sensors (4), and the electronic circuit board (5).

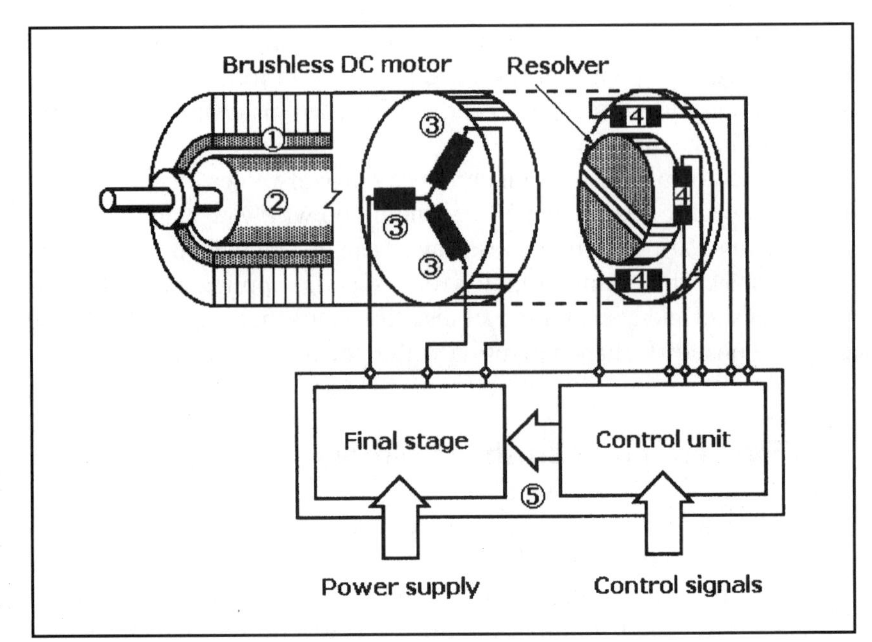

Figure 1-26 A resolver as a rotor position indicator: stationary motor winding (1), permanent-magnet motor rotor (2), three-phase electronically commutated field (3), three magnetic sensors (4), and the electronic circuit board (5).

be used to generate sinusoidal motor currents within the motor controller. The currents through the three motor windings are position independent and respectively 120° phase shifted.

Brushless Motor Advantages

Brushless DC motors have at least four distinct advantages over brush-type DC motors that are attributable to the replacement of mechanical commutation by electronic commutation.

- There is no need to replace brushes or remove the gritty residue caused by brush wear from the motor.
- Without brushes to cause electrical arcing, brushless motors do not present fire or explosion hazards in an environment where flammable or explosive vapors, dust, or liquids are present.
- Electromagnetic interference (EMI) is minimized by replacing mechanical commutation, the source of unwanted radio frequencies, with electronic commutation.
- Brushless motors can run faster and more efficiently with electronic commutation. Speeds of up to 50,000 rpm can be achieved vs. the upper limit of about 5000 rpm for brush-type DC motors.

Brushless DC Motor Disadvantages

There are at least four disadvantages of brushless DC servomotors.

- Brushless PM DC servomotors cannot be reversed by simply reversing the polarity of the power source. The order in which the current is fed to the field coil must be reversed.
- Brushless DC servomotors cost more than comparably rated brush-type DC servomotors.
- Additional system wiring is required to power the electronic commutation circuitry.
- The motion controller and driver electronics needed to operate a brushless DC servomotor are more complex and expensive than those required for a conventional DC servomotor.

Consequently, the selection of a brushless motor is generally justified on a basis of specific application requirements or its hazardous operating environment.

Characteristics of Brushless Rotary Servomotors

It is difficult to generalize about the characteristics of DC rotary servomotors because of the wide range of products available commercially. However, they typically offer continuous torque ratings of 0.62 lb-ft (0.84 N-m) to 5.0 lb-ft (6.8 N-m), peak torque ratings of 1.9 lb-ft (2.6 N-m) to 14 lb-ft (19 N-m), and continuous power ratings of 0.73 hp (0.54 kW) to 2.76 hp (2.06 kW). Maximum speeds can vary from 1400 to 7500 rpm, and the weight of these motors can be from 5.0 lb (2.3 kg) to 23 lb (10 kg). Feedback typically can be either by resolver or encoder.

Linear Servomotors

A linear motor is essentially a rotary motor that has been opened out into a flat plane, but it operates on the same principles. A permanent-magnet DC linear motor is similar to a permanent-magnet rotary motor, and an AC induction squirrel cage motor is similar to an induction linear motor. The same electromagnetic force that produces torque in a rotary motor also produces torque in a linear motor. Linear motors use the same controls and programmable position controllers as rotary motors.

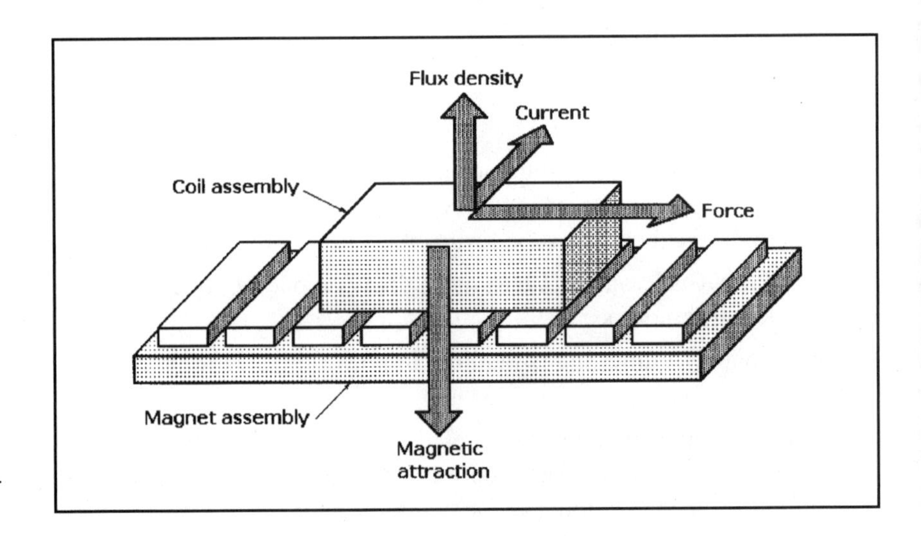

Figure 1-27 Operating principles of a linear servomotor.

Before the invention of linear motors, the only way to produce linear motion was to use pneumatic or hydraulic cylinders, or to translate rotary motion to linear motion with ballscrews or belts and pulleys.

A linear motor consists of two mechanical assemblies: *coil* and *magnet*, as shown in Figure 1-27. Current flowing in a winding in a magnetic flux field produces a force. The copper windings conduct current (I), and the assembly generates magnetic flux density (B). When the current and flux density interact, a force (F) is generated in the direction shown in Figure 1-27, where $F = I \times B$.

Even a small motor will run efficiently, and large forces can be created if a large number of turns are wound in the coil and the magnets are powerful rare-earth magnets. The windings are phased 120 electrical degrees apart, and they must be continually switched or commutated to sustain motion.

Only brushless linear motors for closed-loop servomotor applications are discussed here. Two types of these motors are available commercially—*steel-core* (also called *iron-core*) and *epoxy-core* (also called *ironless*). Each of these linear servomotors has characteristics and features that are optimal in different applications

The coils of steel-core motors are wound on silicon steel to maximize the generated force available with a single-sided magnet assembly or way. Figure 1-28 shows a steel-core brushless linear motor. The steel in these motors focuses the magnetic flux to produce very high force density. The magnet assembly consists of rare-earth bar magnets mounted on the upper surface of a steel base plate arranged to have alternating polarities (i.e., N, S, N, S)

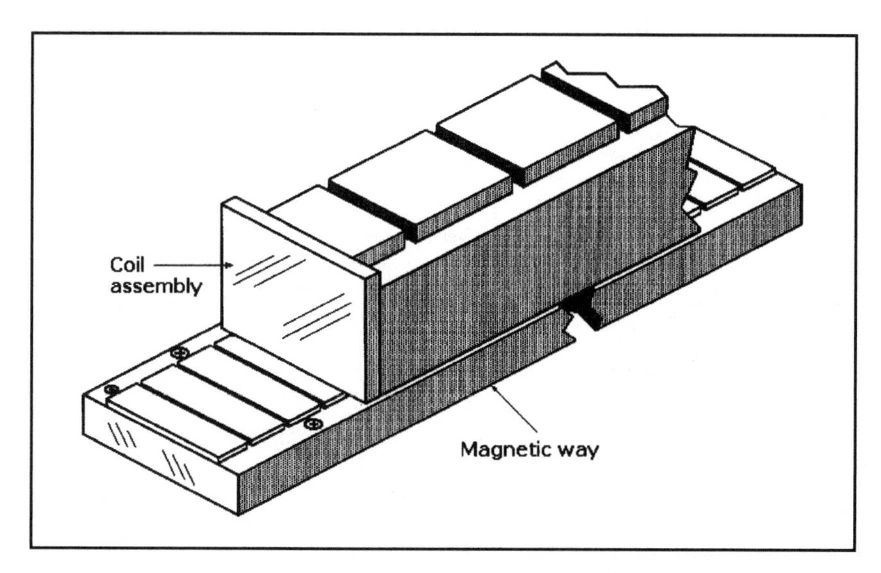

Figure 1-28 A linear iron-core linear servomotor consists of a magnetic way and a mating coil assembly.

The steel in the cores is attracted to the permanent magnets in a direction that is perpendicular (normal) to the operating motor force. The magnetic flux density within the air gap of linear motors is typically several thousand gauss. A constant magnetic force is present whether or not the motor is energized. The normal force of the magnetic attraction can be up to ten times the continuous force rating of the motor. This flux rapidly diminishes to a few gauss as the measuring point is moved a few centimeters away from the magnets.

Cogging is a form of magnetic "detenting" that occurs in both linear and rotary motors when the motor coil's steel laminations cross the alternating poles of the motor's magnets. Because it can occur in steel-core motors, manufacturers include features that minimize cogging. The high thrust forces attainable with steel-core linear motors permit them to accelerate and move heavy masses while maintaining stiffness during machining or process operations.

The features of epoxy-core or ironless-core motors differ from those of the steel-core motors. For example, their coil assemblies are wound and encapsulated within epoxy to form a thin plate that is inserted in the air gap between the two permanent-magnet strips fastened inside the magnet assembly, as shown in Figure 1-29. Because the coil assemblies do not contain steel cores, epoxy-core motors are lighter than steel-core motors and less subject to cogging.

The strip magnets are separated to form the air gap into which the coil assembly is inserted. This design maximizes the generated thrust force and also provides a flux return path for the magnetic circuit. Con-

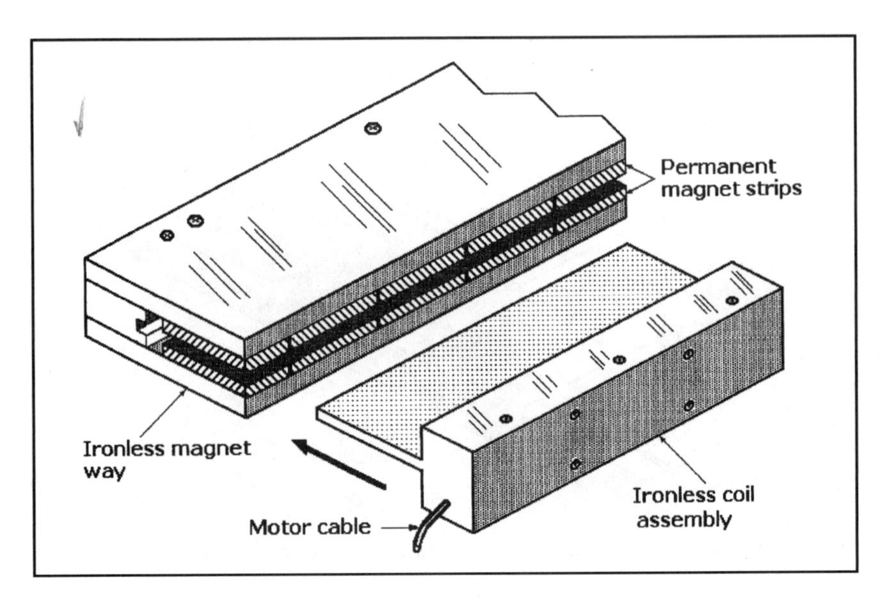

Figure 1-29 A linear ironless servomotor consists of an ironless magnetic way and an ironless coil assembly.

sequently, very little magnetic flux exists outside the motor, thus minimizing residual magnetic attraction.

Epoxy-core motors provide exceptionally smooth motion, making them suitable for applications requiring very low bearing friction and high acceleration of light loads. They also permit constant velocity to be maintained, even at very low speeds.

Linear servomotors can achieve accuracies of 0.1 μm. Normal accelerations are 2 to 3 *g*, but some motors can reach 15 *g*. Velocities are limited by the encoder data rate and the amplifier voltage. Normal peak velocities are from 0.04 in./s (1 mm/s) to about 6.6 ft/s (2 m/s), but the velocity of some models can exceed 26 ft/s (8 m/s).

Ironless linear motors can have continuous force ratings from about 5 to 55 lbf (22 to 245 N) and peak force ratings from about 25 to 180 lbf (110 to 800 N). By contrast, iron-core linear motors are available with continuous force ratings of about 30 to 1100 lbf (130 to 4900 N) and peak force ratings of about 60 to 1800 lbf (270 to 8000 N).

Commutation

The linear motor windings that are phased 120° apart must be continually switched or commutated to sustain motion. There are two ways to commutate linear motors: *sinusoidal* and *Hall-effect device (HED)*, or *trapezoidal*. The highest motor efficiency is achieved with sinusoidal commutation, while HED commutation is about 10 to 15% less efficient.

In sinusoidal commutation, the linear encoder that provides position feedback in the servosystem is also used to commutate the motor. A process called "phase finding" is required when the motor is turned on, and the motor phases are then incrementally advanced with each encoder pulse. This produces extremely smooth motion. In HED commutation a circuit board containing Hall-effect ICs is embedded in the coil assembly. The HED sensors detect the polarity change in the magnet track and switch the motor phases every 60º.

Sinusoidal commutation is more efficient than HED commutation because the coil windings in motors designed for this commutation method are configured to provide a sinusoidally shaped back EMF waveform. As a result, the motors produce a constant force output when the driving voltage on each phase matches the characteristic back EMF waveform.

Installation of Linear Motors

In a typical linear motor application the coil assembly is attached to the moving member of the host machine and the magnet assembly is mounted on the nonmoving base or frame. These motors can be mounted vertically, but if they are they typically require a counterbalance system to prevent the load from dropping if power temporarily fails or is routinely shut off. The counterbalance system, typically formed from pulleys and weights, springs, or air cylinders, supports the load against the force of gravity.

If power is lost, servo control is interrupted. Stages in motion tend to stay in motion while those at rest tend to stay at rest. The stopping time and distance depend on the stage's initial velocity and system friction. The motor's back EMF can provide dynamic braking, and friction brakes can be used to attenuate motion rapidly. However, positive stops and travel limits can be built into the motion stage to prevent damage in situations where power or feedback might be lost or the controller or servo driver fail.

Linear servomotors are supplied to the customer in kit form for mounting on the host machine. The host machine structure must include bearings capable of supporting the mass of the motor parts while maintaining the specified air gap between the assemblies and also resisting the normal force of any residual magnetic attraction.

Linear servomotors must be used in closed loop positioning systems because they do not include built-in means for position sensing. Feedback is typically supplied by such sensors as linear encoders, laser interferometers, LVDTs, or linear Inductosyns.

Advantages of Linear vs. Rotary Servomotors

The advantages of linear servomotors over rotary servomotors include:

- *High stiffness:* The linear motor is connected directly to the moving load, so there is no backlash and practically no compliance between the motor and the load. The load moves instantly in response to motor motion.
- *Mechanical simplicity:* The coil assembly is the only moving part of the motor, and its magnet assembly is rigidly mounted to a stationary structure on the host machine. Some linear motor manufacturers offer modular magnetic assemblies in various modular lengths. This permits the user to form a track of any desired length by stacking the modules end to end, allowing virtually unlimited travel. The force produced by the motor is applied directly to the load without any couplings, bearings, or other conversion mechanisms. The only alignments required are for the air gaps, which typically are from 0.039 in. (1 mm) to 0.020 in. (0.5 mm).
- *High accelerations and velocities:* Because there is no physical contact between the coil and magnet assemblies, high accelerations and velocities are possible. Large motors are capable of accelerations of 3 to 5 *g*, but smaller motors are capable of more than 10 *g*.
- *High velocities:* Velocities are limited by feedback encoder data rate and amplifier bus voltage. Normal peak velocities are up to 6.6 ft/s (2 m/s), although some models can reach 26 ft/s (8 m/s). This compares with typical linear speeds of ballscrew transmissions, which are commonly limited to 20 to 30 in./s (0.5 to 0.7 m/s) because of resonances and wear.
- *High accuracy and repeatability:* Linear motors with position feedback encoders can achieve positioning accuracies of ±1 encoder cycle or submicrometer dimensions, limited only by encoder feedback resolution.
- *No backlash or wear:* With no contact between moving parts, linear motors do not wear out. This minimizes maintenance and makes them suitable for applications where long life and long-term peak performance are required.
- *System size reduction:* With the coil assembly attached to the load, no additional space is required. By contrast, rotary motors typically require ballscrews, rack-and-pinion gearing, or timing belt drives.
- *Clean room compatibility:* Linear motors can be used in clean rooms because they do not need lubrication and do not produce carbon brush grit.

Coil Assembly Heat Dissipation

Heat control is more critical in linear motors than in rotary motors because they do not have the metal frames or cases that can act as large heat-dissipating surfaces. Some rotary motors also have radiating fins on their frames that serve as heatsinks to augment the heat dissipation capability of the frames. Linear motors must rely on a combination of high motor efficiency and good thermal conduction from the windings to a heat-conductive, electrically isolated mass. For example, an aluminum attachment bar placed in close contact with the windings can aid in heat dissipation. Moreover, the carriage plate to which the coil assembly is attached must have effective heat-sinking capability.

Stepper Motors

A *stepper* or *stepping motor* is an AC motor whose shaft is indexed through part of a revolution or *step angle* for each DC pulse sent to it. Trains of pulses provide input current to the motor in increments that can "step" the motor through 360°, and the actual angular rotation of the shaft is directly related to the number of pulses introduced. The position of the load can be determined with reasonable accuracy by counting the pulses entered.

The stepper motors suitable for most open-loop motion control applications have wound stator fields (electromagnetic coils) and iron or permanent magnet (PM) rotors. Unlike PM DC servomotors with mechanical brush-type commutators, stepper motors depend on external controllers to provide the switching pulses for commutation. Stepper motor operation is based on the same electromagnetic principles of attraction and repulsion as other motors, but their commutation provides only the torque required to turn their rotors.

Pulses from the external motor controller determine the amplitude and direction of current flow in the stator's field windings, and they can turn the motor's rotor either clockwise or counterclockwise, stop and start it quickly, and hold it securely at desired positions. Rotational shaft speed depends on the frequency of the pulses. Because controllers can step most motors at audio frequencies, their rotors can turn rapidly.

Between the application of pulses when the rotor is at rest, its armature will not drift from its stationary position because of the stepper motor's inherent holding ability or *detent torque*. These motors generate very little heat while at rest, making them suitable for many different instrument drive-motor applications in which power is limited.

The three basic kinds of stepper motors are *permanent magnet, variable reluctance,* and *hybrid.* The same controller circuit can drive both hybrid and PM stepping motors.

Permanent-Magnet (PM) Stepper Motors

Permanent-magnet stepper motors have smooth armatures and include a permanent magnet core that is magnetized widthwise or perpendicular to its rotation axis. These motors usually have two independent windings, with or without center taps. The most common step angles for PM motors are 45 and 90°, but motors with step angles as fine as 1.8° per step as well as 7.5, 15, and 30° per step are generally available. Armature rotation occurs when the stator poles are alternately energized and deenergized to create torque. A 90° stepper has four poles and a 45° stepper has eight poles, and these poles must be energized in sequence. Permanent-magnet steppers step at relatively low rates, but they can produce high torques and they offer very good damping characteristics.

Variable Reluctance Stepper Motors

Variable reluctance (VR) stepper motors have multitooth armatures with each tooth effectively an individual magnet. At rest these magnets align themselves in a natural detent position to provide larger holding torque than can be obtained with a comparably rated PM stepper. Typical VR motor step angles are 15 and 30° per step. The 30° angle is obtained with a 4-tooth rotor and a 6-pole stator, and the 15° angle is achieved with an 8-tooth rotor and a 12-pole stator. These motors typically have three windings with a common return, but they are also available with four or five windings. To obtain continuous rotation, power must be applied to the windings in a coordinated sequence of alternately deenergizing and energizing the poles.

If just one winding of either a PM or VR stepper motor is energized, the rotor (under no load) will snap to a fixed angle and hold that angle until external torque exceeds the holding torque of the motor. At that point, the rotor will turn, but it will still try to hold its new position at each successive equilibrium point.

Hybrid Stepper Motors

The hybrid stepper motor combines the best features of VR and PM stepper motors. A cutaway view of a typical industrial-grade hybrid stepper

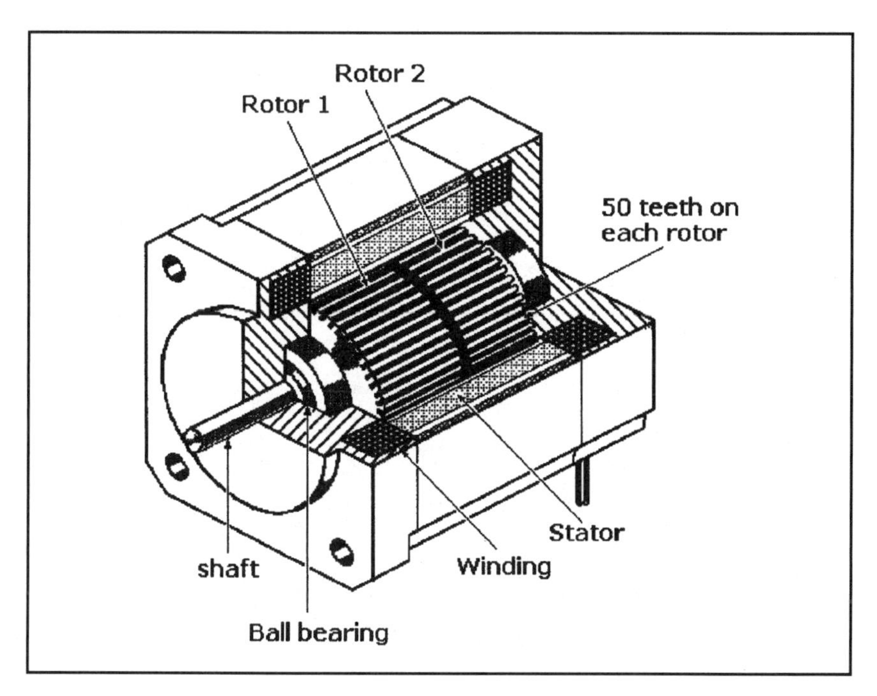

Figure 1-30 Cutaway view of a 5-phase hybrid stepping motor. A permanent magnet is within the rotor assembly, and the rotor segments are offset from each other by 3.5°.

motor with a multitoothed armature is shown in Figure 1-30. The armature is built in two sections, with the teeth in the second section offset from those in the first section. These motors also have multitoothed stator poles that are not visible in the figure. Hybrid stepper motors can achieve high stepping rates, and they offer high detent torque and excellent dynamic and static torque.

Hybrid steppers typically have two windings on each stator pole so that each pole can become either magnetic north or south, depending on current flow. A cross-sectional view of a hybrid stepper motor illustrating the multitoothed poles with dual windings per pole and the multitoothed rotor is illustrated in Figure 1-31. The shaft is represented by the central circle in the diagram.

The most popular hybrid steppers have 3- and 5-phase wiring, and step angles of 1.8 and 3.6° per step. These motors can provide more torque from a given frame size than other stepper types because either all or all but one of the motor windings are energized at every point in the drive cycle. Some 5-phase motors have high resolutions of 0.72° per step (500 steps per revolution). With a compatible controller, most PM and hybrid motors can be run in half-steps, and some controllers are designed to provide smaller fractional steps, or *microsteps*. Hybrid stepper motors capable of a wide range of torque values are available commercially. This range is achieved by scaling length and diameter dimensions.

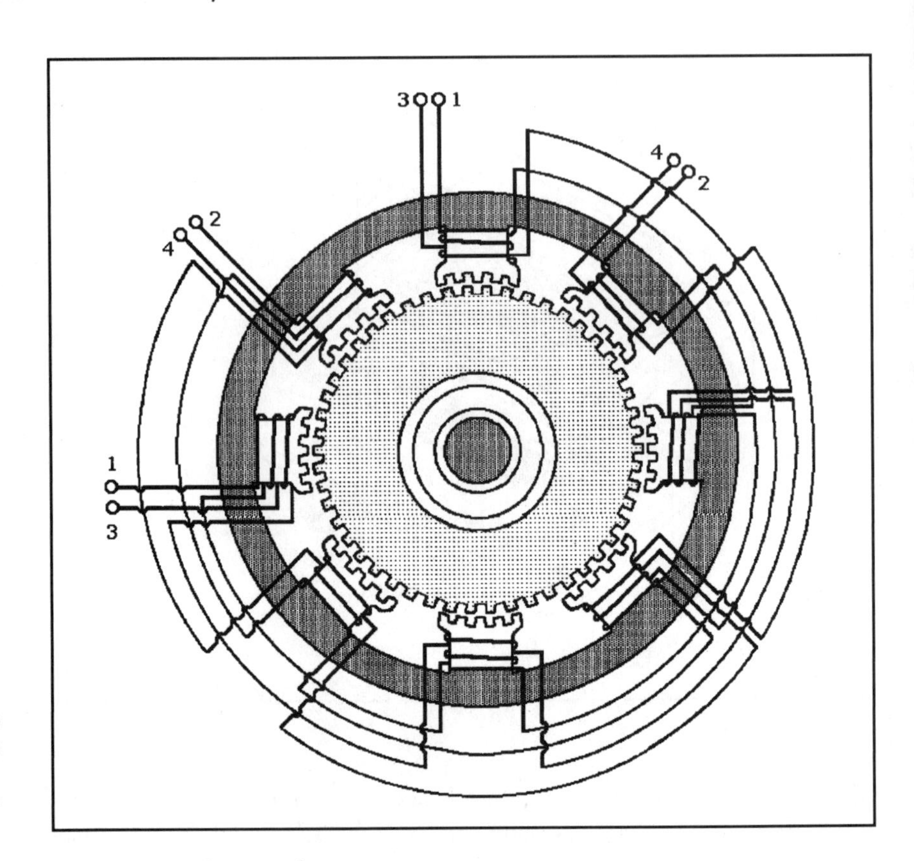

Figure 1-31 Cross-section of a hybrid stepping motor showing the segments of the magnetic-core rotor and stator poles with its wiring diagram.

Hybrid stepper motors are available in NEMA size 17 to 42 frames, and output power can be as high as 1000 W peak.

Stepper Motor Applications

Many different technical and economic factors must be considered in selecting a hybrid stepper motor. For example, the ability of the stepper motor to repeat the positioning of its multitoothed rotor depends on its geometry. A disadvantage of the hybrid stepper motor operating open-loop is that, if overtorqued, its position "memory" is lost and the system must be reinitialized. Stepper motors can perform precise positioning in simple open-loop control systems if they operate at low acceleration rates with static loads. However, if higher acceleration values are required for driving variable loads, the stepper motor must be operated in a closed loop with a position sensor.

DC and AC Motor Linear Actuators

Actuators for motion control systems are available in many different forms, including both linear and rotary versions. One popular configuration is that of a Thomson Saginaw PPA, shown in section view in Figure 1-32. It consists of an AC or DC motor mounted parallel to either a ballscrew or Acme screw assembly through a reduction gear assembly with a slip clutch and integral brake assembly. Linear actuators of this type can perform a wide range of commercial, industrial, and institutional applications.

One version designed for mobile applications can be powered by a 12-, 24-, or 36-VDC permanent-magnet motor. These motors are capable of performing such tasks as positioning antenna reflectors, opening and closing security gates, handling materials, and raising and lowering scissors-type lift tables, machine hoods, and light-duty jib crane arms.

Other linear actuators are designed for use in fixed locations where either 120- or 220-VAC line power is available. They can have either AC or DC motors. Those with 120-VAC motors can be equipped with optional electric brakes that virtually eliminate coasting, thus permitting point-to-point travel along the stroke.

Where variable speed is desired and 120-VAC power is available, a linear actuator with a 90-VDC motor can be equipped with a solid-state rectifier/speed controller. Closed-loop feedback provides speed regulation down to one tenth of the maximum travel rate. This feedback system can maintain its selected travel rate despite load changes.

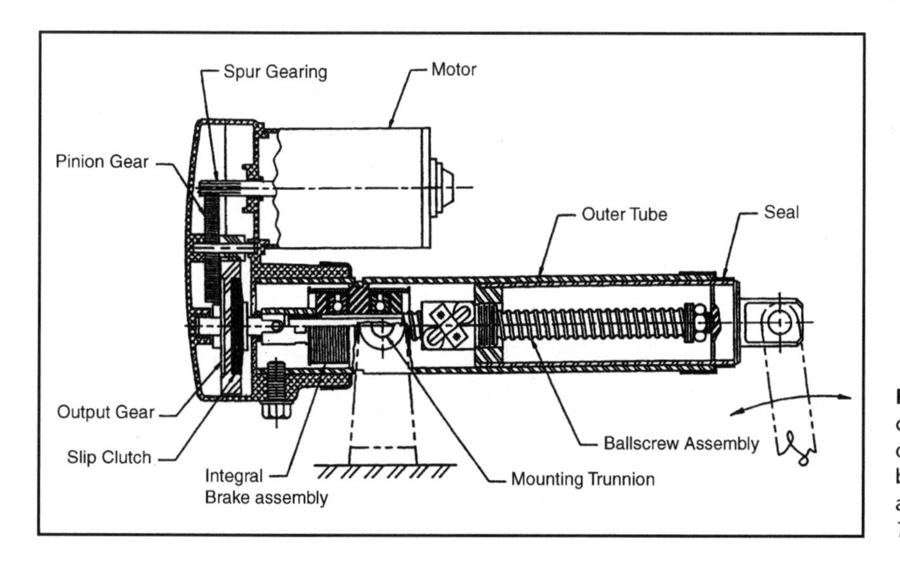

Figure 1-32 This linear actuator can be powered by either an AC or DC motor. It contains ballscrew, reduction gear, clutch, and brake assemblies. *Courtesy of Thomson Saginaw.*

Thomson Saginaw also offers its linear actuators with either Hall-effect or potentiometer sensors for applications where it is necessary or desirable to control actuator positioning. With Hall-effect sensing, six pulses are generated with each turn of the output shaft during which the stroke travels approximately $\frac{1}{32}$ in. (0.033 in. or 0.84 mm). These pulses can be counted by a separate control unit and added or subtracted from the stored pulse count in the unit's memory. The actuator can be stopped at any 0.033-in. increment of travel along the stroke selected by programming. A limit switch can be used together with this sensor.

If a 10-turn, 10,000-ohm potentiometer is used as a sensor, it can be driven by the output shaft through a spur gear. The gear ratio is established to change the resistance from 0 to 10,000 ohms over the length of the actuator stroke. A separate control unit measures the resistance (or voltage) across the potentiometer, which varies continuously and linearly with stroke travel. The actuator can be stopped at any position along its stroke.

Stepper-Motor Based Linear Actuators

Linear actuators are available with axial integral threaded shafts and bolt nuts that convert rotary motion to linear motion. Powered by fractional horsepower permanent-magnet stepper motors, these linear actuators are capable of positioning light loads. Digital pulses fed to the actuator cause the threaded shaft to rotate, advancing or retracting it so that a load coupled to the shaft can be moved backward or forward. The bidirectional digital linear actuator shown in Figure 1-33 can provide linear res-

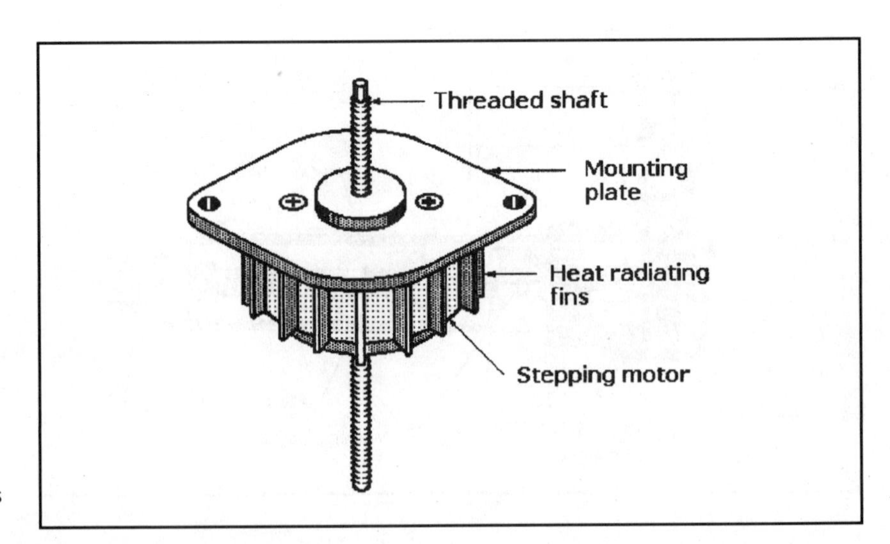

Figure 1-33 This light-duty linear actuator based on a permanent-magnet stepping motor has a shaft that advances or retracts.

olution as fine as 0.001 in. per pulse. Travel per step is determined by the pitch of the leadscrew and step angle of the motor. The maximum linear force for the model shown is 75 oz.

SERVOSYSTEM FEEDBACK SENSORS

A servosystem feedback sensor in a motion control system transforms a physical variable into an electrical signal for use by the motion controller. Common feedback sensors are encoders, resolvers, and linear variable differential transformers (LVDTs) for motion and position feedback, and tachometers for velocity feedback. Less common but also in use as feedback devices are potentiometers, linear velocity transducers (LVTs), angular displacement transducers (ADTs), laser interferometers, and potentiometers. Generally speaking, the closer the feedback sensor is to the variable being controlled, the more accurate it will be in assisting the system to correct velocity and position errors.

For example, direct measurement of the linear position of the carriage carrying the load or tool on a single-axis linear guide will provide more accurate feedback than an indirect measurement determined from the angular position of the guide's leadscrew and knowledge of the drivetrain geometry between the sensor and the carriage. Thus, direct position measurement avoids drivetrain errors caused by backlash, hysteresis, and leadscrew wear that can adversely affect indirect measurement.

Rotary Encoders

Rotary encoders, also called *rotary shaft encoders* or *rotary shaft-angle* encoders, are electromechanical transducers that convert shaft rotation into output pulses, which can be counted to measure shaft revolutions or shaft angle. They provide rate and positioning information in servo feedback loops. A rotary encoder can sense a number of discrete positions per revolution. The number is called *points per revolution* and is analogous to the *steps per revolution* of a stepper motor. The speed of an encoder is in units of counts per second. Rotary encoders can measure the motor-shaft or leadscrew angle to report position indirectly, but they can also measure the response of rotating machines directly.

The most popular rotary encoders are *incremental optical shaft-angle encoders* and the *absolute optical shaft-angle encoders*. There are also *direct contact* or *brush-type* and *magnetic rotary encoders*, but they are not as widely used in motion control systems.

Commercial rotary encoders are available as standard or catalog units, or they can be custom made for unusual applications or survival in extreme environments. Standard rotary encoders are packaged in cylindrical cases with diameters from 1.5 to 3.5 in. Resolutions range from 50 cycles per shaft revolution to 2,304,000 counts per revolution. A variation of the conventional configuration, the *hollow-shaft encoder,* eliminates problems associated with the installation and shaft runout of conventional models. Models with hollow shafts are available for mounting on shafts with diameters of 0.04 to 1.6 in. (1 to 40 mm).

Incremental Encoders

The basic parts of an incremental optical shaft-angle encoder are shown in Figure 1-34. A glass or plastic code disk mounted on the encoder shaft rotates between an internal light source, typically a light-emitting diode (LED), on one side and a mask and matching photodetector assembly on the other side. The incremental code disk contains a pattern of equally spaced opaque and transparent segments or spokes that radiate out from its center as shown. The electronic signals that are generated by the encoder's electronics board are fed into a motion controller that calculates position and velocity information for feedback purposes. An exploded view of an industrial-grade incremental encoder is shown in Figure 1-35.

Glass code disks containing finer graduations capable of 11- to more than 16-bit resolution are used in high-resolution encoders, and plastic (Mylar) disks capable of 8- to 10-bit resolution are used in the more rugged encoders that are subject to shock and vibration.

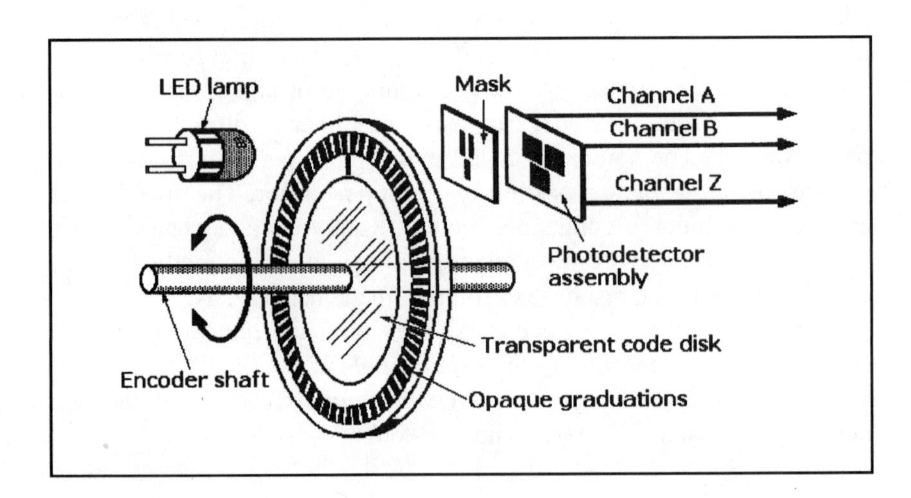

Figure 1-34 Basic elements of an incremental optical rotary encoder.

The quadrature encoder is the most common type of incremental encoder. Light from the LED passing through the rotating code disk and mask is "chopped" before it strikes the photodetector assembly. The output signals from the assembly are converted into two channels of square pulses (A and B) as shown in Figure 1-36. The number of square pulses in each channel is equal to the number of code disk segments that pass the photodetectors as the disk rotates, but the waveforms are 90° out of phase. If, for example, the pulses in channel A lead those in channel B, the disk is rotating in a clockwise direction, but if the pulses in channel A lag those in channel B lead, the disk is rotating counterclockwise. By monitoring both the number of pulses and the relative phases of signals A and B, both position and direction of rotation can be determined.

Many incremental quadrature encoders also include a third output Z channel to obtain a zero reference or index signal that occurs once per revolution. This channel can be gated to the A and B quadrature channels and used to trigger certain events accurately within the system. The signal can also be used to align the encoder shaft to a mechanical reference.

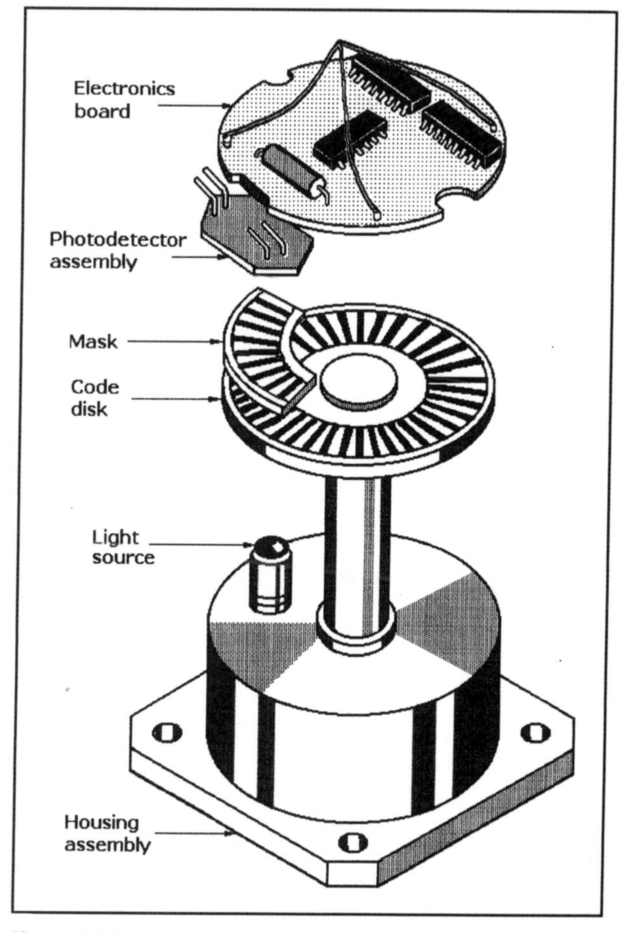

Figure 1-35 Exploded view of an incremental optical rotary encoder showing the stationary mask between the code wheel and the photodetector assembly.

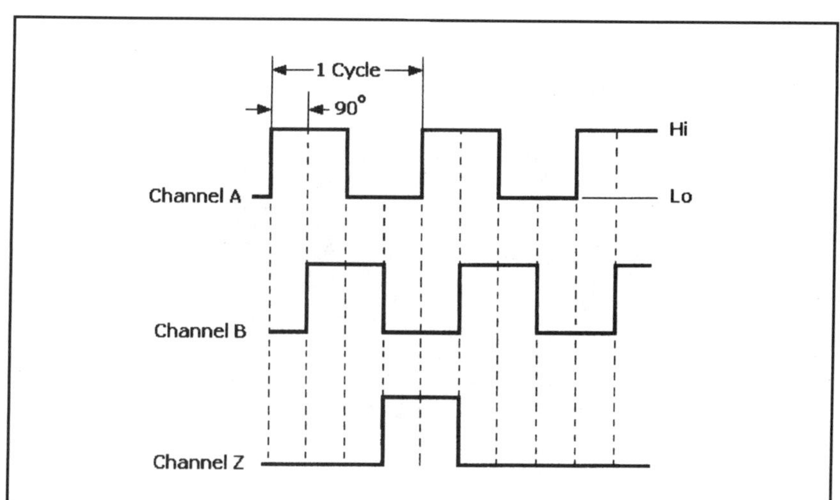

Figure 1-36 Channels A and B provide bidirectional position sensing. If channel A leads channel B, the direction is clockwise; if channel B leads channel A, the direction is counterclockwise. Channel Z provides a zero reference for determining the number of disk rotations.

Absolute Encoders

An *absolute shaft-angle optical encoder* contains multiple light sources and photodetectors, and a code disk with up to 20 tracks of segmented patterns arranged as annular rings, as shown in Figure 1-37. The code disk provides a binary output that uniquely defines each shaft angle, thus providing an absolute measurement. This type of encoder is organized in essentially the same way as the incremental encoder shown in Figure 1-35, but the code disk rotates between linear arrays of LEDs and photodetectors arranged radially, and a LED opposes a photodetector for each track or annular ring.

The arc lengths of the opaque and transparent sectors decrease with respect to the radial distance from the shaft. These disks, also made of glass or plastic, produce either the natural binary or Gray code. Shaft position accuracy is proportional to the number of annular rings or tracks on the disk. When the code disk rotates, light passing through each track or annular ring generates a continuous stream of signals from the detector array. The electronics board converts that output into a binary word. The value of the output code word is read radially from the most significant bit (MSB) on the inner ring of the disk to the least significant bit (LSB) on the outer ring of the disk.

The principal reason for selecting an absolute encoder over an incremental encoder is that its code disk retains the last angular position of the encoder shaft whenever it stops moving, whether the system is shut down deliberately or as a result of power failure. This means that the last readout is preserved, an important feature for many applications.

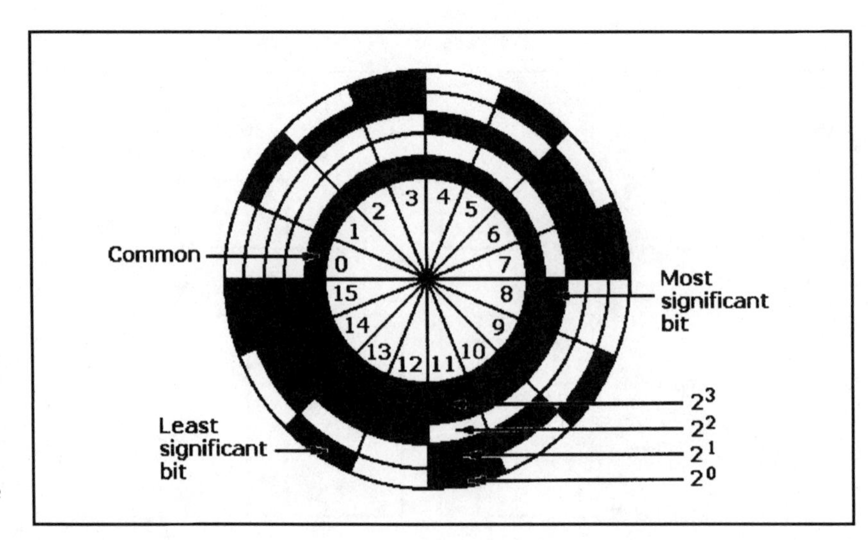

Figure 1-37 Binary-code disk for an absolute optical rotary encoder. Opaque sectors represent a binary value of 1, and the transparent sectors represent binary 0. This four-bit binary-code disk can count from 1 to 15.

Linear Encoders

Linear encoders can make direct accurate measurements of unidirectional and reciprocating motions of mechanisms with high resolution and repeatability. Figure 1-38 illustrates the basic parts of an optical linear encoder. A movable scanning unit contains the light source, lens, graduated glass scanning reticule, and an array of photocells. The scale, typically made as a strip of glass with opaque graduations, is bonded to a supporting structure on the host machine.

A beam of light from the light source passes through the lens, four windows of the scanning reticule, and the glass scale to the array of photocells. When the scanning unit moves, the scale modulates the light beam so that the photocells generate sinusoidal signals.

The four windows in the scanning reticule are each 90° apart in phase. The encoder combines the phase-shifted signal to produce two symmetrical sinusoidal outputs that are phase shifted by 90°. A fifth pattern on the scanning reticule has a random graduation that, when aligned with an identical reference mark on the scale, generates a reference signal.

A fine-scale pitch provides high resolution. The spacing between the scanning reticule and the fixed scale must be narrow and constant to eliminate undesirable diffraction effects of the scale grating. The complete scanning unit is mounted on a carriage that moves on ball bearings along the glass scale. The scanning unit is connected to the host machine

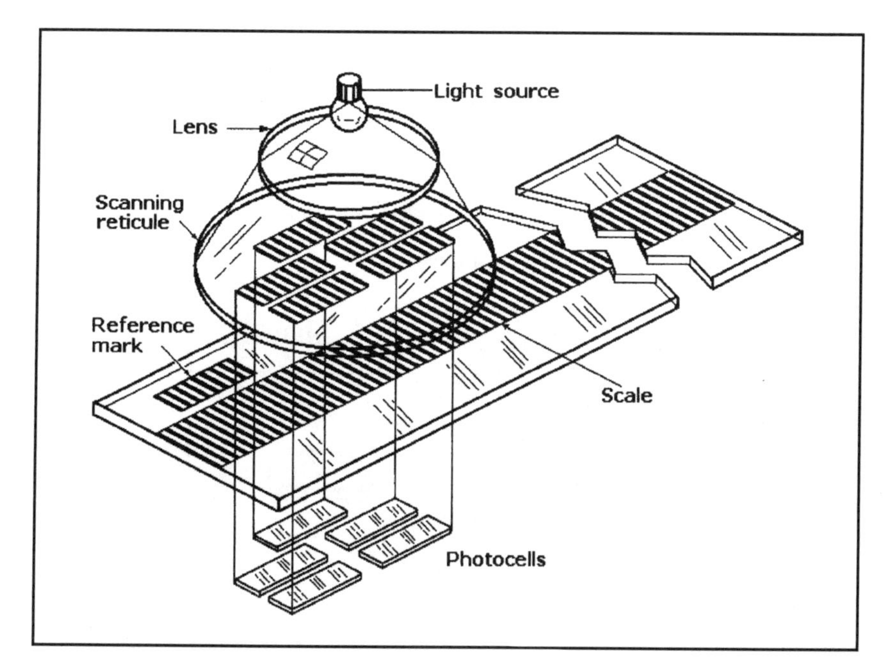

Figure 1-38 Optical linear encoders direct light through a moving glass scale with accurately etched graduations to photocells on the opposite side for conversion to a distance value.

slide by a coupling that compensates for any alignment errors between the scale and the machine guideways.

External electronic circuitry interpolates the sinusoidal signals from the encoder head to subdivide the line spacing on the scale so that it can measure even smaller motion increments. The practical maximum length of linear encoder scales is about 10 ft (3 m), but commercial catalog models are typically limited to about 6 ft (2 m). If longer distances are to be measured, the encoder scale is made of steel tape with reflective graduations that are sensed by an appropriate photoelectric scanning unit.

Linear encoders can make direct measurements that overcome the inaccuracies inherent in mechanical stages due to backlash, hysteresis, and leadscrew error. However, the scale's susceptibility to damage from metallic chips, grit oil, and other contaminants, together with its relatively large space requirements, limits applications for these encoders.

Commercial linear encoders are available as standard catalog models, or they can be custom made for specific applications or extreme environmental conditions. There are both fully enclosed and open linear encoders with travel distances from 2 in. to 6 ft (50 mm to 1.8 m). Some commercial models are available with resolutions down to 0.07 μm, and others can operate at speeds of up to 16.7 ft/s (5 m/s).

Magnetic Encoders

Magnetic encoders can be made by placing a transversely polarized permanent magnet in close proximity to a Hall-effect device sensor. Figure 1-39 shows a magnet mounted on a motor shaft in close proximity to a two-channel HED array which detects changes in magnetic flux density as the magnet rotates. The output signals from the sensors are transmitted to the motion controller. The encoder output, either a square wave or a

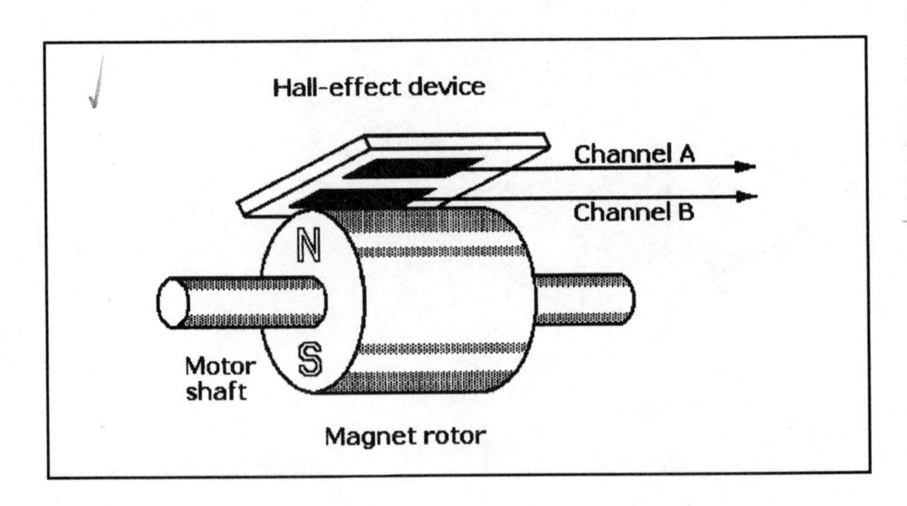

Figure 1-39 Basic parts of a magnetic encoder.

quasi sine wave (depending on the type of magnetic sensing device) can be used to count revolutions per minute (rpm) or determine motor shaft accurately. The phase shift between channels A and B permits them to be compared by the motion controller to determine the direction of motor shaft rotation.

Resolvers

A resolver is essentially a rotary transformer that can provide position feedback in a servosystem as an alternative to an encoder. Resolvers resemble small AC motors, as shown in Figure 1-40, and generate an electrical signal for each revolution of their shaft. Resolvers that sense position in closed-loop motion control applications have one winding on the rotor and a pair of windings on the stator, oriented at 90°. The stator is made by winding copper wire in a stack of iron laminations fastened to the housing, and the rotor is made by winding copper wire in a stack of laminations mounted on the resolver's shaft.

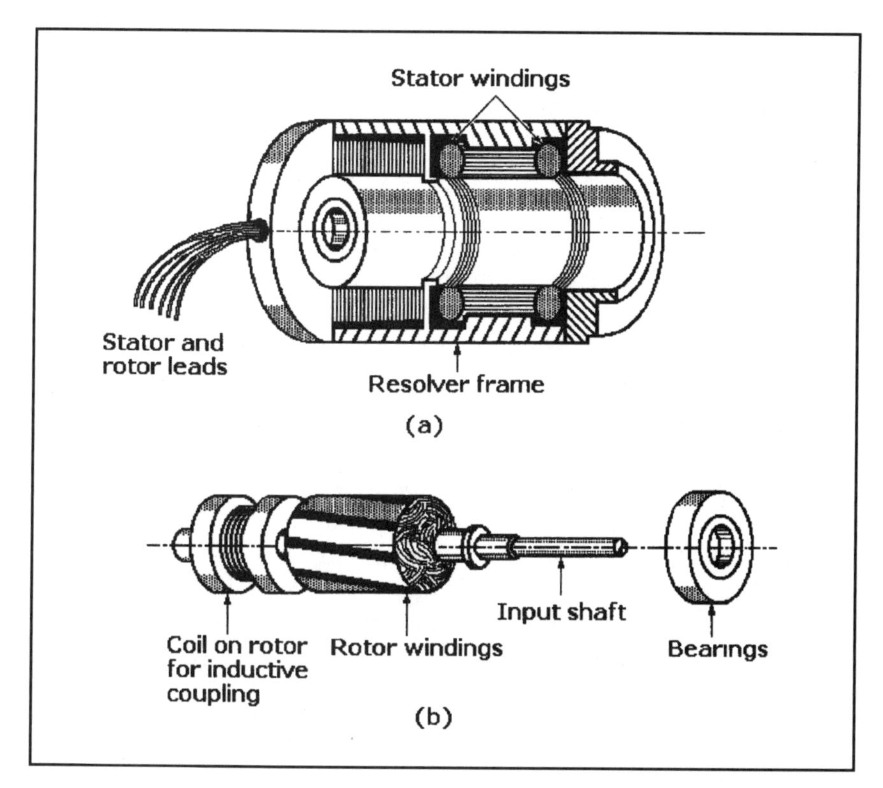

Figure 1-40 Exploded view of a brushless resolver frame (a), and rotor and bearings (b). The coil on the rotor couples speed data inductively to the frame for processing.

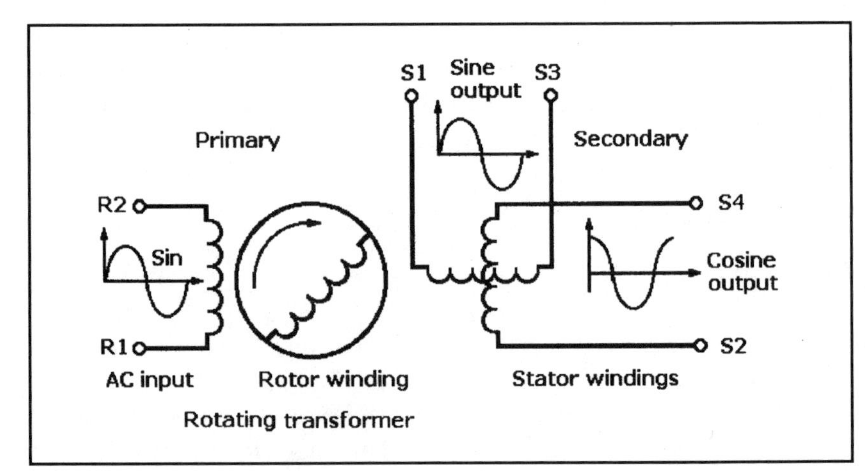

Figure 1-41 Schematic for a resolver shows how rotor position is transformed into sine and cosine outputs that measure rotor position.

Figure 1-41 is an electrical schematic for a brushless resolver showing the single rotor winding and the two stator windings 90° apart. In a servosystem, the resolver's rotor is mechanically coupled to the drive motor and load. When a rotor winding is excited by an AC reference signal, it produces an AC voltage output that varies in amplitude according to the sine and cosine of shaft position. If the phase shift between the applied signal to the rotor and the induced signal appearing on the stator coil is measured, that angle is an analog of rotor position. The absolute position of the load being driven can be determined by the ratio of the sine output amplitude to the cosine output amplitude as the resolver shaft turns through one revolution. (A single-speed resolver produces one sine and one cosine wave as the output for each revolution.)

Connections to the rotor of some resolvers can be made by brushes and slip rings, but resolvers for motion control applications are typically brushless. A rotating transformer on the rotor couples the signal to the rotor inductively. Because brushless resolvers have no slip rings or brushes, they are more rugged than encoders and have operating lives that are up to ten times those of brush-type resolvers. Bearing failure is the most likely cause of resolver failure. The absence of brushes in these resolvers makes them insensitive to vibration and contaminants. Typical brushless resolvers have diameters from 0.8 to 3.7 in. Rotor shafts are typically threaded and splined.

Most brushless resolvers can operate over a 2- to 40-volt range, and their winding are excited by an AC reference voltage at frequencies from 400 to 10,000 Hz. The magnitude of the voltage induced in any stator winding is proportional to the cosine of the angle, q, between the rotor coil axis and the stator coil axis. The voltage induced across any pair of stator terminals will be the vector sum of the voltages across the two connected coils. Accuracies of ±1 arc-minute can be achieved.

In feedback loop applications, the stator's sinusoidal output signals are transmitted to a resolver-to-digital converter (RDC), a specialized analog-to-digital converter (ADC) that converts the signals to a digital representation of the actual angle required as an input to the motion controller.

Tachometers

A tachometer is a DC generator that can provide velocity feedback for a servosystem. The tachometer's output voltage is directly proportional to the rotational speed of the armature shaft that drives it. In a typical servosystem application, it is mechanically coupled to the DC motor and feeds its output voltage back to the controller and amplifier to control drive motor and load speed. A cross-sectional drawing of a tachometer built into the same housing as the DC motor and a resolver is shown in Figure 1-42. Encoders or resolvers are part of separate loops that provide position feedback.

As the tachometer's armature coils rotate through the stator's magnetic field, lines of force are cut so that an electromotive force is induced in each of its coils. This emf is directly proportional to the rate at which

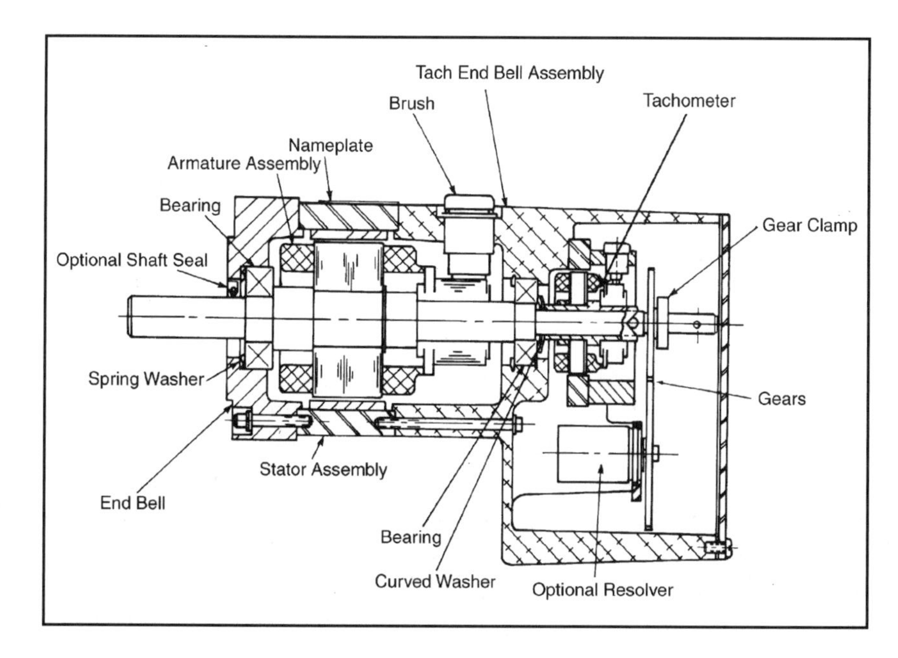

Figure 1-42 Section view of a resolver and tachometer in the same frame as the servomotor.

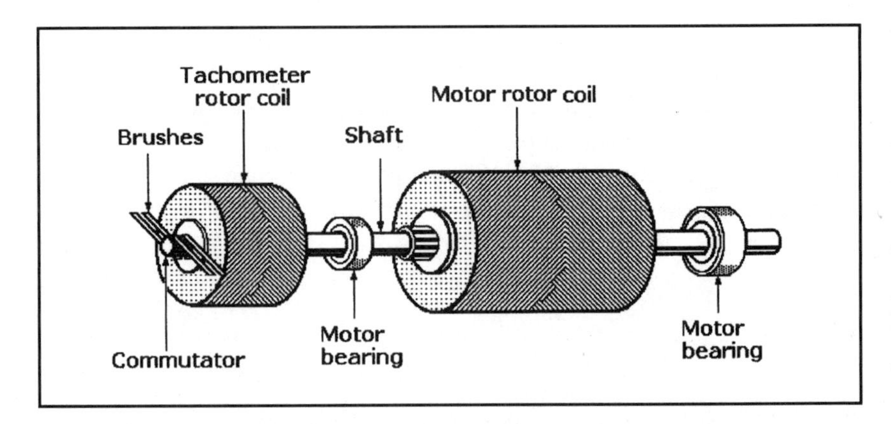

Figure 1-43 The rotors of the DC motor and tachometer share a common shaft.

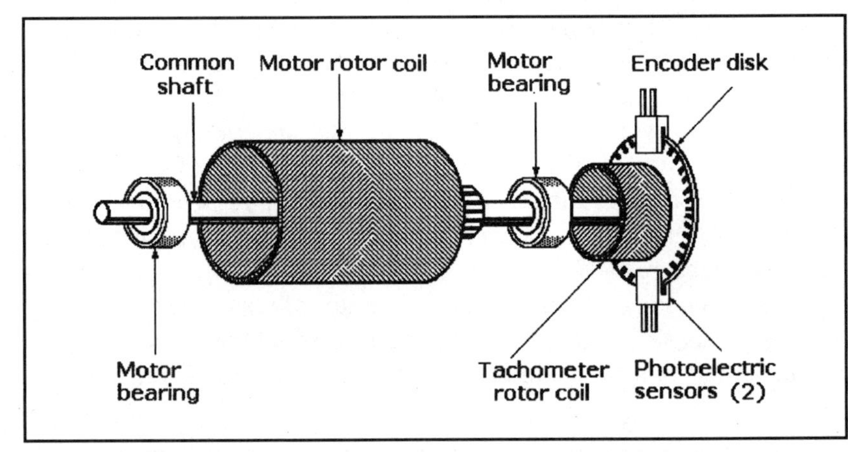

Figure 1-44 This coil-type DC motor obtains velocity feedback from a tachometer whose rotor coil is mounted on a common shaft and position feedback from a two-channel photoelectric encoder whose code disk is also mounted on the same shaft.

the magnetic lines of force are cut as well as being directly proportional to the velocity of the motor's drive shaft. The direction of the emf is determined by Fleming's generator rule.

The AC generated by the armature coil is converted to DC by the tachometer's commutator, and its value is directly proportional to shaft rotation speed while its polarity depends on the direction of shaft rotation.

There are two basic types of DC tachometer: *shunt wound* and *permanent magnet* (PM), but PM tachometers are more widely used in servosystems today. There are also moving-coil tachometers which, like motors, have no iron in their armatures. The armature windings are wound from fine copper wire and bonded with glass fibers and polyester resins into a rigid cup, which is bonded to its coaxial shaft. Because this armature contains no iron, it has lower inertia than conventional copper and iron armatures, and it exhibits low inductance. As a result, the moving-coil tachometer is more responsive to speed changes and provides a DC output with very low ripple amplitudes.

Tachometers are available as standalone machines. They can be rigidly mounted to the servomotor housings, and their shafts can be mechanically coupled to the servomotor's shafts. If the DC servomotor is either a brushless or moving-coil motor, the standalone tachometer will typically be brushless and, although they are housed separately, a common armature shaft will be shared.

A brush-type DC motor with feedback furnished by a brush-type tachometer is shown in Figure 1-43. Both tachometer and motor rotor coils are mounted on a common shaft. This arrangement provides a high resonance frequency. Moreover, the need for separate tachometer bearings is eliminated.

In applications where precise positioning is required in addition to speed regulation, an incremental encoder can be added on the same shaft, as shown in Figure 1-44.

Linear Variable Differential Transformers (LVDTs)

A linear variable differential transformer (LVDT) is a sensing transformer consisting of a primary winding, two adjacent secondary windings, and a ferromagnetic core that can be moved axially within the windings, as shown in the cutaway view Figure 1-45. LVDTs are capable of measuring position, acceleration, force, or pressure, depending on how they are installed. In motion control systems, LVDTs provide position feedback by measuring the variation in mutual inductance between

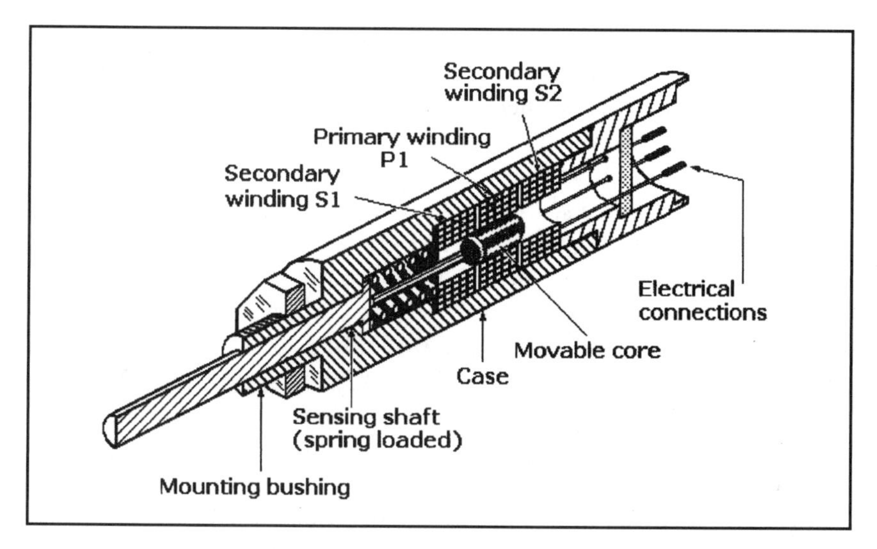

Figure 1-45 Cutaway view of a linear variable displacement transformer (LVDT).

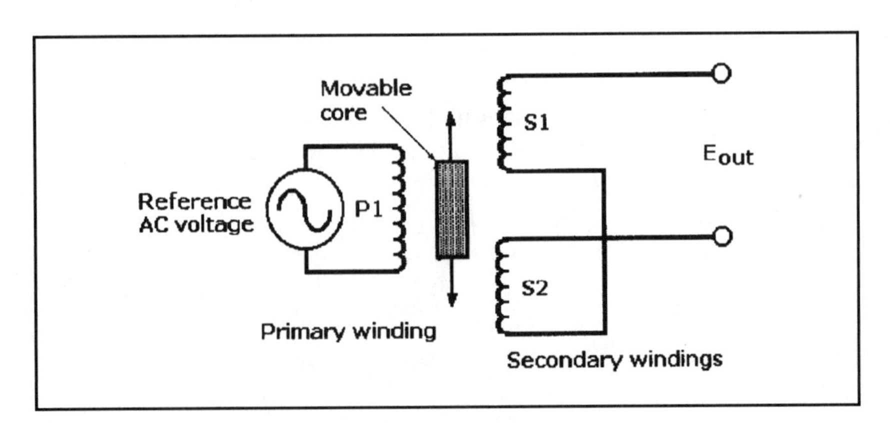

Figure 1-46 Schematic for a linear variable differential transformer (LVDT) showing how the movable core interacts with the primary and secondary windings.

their primary and secondary windings caused by the linear movement of the ferromagnetic core.

The core is attached to a spring-loaded sensing shaft. When depressed, the shaft moves the core axially within the windings, coupling the excitation voltage in the primary (middle) winding P1 to the two adjacent secondary windings S1 and S2.

Figure 1-46 is a schematic diagram of an LVDT. When the core is centered between S1 and S2, the voltages induced in S1 and S2 have equal amplitudes and are 180° out of phase. With a series-opposed connection, as shown, the net voltage across the secondaries is zero because both voltages cancel. This is called the *null position* of the core.

However, if the core is moved to the left, secondary winding S1 is more strongly coupled to primary winding P1 than secondary winding S2, and an output sine wave in phase with the primary voltage is induced. Similarly, if the core is moved to the right and winding S2 is more strongly coupled to primary winding P1, an output sine wave that is 180° out-of-phase with the primary voltage is induced. The amplitudes of the output sine waves of the LVDT vary symmetrically with core displacement, either to the left or right of the null position.

Linear variable differential transformers require signal conditioning circuitry that includes a stable sine wave oscillator to excite the primary winding P1, a demodulator to convert secondary AC voltage signals to DC, a low-pass filter, and an amplifier to buffer the DC output signal. The amplitude of the resulting DC voltage output is proportional to the magnitude of core displacement, either to the left or right of the null position. The phase of the DC voltage indicates the position of the core relative to the null (left or right). An LVDT containing an integral oscillator/demodulator is a DC-to-DC LVDT, also known as a DCDT.

Linear variable differential transformers can make linear displacement (position) measurements as precise as 0.005 in. (0.127 mm).

Output voltage linearity is an important LVDT characteristic, and it can be plotted as a straight line within a specified range. Linearity is the characteristic that largely determines the LVDT's absolute accuracy.

Linear Velocity Transducers (LVTs)

A linear velocity transducer (LVT) consists of a magnet positioned axially within a two wire coils. When the magnet is moved through the coils, it induces a voltage within the coils in accordance with the Faraday and Lenz laws. The output voltage from the coils is directly proportional to the magnet's field strength and axial velocity over its working range.

When the magnet is functioning as a transducer, both of its ends are within the two adjacent coils, and when it is moved axially, its north pole will induce a voltage in one coil and its south pole will induce a voltage in the other coil. The two coils can be connected in series or parallel, depending on the application. In both configurations the DC output voltage from the coils is proportional to magnet velocity. (A single coil would only produce zero voltage because the voltage generated by the north pole would be canceled by the voltage generated by the south pole.)

The characteristics of the LVT depend on how the two coils are connected. If they are connected in series opposition, the output is added and maximum sensitivity is obtained. Also, noise generated in one coil will be canceled by the noise generated in the other coil. However, if the coils are connected in parallel, both sensitivity and source impedance are reduced. Reduced sensitivity improves high-frequency response for measuring high velocities, and the lower output impedance improves the LVT's compatibility with its signal-conditioning electronics.

Angular Displacement Transducers (ATDs)

An angular displacement transducer is an air-core variable differential capacitor that can sense angular displacement. As shown in exploded view Figure 1-47 it has a movable metal rotor sandwiched between a single stator plate and segmented stator plates. When a high-frequency AC signal from an oscillator is placed across the plates, it is modulated by the change in capacitance value due to the position of the rotor with respect to the segmented stator plates. The angular displacement of the rotor can then be determined accurately from the demodulated AC signal.

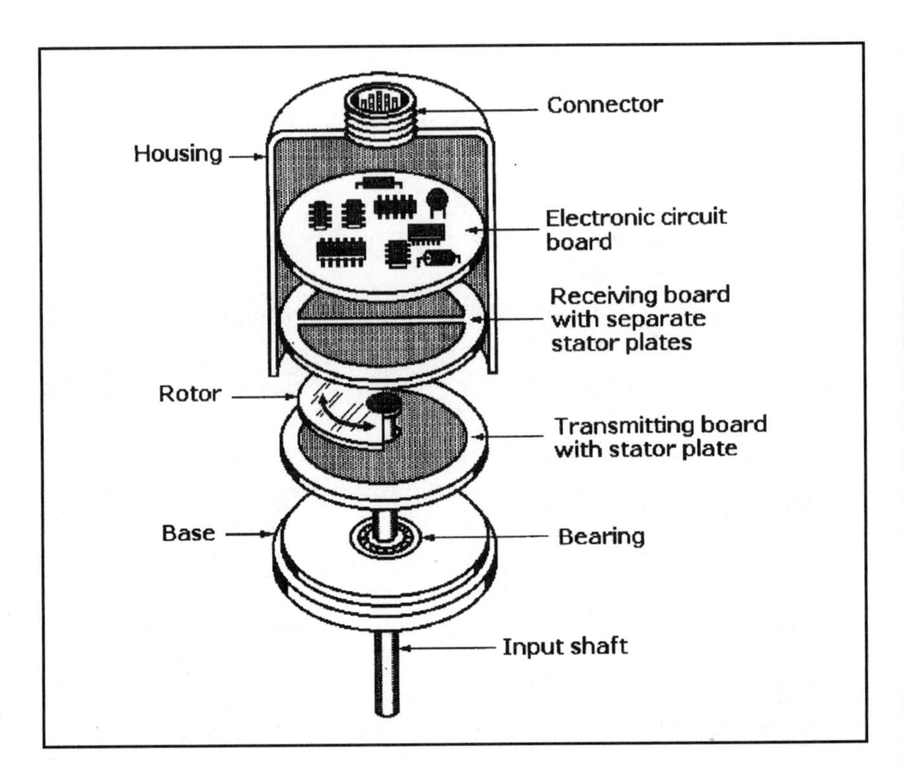

Figure 1-47 Exploded view of an angular displacement transducer (ADT) based on a differential variable capacitor.

The base is the mounting platform for the transducer assembly. It contains the axial ball bearing that supports the shaft to which the rotor is fastened. The base also supports the transmitting board, which contains a metal surface that forms the lower plate of the differential capacitor. The semicircular metal rotor mounted on the shaft is the variable plate or rotor of the capacitor. Positioned above the rotor is the receiving board containing two separate semicircular metal sectors on its lower surface. The board acts as the receiver for the AC signal that has been modulated by the capacitance difference between the plates caused by rotor rotation.

An electronics circuit board mounted on top of the assembly contains the oscillator, demodulator, and filtering circuitry. The ADT is powered by DC, and its output is a DC signal that is proportional to angular displacement. The cup-shaped housing encloses the entire assembly, and the base forms a secure cap.

DC voltage is applied to the input terminals of the ADT to power the oscillator, which generates a 400- to 500-kHz voltage that is applied across the transmitting and receiving stator plates. The receiving plates are at virtual ground, and the rotor is at true ground. The capacitance value between the transmitting and receiving plates remains constant, but the capacitance between the separate receiving plates varies with rotor position.

A null point is obtained when the rotor is positioned under equal areas of the receiving stator plates. In that position, the capacitance between the transmitting stator plate and the receiving stator plates will be equal, and there will be no output voltage. However, as the rotor moves clockwise or counterclockwise, the capacitance between the transmitting plate and one of the receiving plates will be greater than it is between the other receiving plate. As a result, after demodulation, the differential output DC voltage will be proportional to the angular distance the rotor moved from the null point.

Inductosyns

The Inductosyn is a proprietary AC sensor that generates position feedback signals that are similar to those from a resolver. There are rotary and linear Inductosyns. Much smaller than a resolver, a rotary Inductosyn is an assembly of a scale and slider on insulating substrates in a loop. When the scale is energized with AC, the voltage couples into the two slider windings and induces voltages proportional to the sine and cosine of the slider spacing within a cyclic pitch.

An Inductosyn-to-digital (I/D) converter, similar to a resolver-to-digital (R/D) converter, is needed to convert these signals into a digital format. A typical rotary Inductosyn with 360 cyclic pitches per rotation can resolve a total of 1,474,560 sectors for each resolution. This corresponds to an angular rotation of less than 0.9 arc-s. This angular information in a digital format is sent to the motion controller.

Laser Interferometers

Laser interferometers provide the most accurate position feedback for servosystems. They offer very high resolution (to 1.24 nm), noncontact measurement, a high update rate, and intrinsic accuracies of up to 0.02 ppm. They can be used in servosystems either as passive position readouts or as active feedback sensors in a position servo loop. The laser beam path can be precisely aligned to coincide with the load or a specific point being measured, eliminating or greatly reducing Abbe error.

A single-axis system based on the Michaelson interferometer is illustrated in Figure 1-48. It consists of a helium–neon laser, a polarizing beam splitter with a stationary retroreflector, a moving retroreflector that

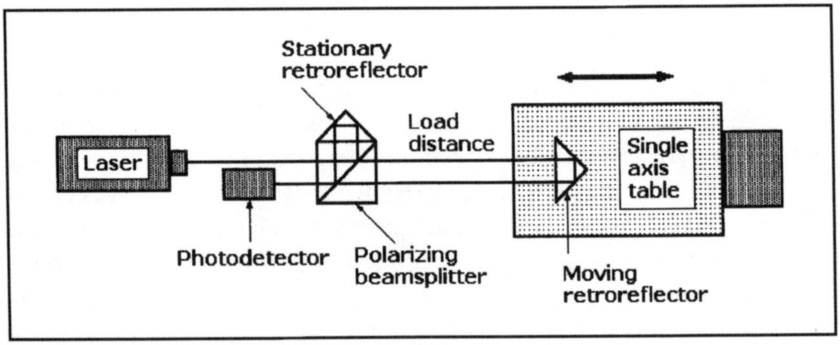

Figure 1-48 Diagram of a laser interferometer for position feedback that combines high resolution with noncontact sensing, high update rates, and accuracies of 0.02 ppm.

can be mounted on the object whose position is to be measured, and a photodetector, typically a photodiode.

Light from the laser is directed toward the polarizing beam splitter, which contains a partially reflecting mirror. Part of the laser beam goes straight through the polarizing beam splitter, and part of the laser beam is reflected. The part that goes straight through the beam splitter reaches the moving reflectometer, which reflects it back to the beam splitter, that passes it on to the photodetector. The part of the beam that is reflected by the beam splitter reaches the stationary retroreflector, a fixed distance away. The retroreflector reflects it back to the beam splitter before it is also reflected into the photodetector.

As a result, the two reflected laser beams strike the photodetector, which converts the combination of the two light beams into an electrical signal. Because of the way laser light beams interact, the output of the detector depends on a *difference* in the distances traveled by the two laser beams. Because both light beams travel the same distance from the laser to the beam splitter and from the beam splitter to the photodetector, these distances are not involved in position measurement. The laser interferometer measurement depends only on the difference in distance between the round trip laser beam travel from the beam splitter to the moving retroreflector and the fixed round trip distance of laser beam travel from the beam splitter to the stationary retroreflector.

If these two distances are exactly the same, the two light beams will recombine in phase at the photodetector, which will produce a high electrical output. This event can be viewed on a video display as a bright *light fringe*. However, if the difference between the distances is as short as one-quarter of the laser's wavelength, the light beams will combine out-of-phase, interfering with each other so that there will be no electrical output from the photodetector and no video output on the display, a condition called a *dark fringe*.

As the moving retroreflector mounted on the load moves farther away from the beam splitter, the laser beam path length will increase and a pattern of light and dark fringes will repeat uniformly. This will result in electrical signals that can be counted and converted to a distance measurement to provide an accurate position of the load. The spacing between the light and dark fringes and the resulting electrical pulse rate is determined by the wavelength of the light from the laser. For example, the wavelength of the light beam emitted by a helium–neon (He–Ne) laser, widely used in laser interferometers, is 0.63 μm, or about 0.000025 in.

Thus the accuracy of load position measurement depends primarily on the known stabilized wavelength of the laser beam. However, that accuracy can be degraded by changes in humidity and temperature as well as airborne contaminants such as smoke or dust in the air between the beam splitter and the moving retroreflector.

Precision Multiturn Potentiometers

The rotary precision multiturn potentiometer shown in the cutaway in Figure 1-49 is a simple, low-cost feedback instrument. Originally developed for use in analog computers, precision potentiometers can provide absolute position data in analog form as a resistance value or voltage. Precise and resettable voltages correspond to each setting of the rotary control shaft. If a potentiometer is used in a servosystem, the analog data

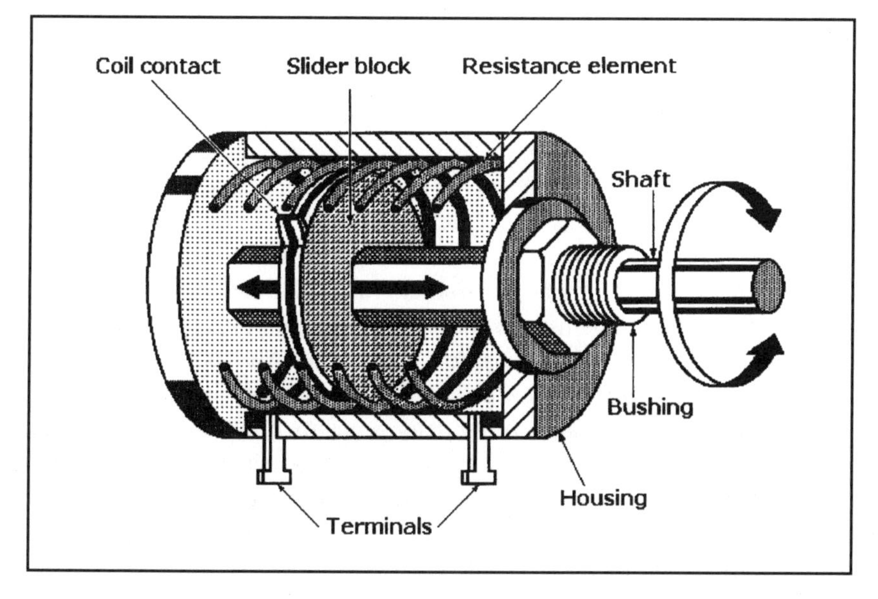

Figure 1-49 A precision potentiometer is a low-cost, reliable feedback sensor for servosystems.

will usually be converted to digital data by an integrated circuit analog-to-digital converter (ADC). Accuracies of 0.05% can be obtained from an instrument-quality precision multiturn potentiometer, and resolutions can exceed 0.005° if the output signal is converted with a 16-bit ADC.

Precision multiturn potentiometers have wirewound or hybrid resistive elements. Hybrid elements are wirewound elements coated with resistive plastic to improve their resolution. To obtain an output from a potentiometer, a conductive wiper must be in contact with the resistive element. During its service life wear on the resistive element caused by the wiper can degrade the precision of the precision potentiometer.

SOLENOIDS AND THEIR APPLICATIONS

Solenoids: An Economical Choice for Linear or Rotary Motion

A solenoid is an electromechanical device that converts electrical energy into linear or rotary mechanical motion. All solenoids include a coil for conducting current and generating a magnetic field, an iron or steel shell or case to complete the magnetic circuit, and a plunger or armature for translating motion. Solenoids can be actuated by either direct current (DC) or rectified alternating current (AC).

Solenoids are built with conductive paths that transmit maximum magnetic flux density with minimum electrical energy input. The mechanical action performed by the solenoid depends on the design of the plunger in a linear solenoid or the armature in a rotary solenoid. Linear solenoid plungers are either spring-loaded or use external methods to restrain axial movement caused by the magnetic flux when the coil is energized and restore it to its initial position when the current is switched off.

Cutaway drawing Figure 1-50 illustrates how pull-in and push-out actions are performed by a linear solenoid. When the coil is energized, the plunger pulls in against the spring, and this motion can be translated into either a "pull-in" or a "push-out" response. All solenoids are basically pull-in-type actuators, but the location of the plunger extension with respect to the coil and spring determines its function. For example, the plunger extension on the left end (end A) provides "push-out" motion against the load, while a plunger extension on the right end terminated by a clevis (end B) provides "pull-in" motion. Commercial solenoids perform only one of these functions. Figure 1-51 is a cross-sectional view of a typical pull-in commercial linear solenoid.

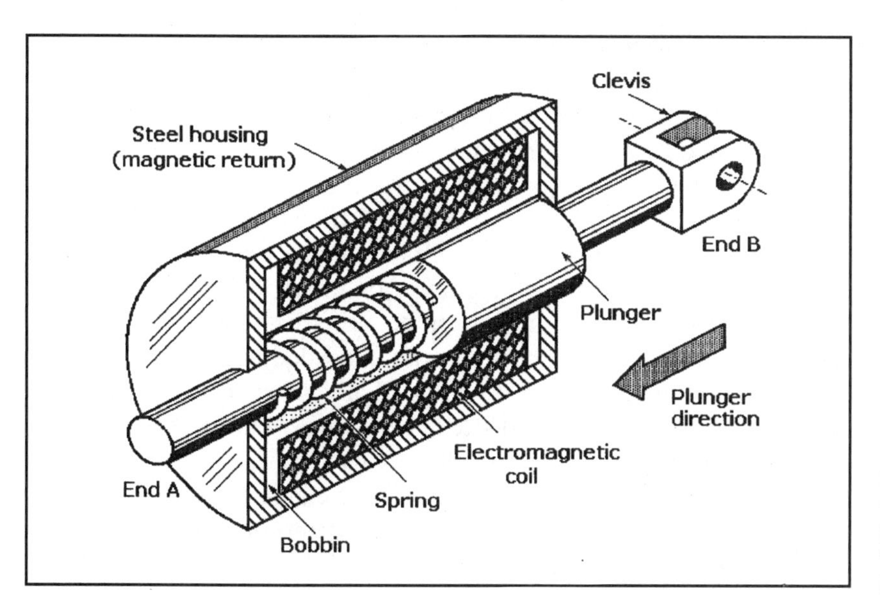

Figure 1-50 The pull-in and push-out functions of a solenoid are shown. End A of the plunger pushes out when the solenoid is energized while the clevis-end B pulls in.

Rotary solenoids operate on the same principle as linear solenoids except that the axial movement of the armature is converted into rotary movement by various mechanical devices. One of these is the use of internal lands or ball bearings and slots or races that convert a pull-in stroke to rotary or twisting motion.

Motion control and process automation systems use many different kinds of solenoids to provide motions ranging from simply turning an event on or off to the performance of extremely complex sequencing. When there are requirements for linear or rotary motion, solenoids should be considered because of their relatively small size and low cost when compared with alternatives such as motors or actuators. Solenoids are easy to install and use, and they are both versatile and reliable.

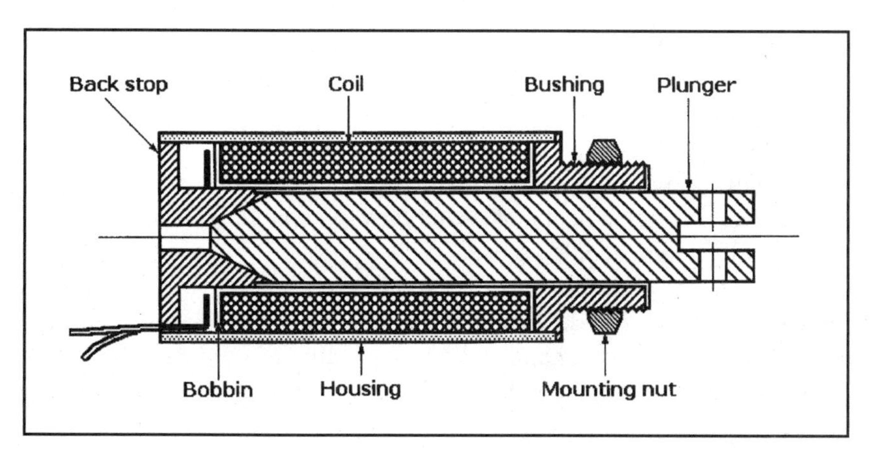

Figure 1-51 Cross-section view of a commercial linear pull-type solenoid with a clevis. The conical end of the plunger increases its efficiency. The solenoid is mounted with its threaded bushing and nut.

Technical Considerations

Important factors to consider when selecting solenoids are their rated torque/force, duty cycles, estimated working lives, performance curves, ambient temperature range, and temperature rise. The solenoid must have a magnetic return path capable of transmitting the maximum amount of magnetic flux density with minimum energy input. Magnetic flux lines are transmitted to the plunger or armature through the bobbin and air gap back through the iron or steel shell. A ferrous metal path is more efficient than air, but the air gap is needed to permit plunger or armature movement. The force or torque of a solenoid is inversely proportional to the square of the distance between pole faces. By optimizing the ferrous path area, the shape of the plunger or armature, and the magnetic circuit material, the output torque/force can be increased.

The torque/force characteristic is an important solenoid specification. In most applications the force can be a minimum at the start of the plunger or armature stroke but must increase at a rapid rate to reach the maximum value before the plunger or armature reaches the backstop.

The magnetizing force of the solenoid is proportional to the number of copper wire turns in its coil, the magnitude of the current, and the permeance of the magnetic circuit. The pull force required by the load must not be greater than the force developed by the solenoid during any portion of its required stroke, or the plunger or armature will not pull in completely. As a result, the load will not be moved the required distance.

Heat buildup in a solenoid is a function of power and the length of time the power is applied. The permissible temperature rise limits the magnitude of the input power. If constant voltage is applied, heat buildup can degrade the efficiency of the coil by effectively reducing its number of ampere turns. This, in turn, reduces flux density and torque/force output. If the temperature of the coil is permitted to rise above the temperature rating of its insulation, performance will suffer and the solenoid could fail prematurely. Ambient temperature in excess of the specified limits will limit the solenoid cooling expected by convection and conduction.

Heat can be dissipated by cooling the solenoid with forced air from a fan or blower, mounting the solenoid on a heat sink, or circulating a liquid coolant through a heat sink. Alternatively, a larger solenoid than the one actually needed could be used.

The heating of the solenoid is affected by the duty cycle, which is specified from 10 to 100%, and is directly proportional to solenoid *on* time. The highest starting and ending torque are obtained with the lowest duty cycle and *on* time. Duty cycle is defined as the ratio of *on* time to the sum of *on* time and *off* time. For example, if a solenoid is energized for 30 s and then turned off for 90 s, its duty cycle is $30/120 = 1/4$, or 25%.

The amount of work performed by a solenoid is directly related to its size. A large solenoid can develop more force at a given stroke than a small one with the same coil current because it has more turns of wire in its coil.

Open-Frame Solenoids

Open-frame solenoids are the simplest and least expensive models. They have open steel frames, exposed coils, and movable plungers centered in their coils. Their simple design permits them to be made inexpensively in high-volume production runs so that they can be sold at low cost. The two forms of open-frame solenoid are the *C-frame solenoid* and the *box-frame solenoid.* They are usually specified for applications where very long life and precise positioning are not critical requirements.

C-Frame Solenoids

C-frame solenoids are low-cost commercial solenoids intended for light-duty applications. The frames are typically laminated steel formed in the shape of the letter C to complete the magnetic circuit through the core, but they leave the coil windings without a complete protective cover. The plungers are typically made as laminated steel bars. However, the coils are usually potted to resist airborne and liquid contaminants. These solenoids can be found in appliances, printers, coin dispensers, security door locks, cameras, and vending machines. They can be powered with either AC or DC current. Nevertheless, C-frame solenoids can have operational lives of millions of cycles, and some standard catalog models are capable of strokes up to 0.5 in. (13 mm).

Box-Frame Solenoids

Box-frame solenoids have steel frames that enclose their coils on two sides, improving their mechanical strength. The coils are wound on phenolic bobbins, and the plungers are typically made from solid bar stock. The frames of some box-type solenoids are made from stacks of thin insulated sheets of steel to control eddy currents as well as keep stray circulating currents confined in solenoids powered by AC. Box-frame solenoids are specified for higher-end applications such as tape decks, industrial controls, tape recorders, and business machines because they offer mechanical and electrical performance that is superior to those of C-frame solenoids. Standard catalog commercial box-frame solenoids can be powered by AC or DC current, and can have strokes that exceed 0.5 in. (13 mm).

Tubular Solenoids

The coils of *tubular solenoids* have coils that are completely enclosed in cylindrical metal cases that provide improved magnetic circuit return and better protection against accidental damage or liquid spillage. These DC solenoids offer the highest volumetric efficiency of any commercial solenoids, and they are specified for industrial and military/aerospace equipment where the space permitted for their installation is restricted. These solenoids are specified for printers, computer disk-and tape drives, and military weapons systems; both pull-in and push-out styles are available. Some commercial tubular linear solenoids in this class have strokes up to 1.5 in. (38 mm), and some can provide 30 lbf (14 kgf) from a unit less than 2.25 in (57 mm) long. Linear solenoids find applications in vending machines, photocopy machines, door locks, pumps, coin-changing mechanisms, and film processors.

Rotary Solenoids

Rotary solenoid operation is based on the same electromagnetic principles as linear solenoids except that their input electrical energy is converted to rotary or twisting rather than linear motion. Rotary actuators should be considered if controlled speed is a requirement in a rotary stroke application. One style of rotary solenoid is shown in the exploded view Figure 1-52. It includes an armature-plate assembly that rotates when it is pulled into the housing by magnetic flux from the coil. Axial stroke is the linear distance that the armature travels to the center of the coil as the solenoid is energized. The three ball bearings travel to the lower ends of the races in which they are positioned.

The operation of this rotary solenoid is shown in Figure 1-53. The rotary solenoid armature is supported by three ball bearings that travel around and down the three inclined ball races. The de-energized state is shown in (a). When power is applied, a linear electromagnetic force pulls in the armature and twists the armature plate, as shown in (b). Rotation

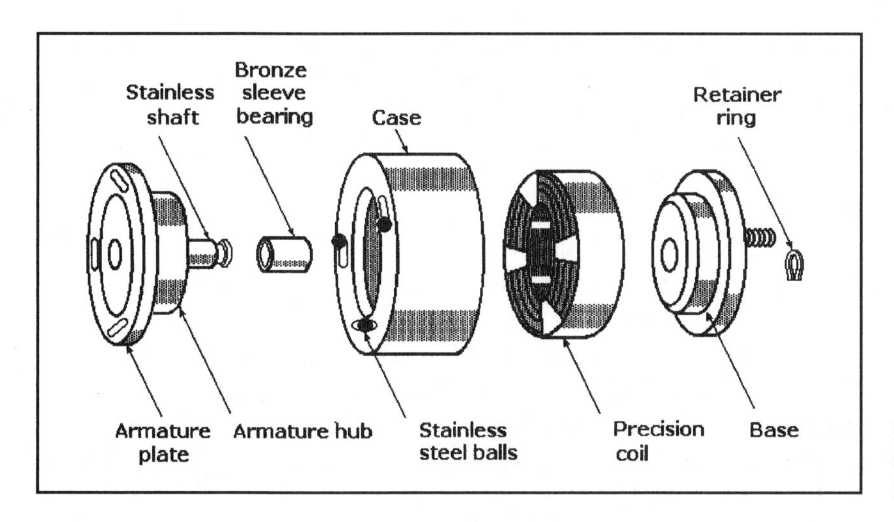

Figure 1-52 Exploded view of a rotary solenoid showing its principal components.

continues until the balls have traveled to the deep ends of the races, completing the conversion of linear to rotary motion.

This type of rotary solenoid has a steel case that surrounds and protects the coil, and the coil is wound so that the maximum amount of copper wire is located in the allowed space. The steel housing provides the high permeability path and low residual flux needed for the efficient conversion of electrical energy to mechanical motion.

Rotary solenoids can provide well over 100 lb-in. (115 kgf-cm) of torque from a unit less than 2.25 in. (57 mm) long. Rotary solenoids are

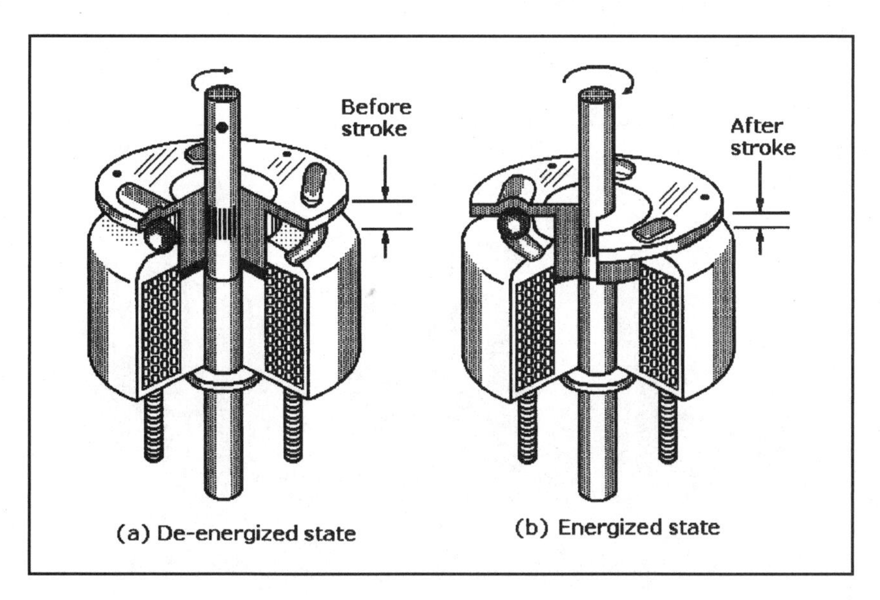

Figure 1-53 Cutaway views of a rotary solenoid de-energized (a) and energized (b). When energized, the solenoid armature pulls in, causing the three ball bearings to roll into the deeper ends of the lateral slots on the faceplate, translating linear to rotary motion.

found in counters, circuit breakers, electronic component pick-and-place machines, ATM machines, machine tools, ticket-dispensing machines, and photocopiers.

Rotary Actuators

The rotary actuator shown in Figure 1-54 operates on the principle of attraction and repulsion of opposite and like magnetic poles as a motor. In this case the electromagnetic flux from the actuator's solenoid interacts with the permanent magnetic field of a neodymium–iron disk magnet attached to the armature but free to rotate.

The patented Ultimag rotary actuator from the Ledex product group of TRW, Vandalia, Ohio, was developed to meet the need for a bidirectional actuator with a limited working stroke of less than 360° but capable of offering higher speed and torque than a rotary solenoid. This fast, short-stroke actuator is finding applications in industrial, office automation, and medical equipment as well as automotive applications

The PM armature has twice as many poles (magnetized sectors) as the stator. When the actuator is not energized, as shown in (a), the armature poles each share half of a stator pole, causing the shaft to seek and hold mid-stroke.

When power is applied to the stator coil, as shown in (b), its associated poles are polarized north above the PM disk and south beneath it. The resulting flux interaction attracts half of the armature's PM poles while repelling the other half. This causes the shaft to rotate in the direction shown.

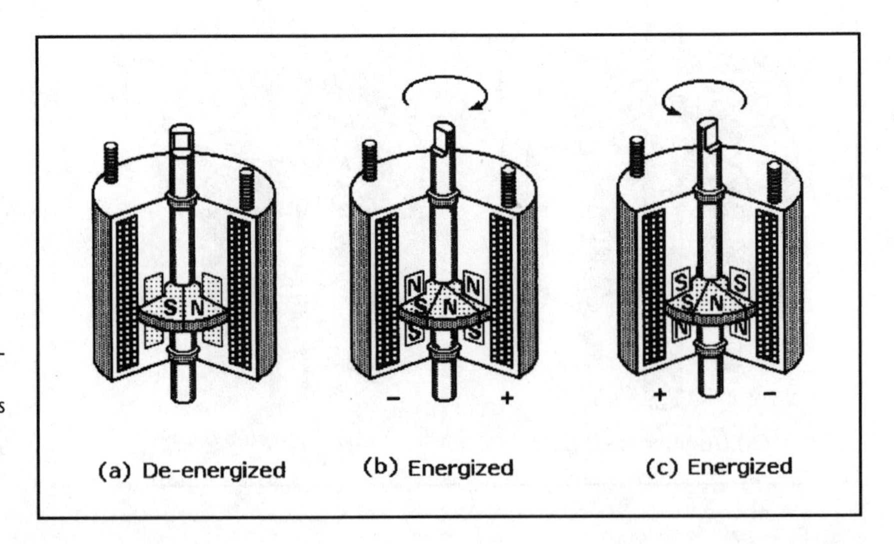

Figure 1-54 This bidirectional rotary actuator has a permanent magnet disk mounted on its armature that interacts with the solenoid poles. When the solenoid is deenergized (a), the armature seeks and holds a neutral position, but when the solenoid is energized, the armature rotates in the direction shown. If the input voltage is reversed, armature rotation is reversed (c).

When the stator voltage is reversed, its poles are reversed so that the north pole is above the PM disk and south pole is below it. Consequently, the opposite poles of the actuator armature are attracted and repelled, causing the armature to reverse its direction of rotation.

According to the manufacturer, Ultimag rotary actuators are rated for speeds over 100 Hz and peak torques over 100 oz-in. Typical actuators offer a 45° stroke, but the design permits a maximum stroke of 160°. These actuators can be operated in an *on/off* mode or proportionally, and they can be operated either open- or closed-loop. Gears, belts, and pulleys can amplify the stroke, but this results in reducing actuator torque.

ACTUATOR COUNT

During the initial design phase of a robot project, it is tempting to add more features and solve mobility or other problems by adding more degrees of freedom (DOF) by adding actuators. This is not always the best approach. The number of actuators in any mechanical device has a direct impact on debugging, reliability, and cost. This is especially true with mobile robots, whose interactions between sensors and actuators must be carefully integrated, first one set at a time, then in the whole robot. Adding more actuators extends this process considerably and increases the chance that problems will be overlooked.

Debugging

Debugging effort, the process of testing, discovering problems, and working out fixes, is directly related to the number of actuators. The more actuators there are, the more problems there are, and each has to be debugged separately. Frequently the actuators have an affect on each other or act together and this in itself adds to the debugging task. This is good reason to keep the number of actuators to a minimum.

Debugging a robot happens in many stages, and is often an iterative process. Each engineering discipline builds (or simulates), tests, and debugs their own piece of the puzzle. The pieces are assembled into larger blocks of the robot and tests and debugging are done on those subassemblies, which may be just breadboard electronics with some control software, or perhaps electronics controlling some test motors. The subassemblies are put together, tested, and debugged in the assembled robot.

This is when the number of actuators has a large affect on debug complexity and time. Each actuator must be controlled with some piece of electronics, which is, in turn, controlled by the software, which takes inputs from the sensors to make its decisions. The relationship between the sensors and actuators is much more complicated than just one sensor connected through software to one actuator. The sensors work sometimes individually and sometimes as a group. The control software must look at the inputs from the all sensors, make intelligent decisions based on that information, and then send commands to one, or many of the actuators. Bugs will be found at any point in this large number of combinations of sensors and actuators.

Mechanical bugs, electronic bugs, software bugs, and bugs caused by interactions between those engineering disciplines will appear and solutions must be found for them. Every actuator adds a whole group of relationships, and therefore the potential for a whole group of bugs.

Reliability

For much the same reasons, reliability is also affected by actuator count. There are simply more things that can go wrong, and they will. Every moving part has a limited lifetime, and every piece of the robot has a chance of being made incorrectly, assembled incorrectly, becoming loose from vibration, being damaged by something in the environment, etc. A rule of thumb is that every part added potentially decreases reliability.

Cost

Cost should also be figured in when working on the initial phases of design, though for some applications cost is less important. Each actuator adds its own cost, its associated electronics, the parts that the actuator moves or uses, and the cost of the added debug time. The designer or design team should seriously consider having a slightly less capable platform or manipulator and leave out one or two actuators, for a significant increase in reliability, greatly reduced debug time, and reduced cost.

Chapter 2 Indirect Power Transfer Devices

As mentioned in Chapter One, electric motors suffer from a problem that must be solved if they are to be used in robots. They turn too fast with too little torque to be very effective for many robot applications, and if slowed down to a useable speed by a motor speed controller their efficiency drops, sometimes drastically. Stepper motors are the least prone to this problem, but even they loose some system efficiency at very low speeds. Steppers are also less volumetrically efficient, they require special drive electronics, and do not run as smoothly as simple permanent magnet (PMDC) motors. The solution to the torque problem is to attach the motor to some system that changes the high speed/low torque on the motor output shaft into the low speed/high torque required for most applications in mobile robots.

Fortunately, there are many mechanisms that perform this transformation of speed to torque. Some attach directly to the motor and essentially make it a bigger and heavier but more effective motor. Others require separate shafts and mounts between the motor and the output shaft; and still others directly couple the motor to the output shaft, deal with any misalignment, and exchange speed for torque all in one mechanism. Power transfer mechanisms are normally divided into five general categories:

1. belts (flat, round, V-belts, timing)
2. chain (roller, ladder, timing)
3. plastic-and-cable chain (bead, ladder, pinned)
4. friction drives
5. gears (spur, helical, bevel, worm, rack and pinion, and many others)

Some of these, like V-belts and friction drives, can be used to provide the further benefit of mechanically varying the output speed. This ability is not usually required on a mobile robot, indeed it can cause control problems in certain cases because the computer does not have direct control over the actual speed of the output shaft. Other power transfer devices like timing belts, plastic-and-cable chain, and all types of steel chain connect the input to the output mechanically by means of teeth just like gears. These devices could all be called synchronous because they keep the input and output shafts in synch, but roller chain is usually left out of this category because the rollers allow some relative motion between the chain and the sprocket. The term *synchronous* is usually applied only to toothed belts which fit on their sprockets much tighter than roller chain.

For power transfer methods that require attaching one shaft to another, like motor-mounted gearboxes driving a separate output shaft, a method to deal with misalignment and vibration should be incorporated. This is done with shaft couplers and flexible drives. In some cases where shock loads might be high, a method of protecting against overloading and breaking the power transfer mechanism should be included. This is done with torque limiters and clutches.

Let's take a look at each method. We'll start with mechanisms that transfer power between shafts that are not inline, then look at couplers and torque limiters. Each section has a short discussion on how well that method applies to mobile robots.

BELTS

Belts are available in at least 4 major variations and many smaller variations. They can be used at power levels from fractional horsepower to tens of horsepower. They can be used in variable speed drives, remembering that this may cause control problems in an autonomous robot. They are durable, in most cases quiet, and handle some misalignment. The four variations are

- flat belt
- O-ring belt
- V-belt
- timing belt

There are many companies that make belts, many of which have excellent web sites on the world wide web. Their web sites contain an enormous amount of information about belts of all types.

- V-belt.com
- fennerprecision.com
- brecoflex.com
- gates.com

- intechpower.com
- mectrol.com
- dodge-pt.com

Flat Belts

Flat belts are an old design that has only limited use today. The belt was originally made flat primarily because the only available durable belt material was leather. In the late 18th and early 19th centuries, it was used extensively in just about every facility that required moving rotating power from one place to another. There are examples running in museums and some period villages, but for the most part flat belts are obsolete. Leather flat belts suffered from relatively short life and moderate efficiency.

Having said all that, they are still available for low power devices with the belts now being made of more durable urethane rubber, sometimes reinforced with nylon, kevlar, or polyester tension members. They require good alignment between the driveR and driveN pulleys and the pulleys themselves are not actually flat, but slightly convex. While they do work, there are better belt styles to use for most applications. They are found in some vacuum cleaners because they are resistant to dirt buildup.

O-Ring Belts

O-ring belts are used in some applications mostly because they are extremely cheap. They too suffer from moderate efficiency, but their cost is so low that they are used in toys and low power devices like VCRs etc. They are a good choice in their power range, but require proper tension and alignment for good life and efficiency.

V-Belts

V-belts get their name from the shape of a cross section of the belt, which is similar to a V with the bottom chopped flat. Their design relies on friction, just like flat belts and O-ring belts, but they have the advantage that the V shape jams in a matching V shaped groove in the pulley. This increases the friction force because of the steep angle of the V and therefore increases the transmittable torque under the same tension as is required for flat or O-ring belts. V-belts are also very quiet, allow some misalignment, and are surprisingly efficient. They are a good choice for power levels from fractional to tens of horsepower. Their only draw-

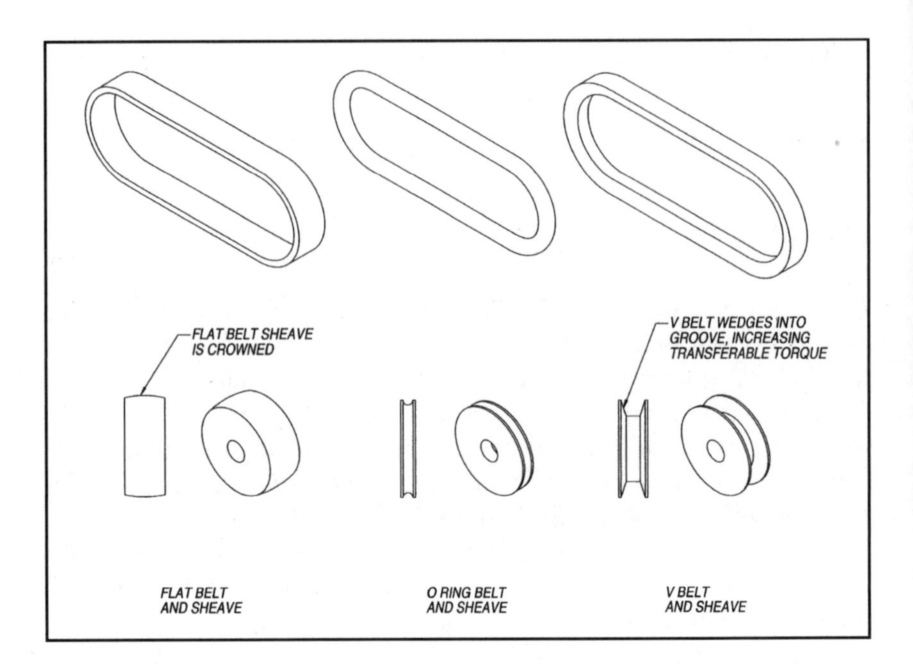

Figure 2-1 Flat, O-ring, and V-belt profiles and pulleys

back is a slight tendency to slip over time. This slip means the computer has no precise control of the orientation of the output shaft, unless a feedback device is on the driveN pulley. There are several applications, however, where some slip is not much of a problem, like in some wheel and track drives. Figure 2-1 shows the cross sectional shape of each belt.

In spite of the warnings on the possibility of problems using variable speed drives, here are some examples of methods of varying the speed and torque by using variable diameter sheaves. Figure 2-2 (from *Mechanisms and Mechanical Devices Sourcebook*, as are many of the figures in this book) shows how variable speed drives work. They may have some applications, especially in teleoperated vehicles.

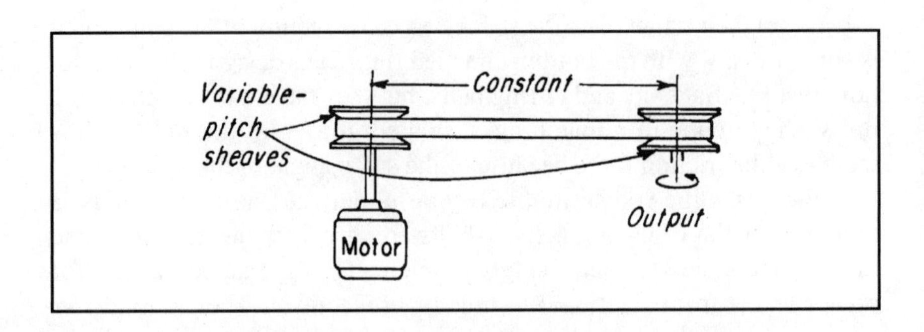

Figure 2-2 Variable Belt

SMOOTHER DRIVE WITHOUT GEARS

The transmission in the motor scooter in Figure 2-3 is torque-sensitive; motor speed controls the continuously variable drive ratio. The operator merely works the throttle and brake.

Variable-diameter V-belt pulleys connect the motor and chain drive sprocket to give a wide range of speed reduction. The front pulley incorporates a three-ball centrifugal clutch which forces the flanges together when the engine speeds up. At idle speed the belt rides on a ballbearing between the retracted flanges of the pulley. During starting and warmup, a lockout prevents the forward clutch from operating.

Upon initial engagement, the overall drive ratio is approximately 18:1. As engine speed increases, the belt rides higher up on the forward-pulley flanges until the overall drive ratio becomes approximately 6:1. The resulting variations in belt tension are absorbed by the spring-loaded flanges of the rear pulley. When a clutch is in an idle position, the V-belt is forced to the outer edge of the rear pulley by a spring force. When the clutch engages, the floating half of the front pulley moves inward, increasing its effective diameter and pulling the belt down between the flanges of the rear pulley.

The transmission is torque-responsive. A sudden engine acceleration increases the effective diameter of the rear pulley, lowering the drive ratio. It works this way: An increase in belt tension rotates the floating flange ahead in relation to the driving flange. The belt now slips slightly on its driver. At this time nylon rollers on the floating flange engage cams on the driving flange, pulling the flanges together and increasing the effective diameter of the pulley.

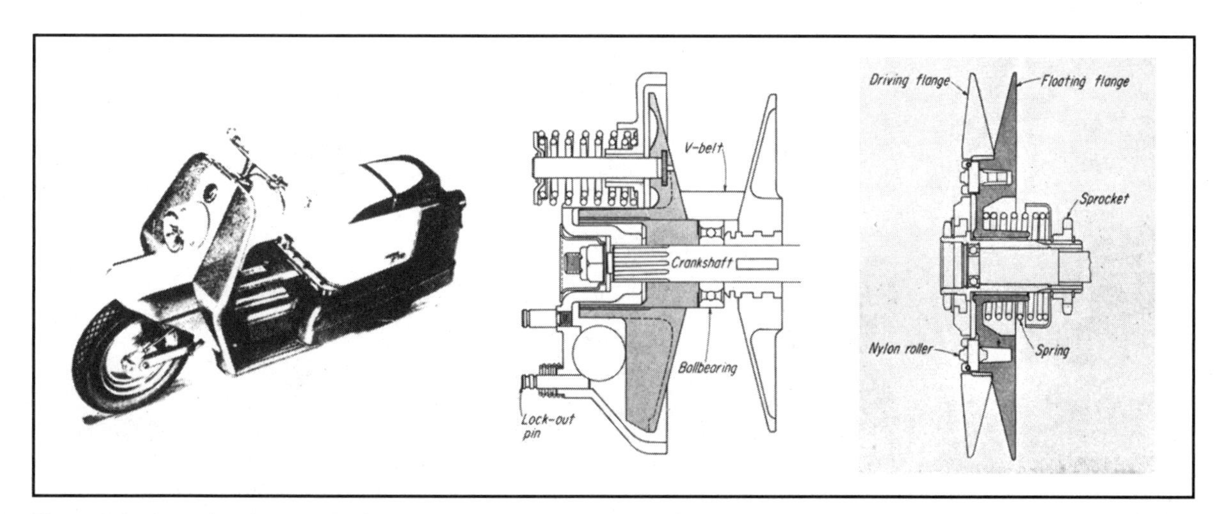

Figure 2-3 Smoother Drive Without Gears

Timing Belts

Timing belts solve the slip problems of flat, O ring, and V belts by using a flexible tooth, molded to a belt that has tension members built in. The teeth are flexible allowing the load to be spread out over all the teeth in contact with the pulley. Timing belts are part of a larger category of power transmission devices called *synchronous drives*. These belt or cable-based drives have the distinct advantage of not slipping, hence the name synchronous. Synchronous or *positive drive* also means these belts can even be used in wet conditions, provided the pulleys are stainless steel or plastic to resist corrosion.

Timing belts come in several types, depending on their tooth profile and manufacturing method. The most common timing belt has a trapezoidal shaped tooth. This shape has been the standard for many years, but it does have drawbacks. As each tooth comes in contact with the mating teeth on a pulley, the tooth tends to be deflected by the cantilever force, deforming the belt's teeth so that only the base of the tooth remains in contact. This bending and deformation wastes energy and also can make the teeth ride up pulley's teeth and skip teeth. The deformation also increases wear of the tooth material and causes the timing belt drive to be somewhat noisy.

Several other shapes have been developed to improve on this design, the best of which is the curved tooth profile. A trade name for this shape is HTD for High Torque Design. Timing belts can be used at very low rpm, high torque, and at power levels up to 250 horsepower. They are an excellent method of power transfer, but for a slightly higher price than chain or plastic-and-cable chain discussed later in this chapter.

Table 2-1 Timing Belts

Table 2-1 Timing Belts

Type	Circular pitch, in.	Wkg. tension lb/in. width	Centr. loss const., K_c
Standard			
MXL	0.080	32	10×10^{-9}
XL	0.200	41	27×10^{-9}
L	0.375	55	38×10^{-9}
H	0.500	140	53×10^{-9}
40DP	0.0816	13	—
High-torque			
3 mm	0.1181	60	15×10^{-9}
5 mm	0.1968	100	21×10^{-9}
8 mm	0.3150	138	34×10^{-9}

Courtesy Stock Drive Products

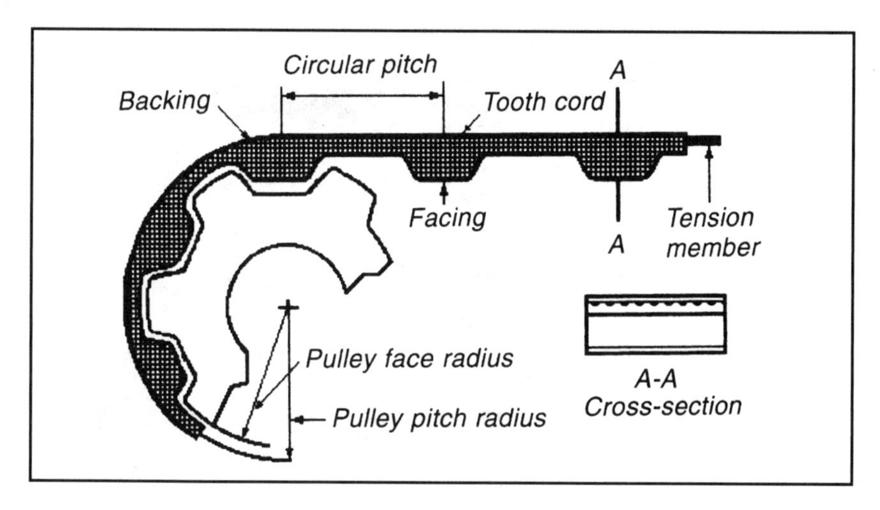

Figure 2-4 Trapezoidal Tooth Timing Belt

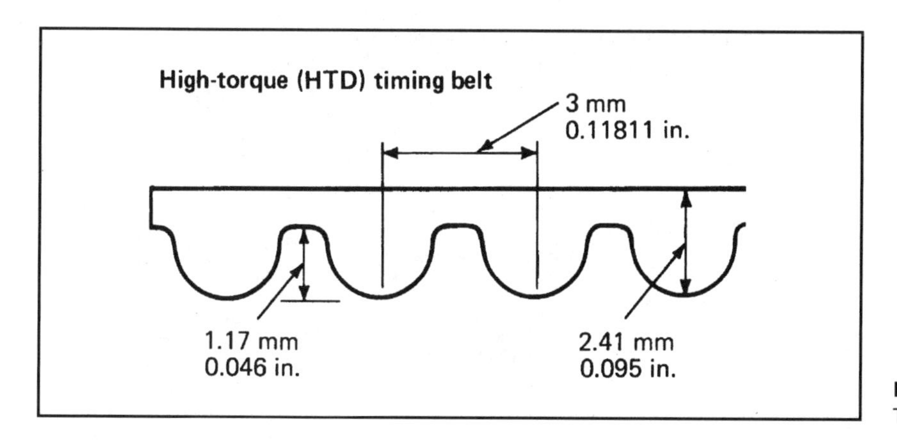

Figure 2-5 HTD Timing Belt Tooth Profile

Plastic-and-Cable Chain

The other type of synchronous drive is based on a steel cable core. It is actually the reverse of belt construction where the steel or synthetic cable is molded into the rubber or plastic belt. Plastic-and-cable chain starts with the steel cable and over-molds plastic or hard rubber teeth onto the cable. The result appears almost like a roller chain. This style is sometimes called Posi-drive, plastic-and-cable, or cable chain. It is made in three basic forms.

The simplest is molding beads onto the cable as shown in Figure 2-6. Figure 2-7 shows a single cable form where the plastic teeth protrude out of both sides of the cable, or even 4 sides of the cable. The third form is

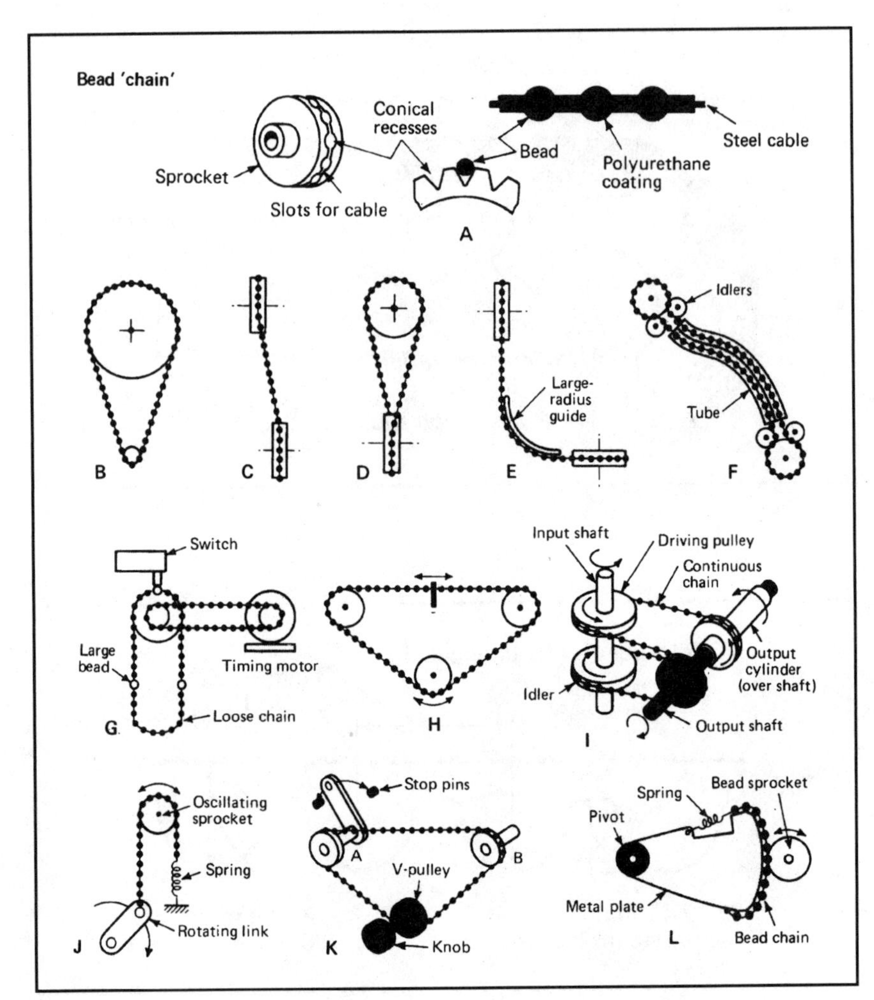

Figure 2-6 Polyurethane-coated steel-cable "chains"—both beaded and 4-pinned—can cope with conditions unsuitable for most conventional belts and chains.

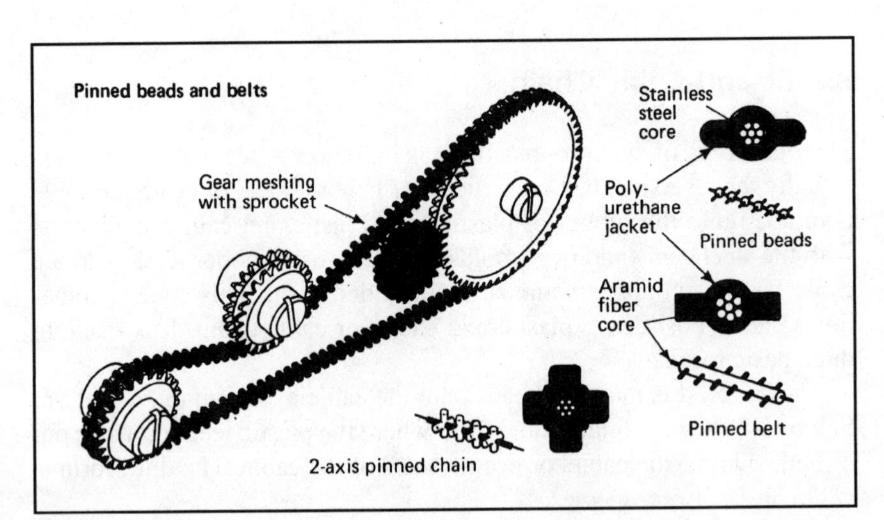

Figure 2-7 Plastic pins eliminate the bead chain's tendency to cam out of pulley recesses, and permit greater precision in angular transmission.

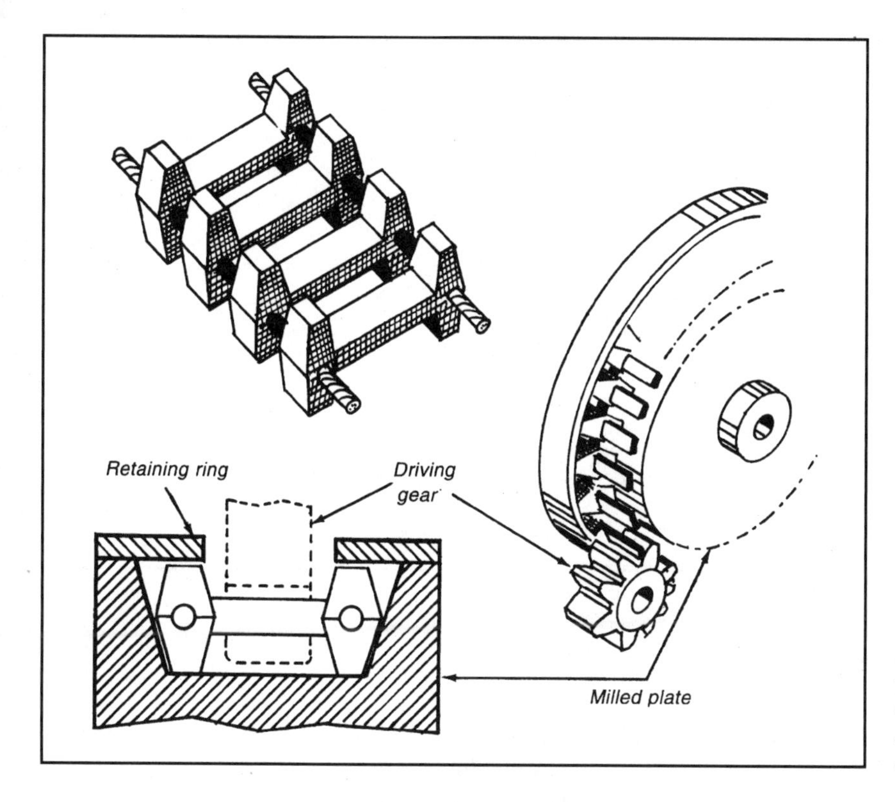

Figure 2-8 A gear chain can function as a ladder chain, as a wide V-belt, or, as here, a gear surrogate meshing with a standard pinion.

shown in Figure 2-8. This is sometimes called plastic ladder chain. It is a double cable form and is the kind that looks like a roller chain, except the rollers are replaced with non-rolling plastic cross pieces. These teeth engage a similar shape profile cut in the mating pulleys.

Another form of the cable-based drive wraps a spiral of plastic coated steel cable around the base cable. The pulleys for this form have a matching spiral-toothed groove. This type can bend in any direction, allowing it to be used to change drive planes. Both of these synchronous drive types are cheap and functional for low power applications.

CHAIN

Chain comes in three basic types.

- Ladder chain, generally used for power levels below 1/4 horsepower
- Roller chain, for fractional to hundreds of horsepower
- Timing chain, also called silent chain, for power levels in the tens to hundreds of horsepower

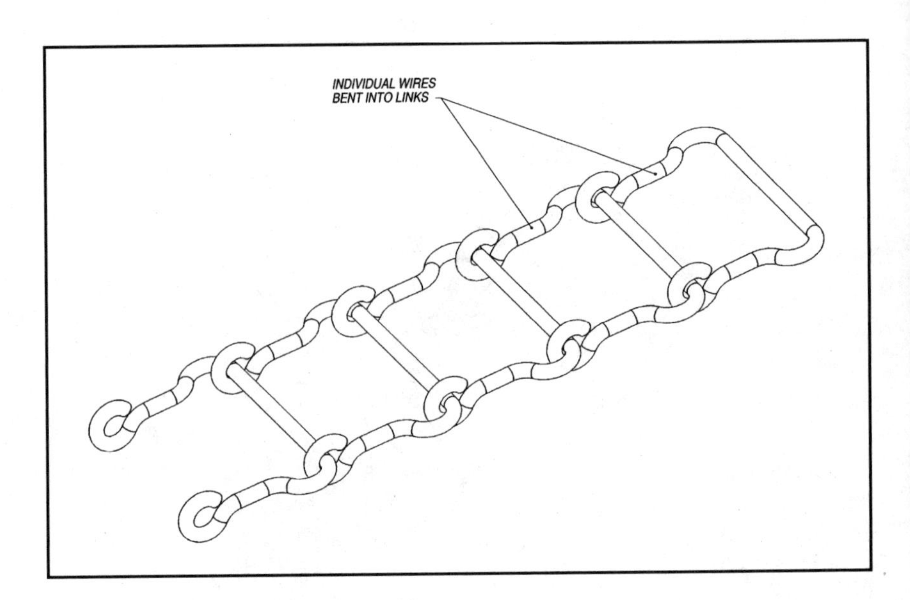

Figure 2-9 Ladder Chain

Ladder Chain

Ladder chain is so named because it looks like a very small ladder. Its construction is extremely simple and inexpensive. A short piece of wire is bent into a U shape and looped over the next U in the chain. See Figure 2-9. This construction is not very strong so this chain is used mainly where low cost is paramount and the power being transferred is less than 1/4 horsepower.

Roller Chain

Roller chain is an efficient power transfer method. It is called roller chain because it has steel rollers turning on pins held together by links. Roller chain is robust and can handle some misalignment between the driveR and driveN gears, and in many applications does not require precise pre-tensioning of the pulleys. It has two minor weaknesses.

1. It doesn't tolerate sand or abrasive environments very well.
2. It can be noisy.

Roller chain can be used for single stage reductions of up to 6:1 with careful attention to pulley spacing, making it a simple way to get an efficient, high reduction system. It is also surprisingly strong. The most common size chain, #40 (the distance from one roller to the next is .4") can transfer up to 2 horsepower at 300 rpm without special lubrication. Even the smallest size, #25, can transmit more than 5 horsepower at 3000 rpm with adequate forced lubrication and sufficiently large pulleys. There are several good references online that give much more detail than is within the scope of this book—they are

- americanchainassn.org
- bostongear.com
- diamondchain.com
- ramseychain.com
- ustsubaki.com

As shown in Figure 2-10 (a–d), roller chain comes in many sizes and styles, some of which are useful for things other than simply transferring power from one pulley to another.

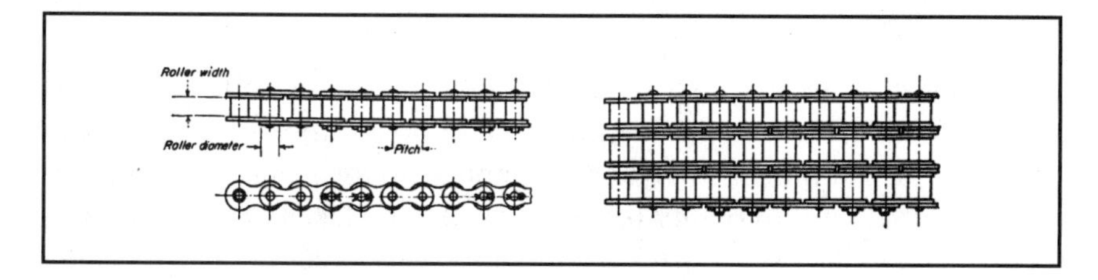

Figure 2-10a Standard roller chain—for power transmission and conveying.

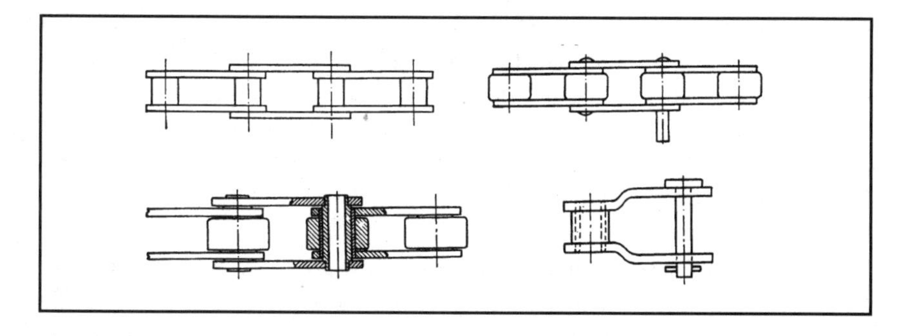

Figure 2-10b Extended pitch chain—for conveying

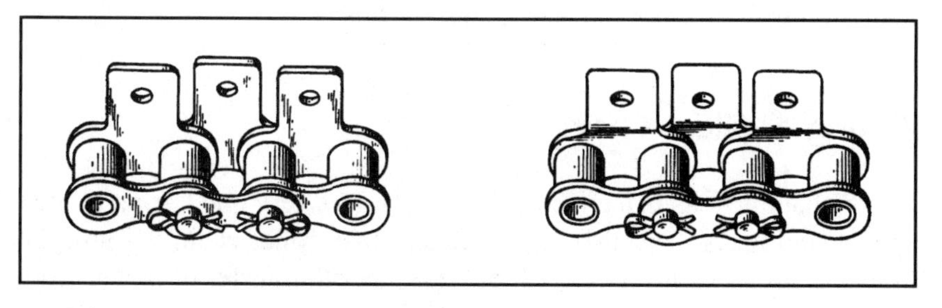

Figure 2-10c Standard pitch adaptations

Figure 2-10d Extended pitch adaptations

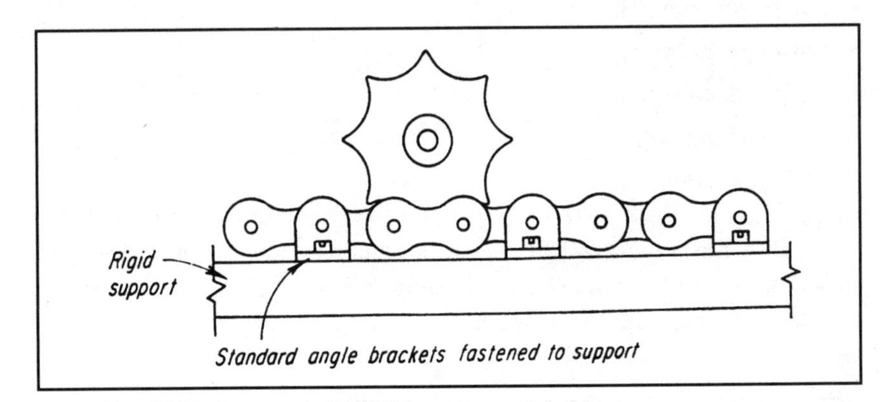

Figure 2-11 Bent lug roller chain used for rack and pinion linear actuator.

A clever, commercially available modification of roller chain has extended and bent lugs. These lugs can be bolted directly to pads and used for tracks on tracked vehicles, simplifying this sometimes complicated part of a high-mobility robot. Care must be taken to keep the pads as thin as possible, or to space them out to every other bent lug because debris can jam between the pads and cause problems. This is why tracks on excavators and military tanks are specially designed with the pivot point as close to the ground as possible. Other than that small issue, however, this chain can be and has been used as the backbone for tracks.

Rack and Pinion Chain Drive

Bent lug roller chain can also be used as a low cost rack and pinion drive to get linear motion from rotary motion. Though crude, this system works well if noise and a slightly non-smooth linear motion can be tolerated. Figure 2-11 shows a basic layout for this concept.

Timing or Silent Chain

Silent chain gets its name from the fact that it is very quiet, even at high speeds and loads. It is also more efficient than roller chain because the clever shape of its inverted teeth provide smooth transfer of power from

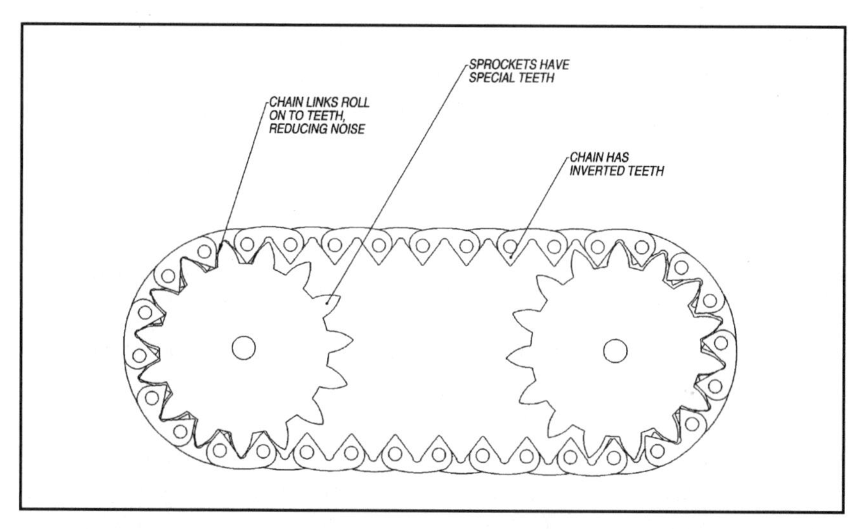

Figure 2-12 Silent chain tooth profile

the chain to the pulley. It is intolerant of grit, is somewhat more expensive, and requires more precision in alignment between the driveR and driveN pulleys than a roller chain.

It is a very good choice for transmitting high horsepower at thousands of rpm from an electric motor or an internal combustion engine to the transmission of large vehicles. It was used in the Oldsmobile Toranado automobile in the late 1970s, where it transmitted several hundred horsepower from the engine to the transmission. It is not made in small sizes because of the special shape of its teeth (Figure 2-12) and is designed mainly for power ranges from tens to hundreds of horsepower.

With proper design and simple maintenance, a silent chain drive will last for thousands of hours. If high efficiency and high power are required with operation in a clean environment, and the higher price can be afforded, silent chain is the best choice of any power transfer device in this book.

FRICTION DRIVES

Power can be transferred by friction alone. This technique is usually reserved for special cases, where its short life is acceptable. Its claim to fame is its high efficiency and ability to vary speed. The usual layout for a variable speed friction drive is a hardened steel wheel mounted on the input shaft, which is pushed very hard against a steel disk mounted on the output shaft. Efficiencies can be high, but the high forces required to carry the torque through only friction wear out the mating surfaces at a high rate. This drive has been used with some success in walk-behind lawn mowers, but its life in that application is usually only a couple seasons. Figure 2-13 shows one of several versions of a friction drive.

CONE DRIVE NEEDS NO GEARS OR PULLEYS

A variable-speed-transmission cone drive operates without gears or pulleys. The drive unit has its own limited slip differential and clutch.

As the drawing shows, two cones made of brake lining material are mounted on a shaft directly connected to the engine. These drive two larger steel conical disks mounted on the output shaft. The outer disks are mounted on pivoting frames that can be moved by a simple control rod.

To center the frames and to provide some resistance when the outer disks are moved, two torsion bars attached to the main frame connect and support the disk-support frames. By altering the position of the frames relative to the driving cones, the direction of rotation and speed can be varied.

The unit was invented by Marion H. Davis of Indiana.

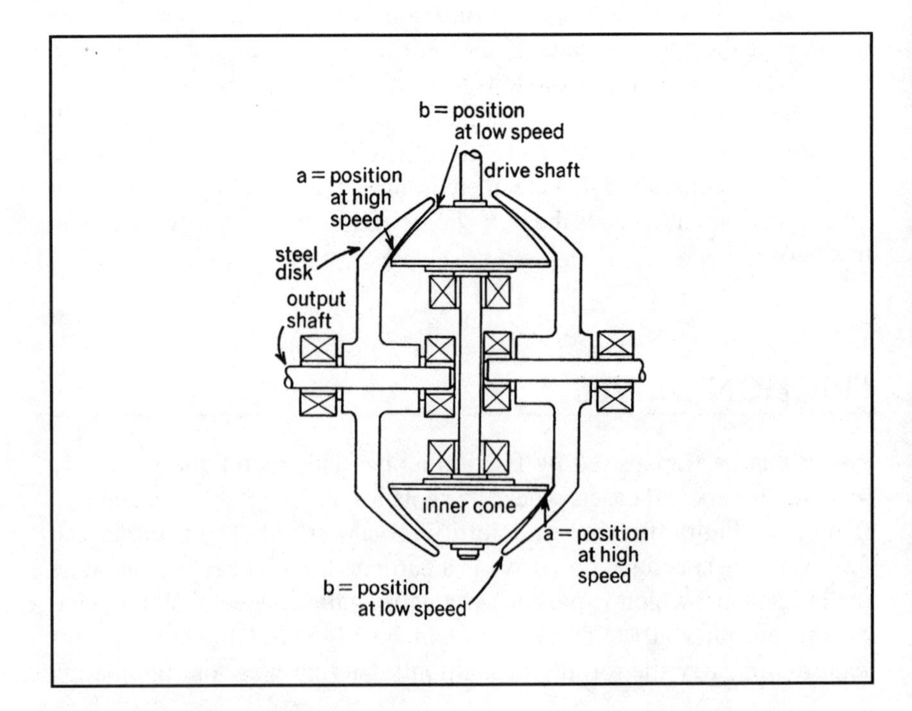

Figure 2-13 Cone drive operates without lubrication.

GEARS

Gears are the most common form of power transmission for several reasons. They can be scaled to transmit power from small battery powered watch motors (or even microscopic), up to the power from thousand horsepower gas turbine engines. Properly mounted and lubricated, they transmit power efficiently, smoothly, and quietly. They can transmit power between shafts that are parallel, intersecting, or even skew. For all their pluses, there are a few important things to remember about gears. To be efficient and quiet, they require high precision, both in the shape of the teeth and the distance between one gear and its mating gear. They do not tolerate dirt and must be enclosed in a sealed case that keeps the teeth clean and contains the required lubricating oil or grease. In general, gears are an excellent choice for the majority of power transmission applications.

Gears come in many forms and standard sizes, both inch and metric. They vary in diameter, tooth size, face width (the width of the gear), and tooth shape. Any two gears with the same tooth size can be used together, allowing very large ratios in a single stage. Large ratios between a single pair of gears cause problems with tooth wear and are usually obtained by using cluster gears to reduce the gearbox's overall size. Figure 2-14 shows an example of a cluster gear. Cluster gears reduce the size of a gearbox by adding an interim stage of gears. They are ubiquitous in practically every gearbox with a gear ratio of more than 5:1, with the exception of planetary and worm gearboxes.

Gears are available as spur, internal, helical, double helical (herringbone), bevel, spiral bevel, miter, face, hypoid, rack, straight worm, double enveloping worm, and harmonic. Each type has its own pros and cons, including differences in efficiency, allowable ratios, mating shaft angles, noise, and cost. Figure 2-15 shows the basic tooth profile of a spur gear.

Gears are versatile mechanical components capable of performing many different kinds of power transmission or motion control. Examples of these are

- Changing rotational speed.
- Changing rotational direction.
- Changing the angular orientation of rotational motion.
- Multiplication or division of torque or magnitude of rotation.
- Converting rotational to linear motion and its reverse.
- Offsetting or changing the location of rotating motion.

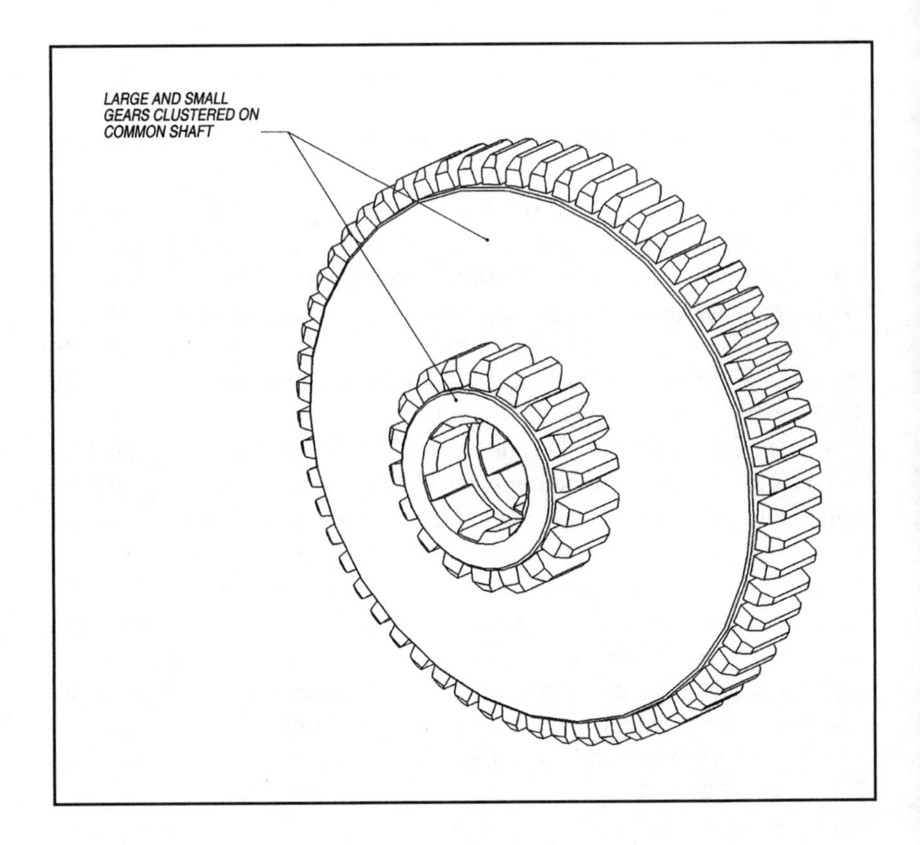

Figure 2-14 Cluster gear

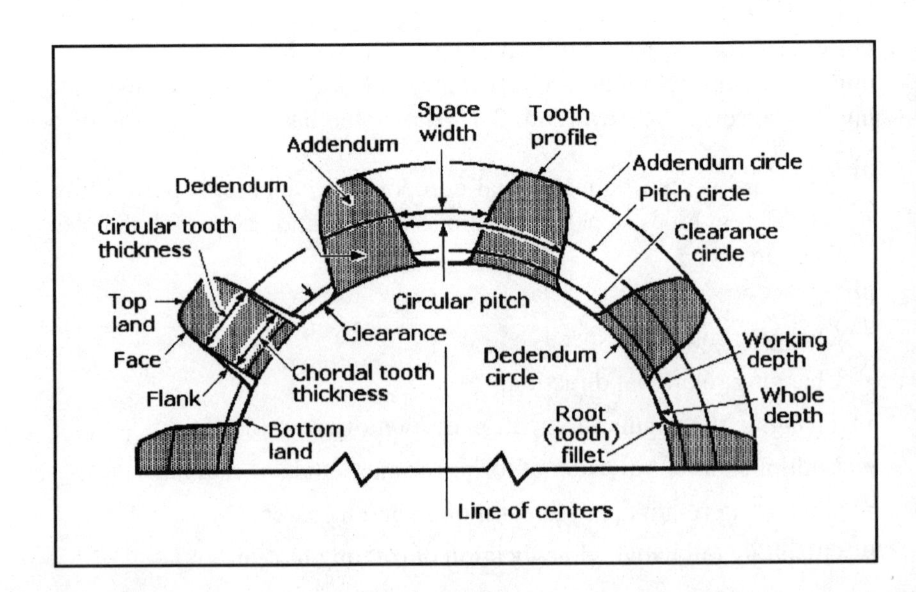

Figure 2-15 Gear Tooth Terminology

Gear Tooth Geometry: This is determined primarily by pitch, depth, and pressure angle.

Gear Terminology

addendum: The radial distance between the *top land* and the *pitch circle.*

addendum circle: The circle defining the outer diameter of the gear.

circular pitch: The distance along the pitch circle from a point on one tooth to a corresponding point on an adjacent tooth. It is also the sum of the *tooth thickness* and the space width, measured in inches or millimeters.

clearance: The radial distance between the *bottom land* and the *clearance circle.*

contact ratio: The ratio of the number of teeth in contact to the number of those not in contact.

dedendum circle: The theoretical circle through the *bottom lands* of a gear.

dedendum: The radial distance between the *pitch circle* and the *dedendum circle.*

depth: A number standardized in terms of pitch. Full-depth teeth have *a working depth of 2/P*. If the teeth have equal *addenda* (as in standard interchangeable gears), the addendum is 1/P. Full-depth gear teeth have a larger contact ratio than stub teeth, and their working depth is about 20% more than that of stub gear teeth. Gears with a small number of teeth might require *undercutting* to prevent one interfering with another during engagement.

diametral pitch (P): The ratio of the number of teeth to the *pitch diameter*. A measure of the coarseness of a gear, it is the index of tooth size when U.S. units are used, expressed as teeth per inch.

pitch: A standard pitch is typically a whole number when measured as *a diametral pitch (P)*. *Coarse-pitch gears* have teeth larger than a diametral pitch of 20 (typically 0.5 to 19.99). *Fine-pitch gears* usually have teeth of diametral pitch greater than 20. The usual maximum fineness is 120 diametral pitch, but involute-tooth gears can be made with diametral pitches as fine as 200, and cycloidal tooth gears can be made with diametral pitches to 350.

pitch circle: A theoretical circle upon which all calculations are based.

pitch diameter: The diameter of the *pitch circle*, the imaginary circle that rolls without slipping with the pitch circle of the mating gear, measured in inches or millimeters.

pressure angle: The angle between the *tooth profile* and a line perpendicular to the *pitch circle*, usually at the point where the pitch circle and the tooth profile intersect. Standard angles are 20 and 25°. The pressure angle affects the force that tends to separate mating gears. A high pressure angle decreases the *contact ratio*, but it permits the teeth to have higher capacity and it allows gears to have fewer teeth without *undercutting*.

Gear Dynamics Terminology

backlash: The amount by which the width of a tooth space exceeds the thickness of the engaging tooth measured on the pitch circle. It is the shortest distance between the noncontacting surfaces of adjacent teeth.

gear efficiency: The ratio of output power to input power, taking into consideration power losses in the gears and bearings and from windage and churning of lubricant.

gear power: A gear's load and speed capacity, determined by gear dimensions and type. Helical and helical-type gears have capacities to approximately 30,000 hp, spiral bevel gears to about 5000 hp, and worm gears to about 750 hp.

gear ratio: The number of teeth in the gear (larger of a pair) divided by the number of teeth in the *pinion* (smaller of a pair). Also, the ratio of the speed of the pinion to the speed of the gear. In reduction gears, the ratio of input to output speeds.

gear speed: A value determined by a specific pitchline velocity. It can be increased by improving the accuracy of the gear teeth and the balance of rotating parts.

undercutting: Recessing in the bases of gear tooth flanks to improve clearance.

Gear Classification

External gears have teeth on the outside surface of a disk or wheel.

Internal gears have teeth on the inside surface of a cylinder.

Spur gears are cylindrical gears with teeth that are straight and parallel to the axis of rotation. They are used to transmit motion between parallel shafts.

Rack gears have teeth on a flat rather than a curved surface that provide straight-line rather than rotary motion.

Helical gears have a cylindrical shape, but their teeth are set at an angle to the axis. They are capable of smoother and quieter action than spur gears. When their axes are parallel, they are called *parallel helical gears*, and when they are at right angles they are called helical gears. Herringbone and worm gears are based on helical gear geometry.

Herringbone gears are double helical gears with both right-hand and left-hand helix angles side by side across the face of the gear. This geometry neutralizes axial thrust from helical teeth.

Worm gears are crossed-axis helical gears in which the helix angle of one of the gears (the worm) has a high helix angle, resembling a screw.

Pinions are the smaller of two mating gears; the larger one is called the *gear* or *wheel*.

Bevel gears have teeth on a conical surface that mate on axes that intersect, typically at right angles. They are used in applications where there are right angles between input and output shafts. This class of gears includes the most common straight and spiral bevel as well as the miter and hypoid.

Straight bevel gears are the simplest bevel gears. Their straight teeth produce instantaneous line contact when they mate. These gears provide moderate torque transmission, but they are not as smooth running or quiet as spiral bevel gears because the straight teeth engage with full-line contact. They permit medium load capacity.

Spiral bevel gears have curved oblique teeth. The spiral angle of curvature with respect to the gear axis permits substantial tooth overlap. Consequently, teeth engage gradually and at least two teeth are in contact at the same time. These gears have lower tooth loading than straight bevel gears, and they can turn up to eight times faster. They permit high load capacity.

Miter gears are mating bevel gears with equal numbers of teeth and with their axes at right angles.

Hypoid gears are spiral bevel gears with offset intersecting axes.

Face gears have straight tooth surfaces, but their axes lie in planes perpendicular to shaft axes. They are designed to mate with instantaneous point contact. These gears are used in right-angle drives, but they have low load capacities.

Designing a properly sized gearbox is not a simple task and tables or manufacturer's recommendations are usually the best place to look for help. The amount of power a gearbox can transmit is affected by gear size, tooth size, rpm of the faster shaft, lubrication method, available cooling method (everything from nothing at all to forced air), gear materials, bearing types, etc. All these variables must be taken into account to come up with an effectively sized gearbox. Don't be daunted by this. In most cases the gearbox is not designed at all, but easily selected from a large assortment of off-the-shelf gearboxes made by one of many manufacturers. Let's now turn our attention to more complicated gearboxes that do more than just exchange speed for torque.

Worm Gears

Worm gear drives get their name from the unusual input gear which looks vaguely like a worm wrapped around a shaft. They are used primarily for high reduction ratios, from 5:1 to 100s:1. Their main disadvantage is inefficiency caused by the worm gear's sliding contact with the worm wheel. In larger reduction ratios, they can be self locking, meaning when the input power is turned off, the output cannot be rotated. The following section discusses an unusual double enveloping, internally-lubricated worm gear layout that is an attempt to increase efficiency and the life of the gearbox.

WORM GEAR WITH HYDROSTATIC ENGAGEMENT

Friction would be reduced greatly.

Lewis Research Center, Cleveland, Ohio

In a proposed worm-gear transmission, oil would be pumped at high pressure through the meshes between the teeth of the gear and the worm coil (Figure 2-16). The pressure in the oil would separate the meshing surfaces slightly, and the oil would reduce the friction between these sur-

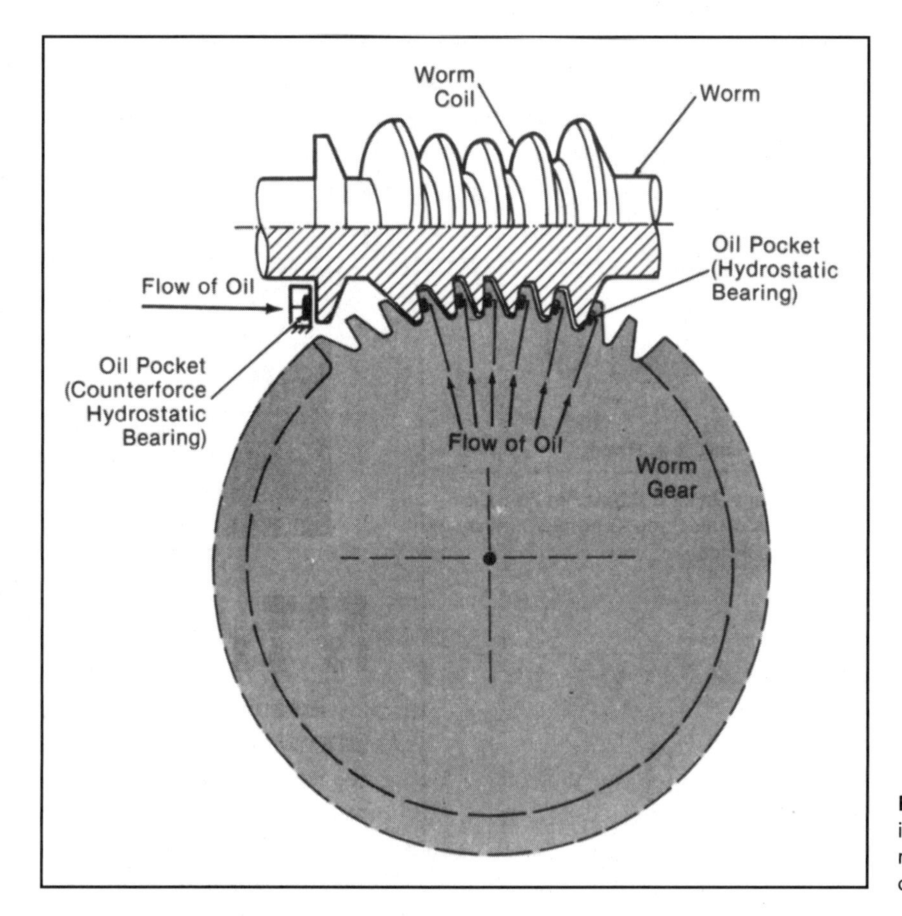

Figure 2-16 Oil would be injected at high pressure to reduce friction in critical areas of contact

faces. Each of the separating forces in the several meshes would contribute to the torque on the gear and to an axial force on the worm. To counteract this axial force and to reduce the friction that it would otherwise cause, oil would also be pumped under pressure into a counterforce hydrostatic bearing at one end of the worm shaft.

This type of worm-gear transmission was conceived for use in the drive train between the gas-turbine engine and the rotor of a helicopter and might be useful in other applications in which weight is critical. Worm gear is attractive for such weight-critical applications because (1) it can transmit torque from a horizontal engine (or other input) shaft to a vertical rotor (or other perpendicular output) shaft, reducing the speed by the desired ratio in one stage, and (2) in principle, a one-stage design can be implemented in a gearbox that weighs less than does a conventional helicopter gearbox.

Heretofore, the high sliding friction between the worm coils and the gear teeth of worm-gear transmissions has reduced efficiency so much

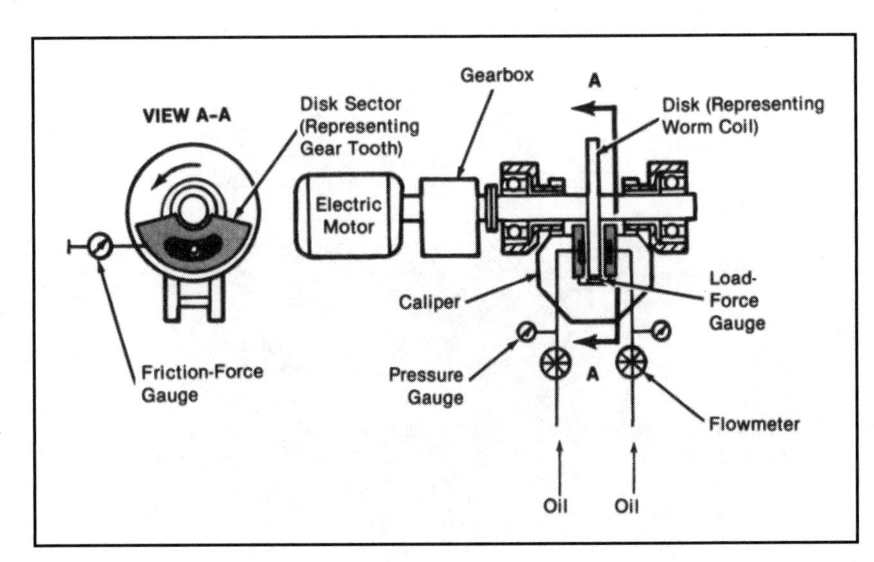

Figure 2-17 This test apparatus simulates and measures some of the loading conditions of the proposed worm gear with hydrostatic engagement. The test data will be used to design efficient worm-gear transmissions.

that such transmissions could not be used in helicopters. The efficiency of the proposed worm-gear transmission with hydrostatic engagement would depend partly on the remaining friction in the hydrostatic meshes and on the power required to pump the oil. Preliminary calculations show that the efficiency of the proposed transmission could be the same as that of a conventional helicopter gear train.

Figure 2-17 shows an apparatus that is being used to gather experimental data pertaining to the efficiency of a worm gear with hydrostatic engagement. Two stationary disk sectors with oil pockets represent the gear teeth and are installed in a caliper frame. A disk that represents the worm coil is placed between the disk sectors in the caliper and is rotated rapidly by a motor and gearbox. Oil is pumped at high pressure through the clearances between the rotating disk and the stationary disk sectors. The apparatus is instrumented to measure the frictional force of meshing and the load force.

The stationary disk sectors can be installed with various clearances and at various angles to the rotating disk. The stationary disk sectors can be made in various shapes and with oil pockets at various positions. A flowmeter and pressure gauge will measure the pump power. Oils of various viscosities can be used. The results of the tests are expected to show the experimental dependences of the efficiency of transmission on these factors.

It has been estimated that future research and development will make it possible to make worm-gear helicopter transmission that weigh half as much as conventional helicopter transmissions do. In addition, the new hydrostatic meshes would offer longer service life and less noise. It might even be possible to make the meshing worms and gears, or at least parts of them, out of such lightweight materials as titanium, aluminum, and composites.

This work was done by Lev. I. Chalko of the U.S. Army Propulsion Directorate (AVSCOM) for **Lewis Research Center.**

CONTROLLED DIFFERENTIAL DRIVES

By coupling a differential gear assembly to a variable speed drive, a drive's horsepower capacity can be increased at the expense of its speed range. Alternatively, the speed range can be increased at the expense of the horsepower range. Many combinations of these variables are possible. The features of the differential depend on the manufacturer. Some systems have bevel gears, others have planetary gears. Both single and double differentials are employed. Variable-speed drives with differential gears are available with ratings up to 30 hp.

Horsepower-increasing differential. The differential is coupled so that the output of the motor is fed into one side and the output of the speed variator is fed into the other side. An additional gear pair is employed as shown in Figure 2-18.

Output speed

$$n_4 = \frac{1}{2}\left(n_1 + \frac{n_2}{R}\right)$$

Output torque

$$T_4 = 2T_3 = 2RT_2$$

Output hp

$$\text{hp} = \left(\frac{Rn_1 + n_2}{63{,}025}\right)T_2$$

hp increase

$$\Delta\text{hp} = \left(\frac{Rn_1}{63{,}025}\right)T_2$$

Speed variation

$$n_{4\,\max} - n_{4\,\min} = \frac{1}{2R}(n_{2\,\max} - n_{2\,\min})$$

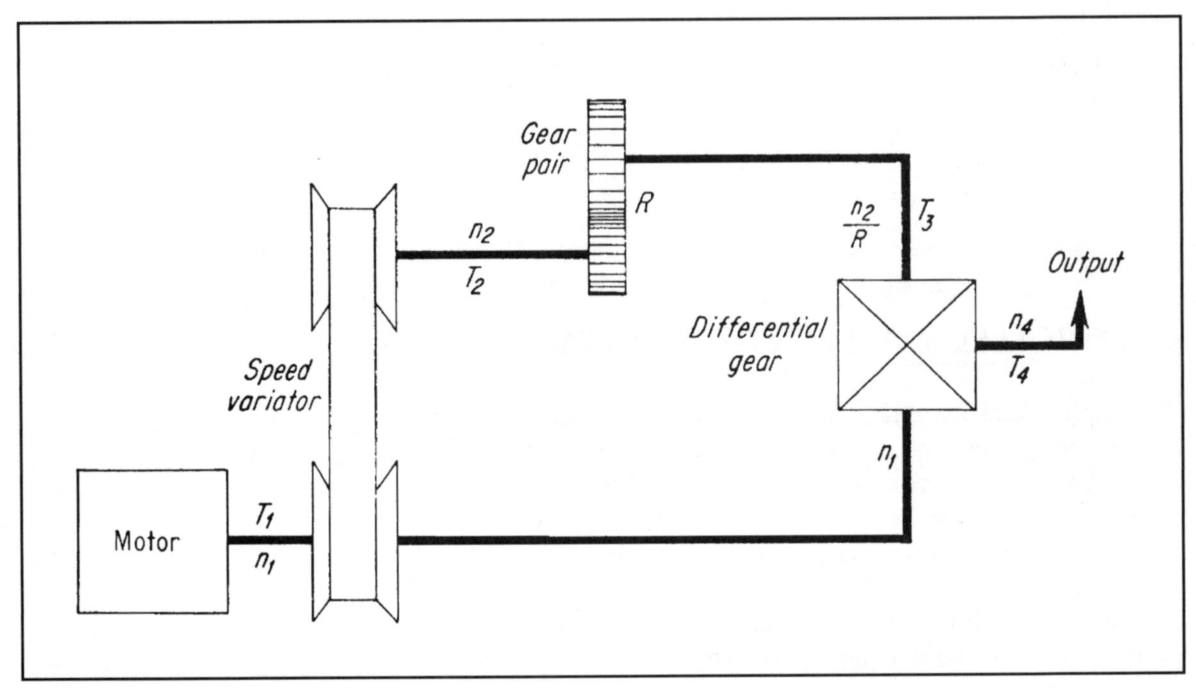

Figure 2-18

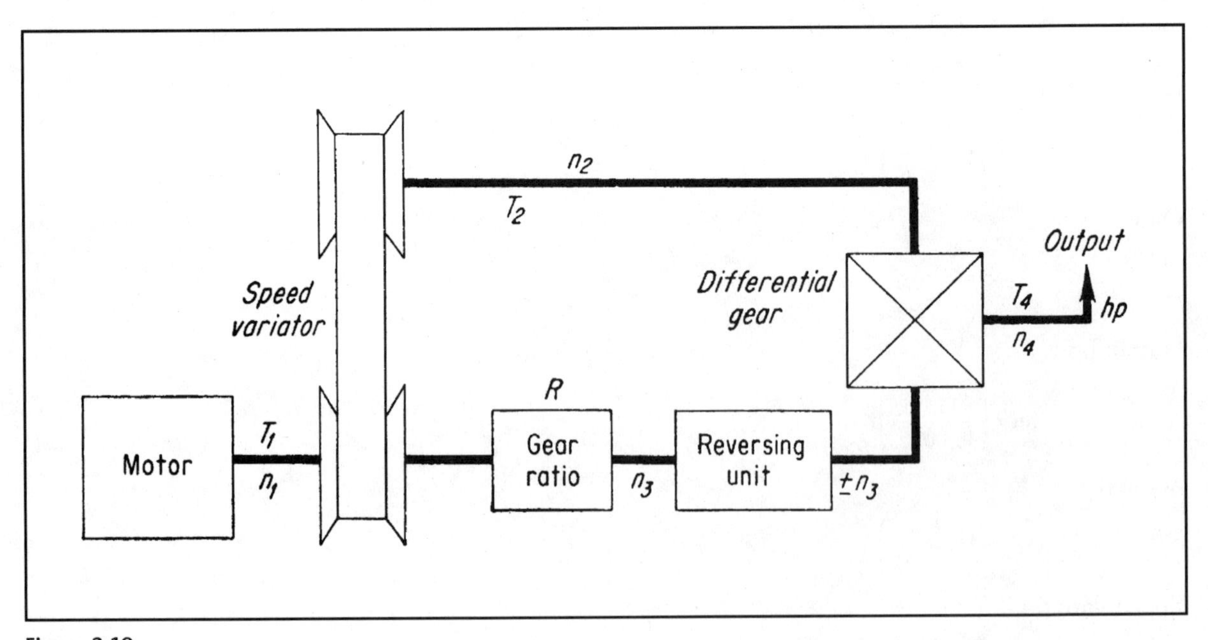

Figure 2-19

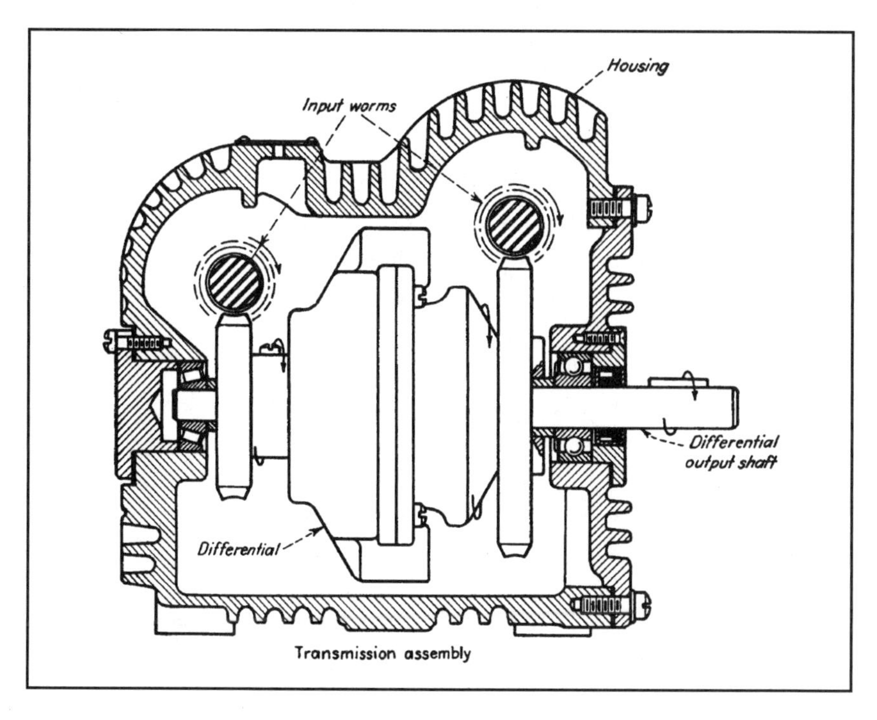

Figure 2-20 A variable-speed transmission consists of two sets of worm gears feeding a differential mechanism. The output shaft speed depends on the difference in rpm between the two input worms. When the worm speeds are equal, output is zero. Each worm shaft carries a cone-shaped pulley. These pulley are mounted so that their tapers are in opposite directions. Shifting the position of the drive belt on these pulleys has a compound effect on their output speed.

Speed range increase differential (Figure 2-19). This arrangement achieves a wide range of speed with the low limit at zero or in the reverse direction.

TWIN-MOTOR PLANETARY GEARS PROVIDE SAFETY PLUS DUAL-SPEED

Many operators and owners of hoists and cranes fear the possible catastrophic damage that can occur if the driving motor of a unit should fail for any reason. One solution to this problem is to feed the power of two motors of equal rating into a planetary gear drive.

Power supply. Each of the motors is selected to supply half the required output power to the hoisting gear (see Figure 2-21). One motor drives the ring gear, which has both external and internal teeth. The second motor drives the sun gear directly.

Both the ring gear and sun gear rotate in the same direction. If both gears rotate at the same speed, the planetary cage, which is coupled to

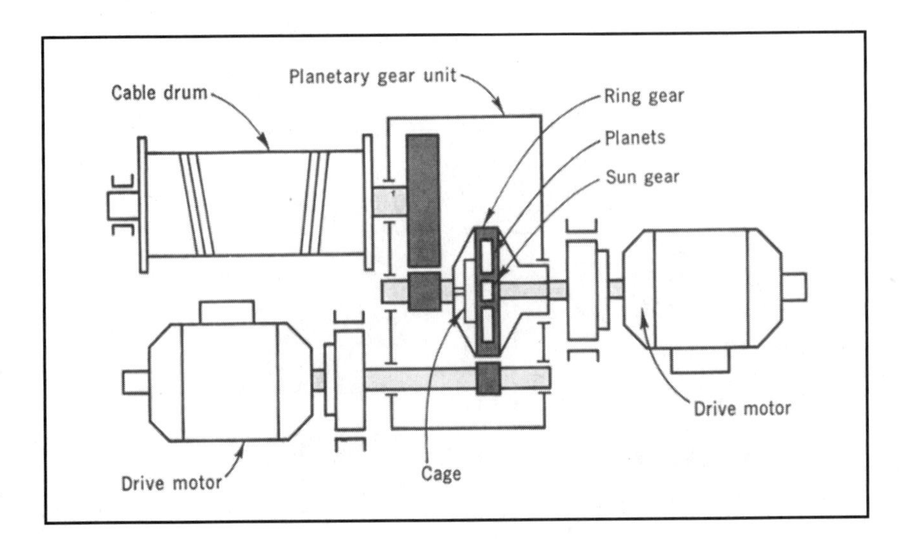

Figure 2-21 Power flow from two motors combine in a planetary that drives the cable drum.

the output, will also revolve at the same speed (and in the same direction). It is as if the entire inner works of the planetary were fused together. There would be no relative motion. Then, if one motor fails, the cage will revolve at half its original speed, and the other motor can still lift with undiminished capacity. The same principle holds true when the ring gear rotates more slowly than the sun gear.

No need to shift gears. Another advantage is that two working speeds are available as a result of a simple switching arrangement. This makes is unnecessary to shift gears to obtain either speed.

The diagram shows an installation for a steel mill crane.

HARMONIC-DRIVE SPEED REDUCERS

The harmonic-drive speed reducer was invented in the 1950s at the Harmonic Drive Division of the United Shoe Machinery Corporation, Beverly, Massachusetts. These drives have been specified in many high-performance motion-control applications. Although the Harmonic Drive Division no longer exists, the manufacturing rights to the drive have been sold to several Japanese manufacturers, so they are still made and sold. Most recently, the drives have been installed in industrial robots, semiconductor manufacturing equipment, and motion controllers in military and aerospace equipment.

The history of speed-reducing drives dates back more than 2000 years. The first record of reducing gears appeared in the writings of the Roman engineer Vitruvius in the first century B.C. He described wooden-

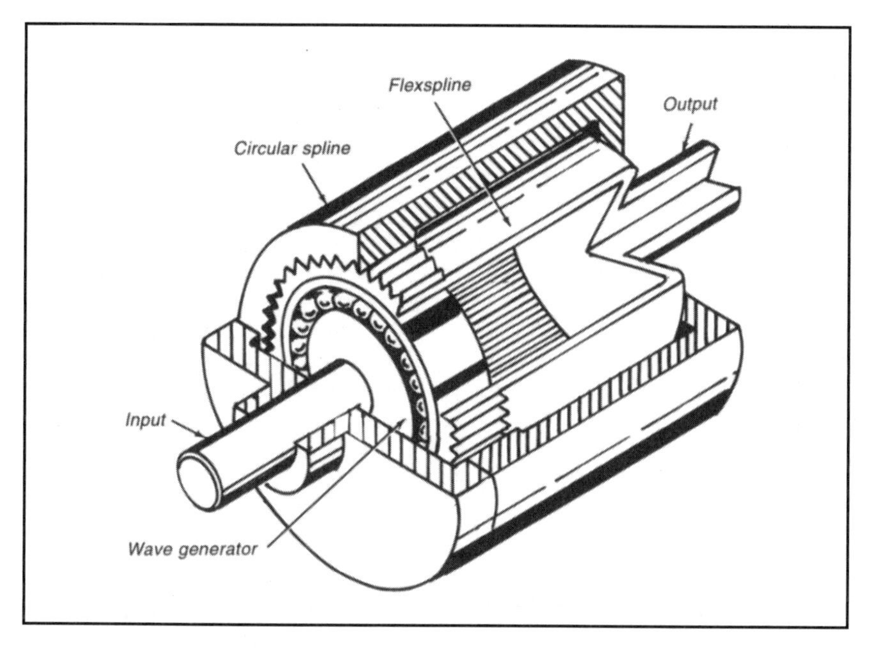

Figure 2-22 Exploded view of a typical harmonic drive showing its principal parts. The flexspline has a smaller outside diameter than the inside diameter of the circular spline, so the elliptical wave generator distorts the flexspline so that its teeth, 180° apart, mesh.

tooth gears that coupled the power of water wheel to millstones for grinding corn. Those gears offered about a 5 to 1 reduction. In about 300 B.C., Aristotle, the Greek philosopher and mathematician, wrote about toothed gears made from bronze.

In 1556, the Saxon physician, Agricola, described geared, horse-drawn windlasses for hauling heavy loads out of mines in Bohemia. Heavy-duty cast-iron gear wheels were first introduced in the mid-eighteenth century, but before that time gears made from brass and other metals were included in small machines, clocks, and military equipment.

The harmonic drive is based on a principle called *strain-wave gearing,* a name derived from the operation of its primary torque-transmitting element, the flexspline. Figure 2-22 shows the three basic elements of the harmonic drive: the rigid circular spline, the fliexible flexspline, and the ellipse-shaped wave generator.

The *circular spline* is a nonrotating, thick-walled, solid ring with internal teeth. By contrast, a *flexspline* is a thin-walled, flexible metal cup with external teeth. Smaller in external diameter than the inside diameter of the circular spline, the flexspline must be deformed by the wave generator if its external teeth are to engage the internal teeth of the circular spline.

When the *elliptical cam wave generator* is inserted into the bore of the flexspline, it is formed into an elliptical shape. Because the major axis of the wave generator is nearly equal to the inside diameter of the circular

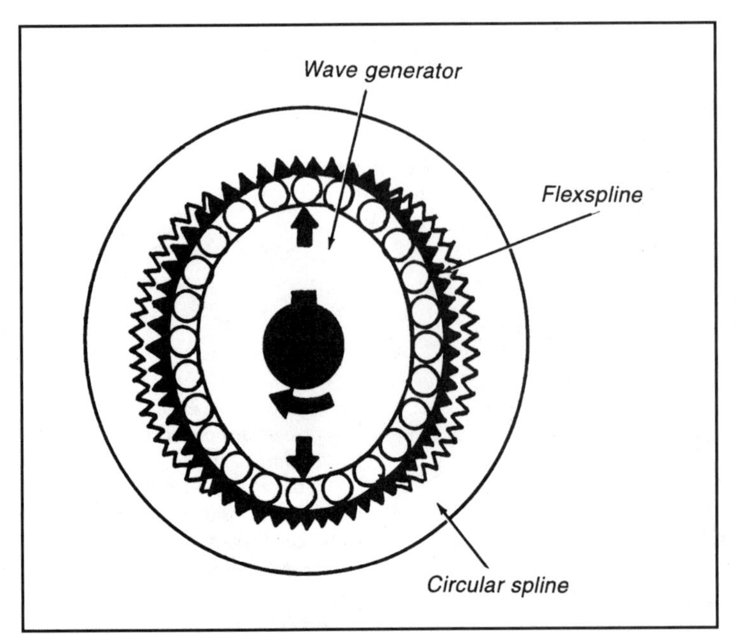

Figure 2-23 Schematic of a typical harmonic drive showing the mechanical relationship between the two splines and the wave generator.

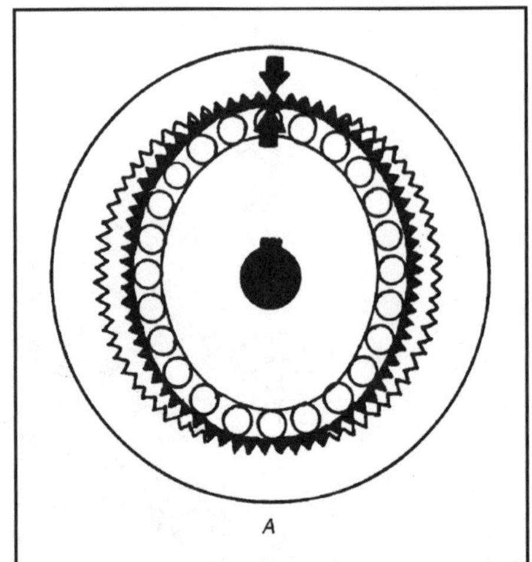

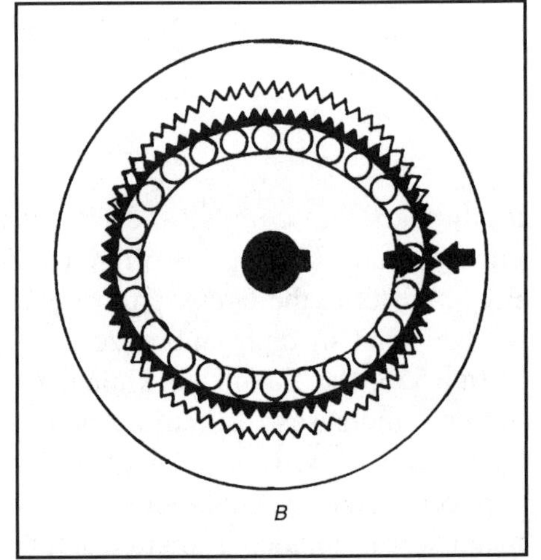

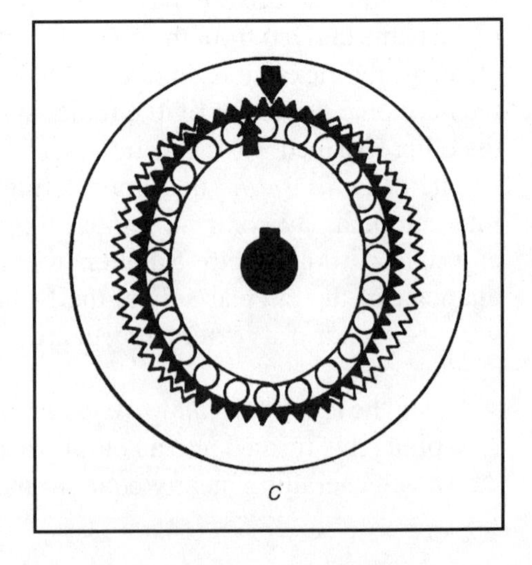

spline, external teeth of the flexspline that are 180° apart will engage the internal circular-spline teeth.

Modern wave generators are enclosed in a ball-bearing assembly that functions as the rotating input element. When the wave generator transfers its elliptical shape to the flexspline and the external circular spline teeth have engaged the internal circular spline teeth at two opposing locations, a positive gear mesh occurs at those engagement points. The shaft attached to the flexspline is the rotating output element.

Figure 2-23 is a schematic presentation of harmonic gearing in a section view. The flexspline typically has two fewer external teeth than the number of internal teeth on the circular spline. The keyway of the input shaft is at its zero-degree or 12 o'clock position. The small circles around the shaft are the ball bearings of the wave generator.

Figure 2-24 is a schematic view of a harmonic drive in three operating positions. In Figure 2-24A, the inside and outside arrows are aligned. The inside arrow indicates that the wave generator is in its 12 o'clock position with respect to the circular spline, prior to its clockwise rotation.

Figure 2-24 Three positions of the wave generator: (A) the 12 o'clock or zero degree position; (B) the 3 o'clock or 90° position; and (C) the 360° position showing a two-tooth displacement.

Because of the elliptical shape of the wave generator, full tooth engagement occurs only at the two areas directly in line with the major axis of the ellipse (the vertical axis of the diagram). The teeth in line with the minor axis are completely disengaged.

As the wave generator rotates 90° clockwise, as shown in Figure 2-24B, the inside arrow is still pointing at the same flexspline tooth, which has begun its counterclockwise rotation. Without full tooth disengagement at the areas of the minor axis, this rotation would not be possible.

At the position shown in Figure 2-24C, the wave generator has made one complete revolution and is back at its 12 o'clock position. The inside arrow of the flexspline indicates a two-tooth per revolution displacement counterclockwise. From this one revolution motion the reduction ratio equation can be written as:

$$GR = \frac{FS}{CS - FS}$$

where:
GR =gear ratio
FS = number of teeth on the flexspline
CS = number of teeth on the circular spline
Example:
$FS = 200$ teeth
$CS = 202$ teeth

$$GR = \frac{200}{202 - 200} = 100 : 1 \text{ reduction}$$

As the wave generator rotates and flexes the thin-walled spline, the teeth move in and out of engagement in a rotating wave motion. As might be expected, any mechanical component that is flexed, such as the flexspline, is subject to stress and strain.

Advantages and Disadvantages

The harmonic drive was accepted as a high-performance speed reducer because of its ability to position moving elements precisely. Moreover, there is no backlash in a harmonic drive reducer. Therefore, when positioning inertial loads, repeatability and resolution are excellent (one arc-minute or less).

Because the harmonic drive has a concentric shaft arrangement, the input and output shafts have the same centerline. This geometry contributes to its compact form factor. The ability of the drive to provide high reduction ratios in a single pass with high torque capacity recommends it for many machine designs. The benefits of high mechanical efficiency are high torque capacity per pound and unit of volume, both attractive performance features.

One disadvantage of the harmonic drive reducer has been its wind-up or torsional spring rate. The design of the drive's tooth form necessary for the proper meshing of the flexspline and the circular spline permits only one tooth to be completely engaged at each end of the major elliptical axis of the generator. This design condition is met only when there is no torsional load. However, as torsional load increases, the teeth bend slightly and the flexspline also distorts slightly, permitting adjacent teeth to engage.

Paradoxically, what could be a disadvantage is turned into an advantage because more teeth share the load. Consequently, with many more teeth engaged, torque capacity is higher, and there is still no backlash. However, this bending and flexing causes torsional wind-up, the major contributor to positional error in harmonic-drive reducers.

At least one manufacturer claims to have overcome this problem with redesigned gear teeth. In a new design, one company replaced the original involute teeth on the flexspline and circular spline with noninvolute teeth. The new design is said to reduce stress concentration, double the fatigue limit, and increase the permissible torque rating.

The new tooth design is a composite of convex and concave arcs that match the loci of engagement points. The new tooth width is less than the width of the tooth space and, as a result of these dimensions and proportions, the root fillet radius is larger.

FLEXIBLE FACE-GEARS MAKE EFFICIENT HIGH-REDUCTION DRIVES

A system of flexible face-gearing provides designers with a means for obtaining high-ratio speed reductions in compact trains with concentric input and output shafts.

With this approach, reduction ratios range from 10:1 to 200:1 for single-stage reducers, whereas ratios of millions to one are possible for multi-stage trains. Patents on the flexible face-gear reducers were held by Clarence Slaughter of Grand Rapids, Michigan.

Building blocks. Single-stage gear reducers consist of three basic parts: a flexible face-gear (Figure 2-25) made of plastic or thin metal; a solid, non-flexing face-gear; and a wave former with one or more sliders and rollers to force the flexible gear into mesh with the solid gear at points where the teeth are in phase.

The high-speed input to the system usually drives the wave former. Low-speed output can be derived from either the flexible or the solid face gear; the gear not connected to the output is fixed to the housing.

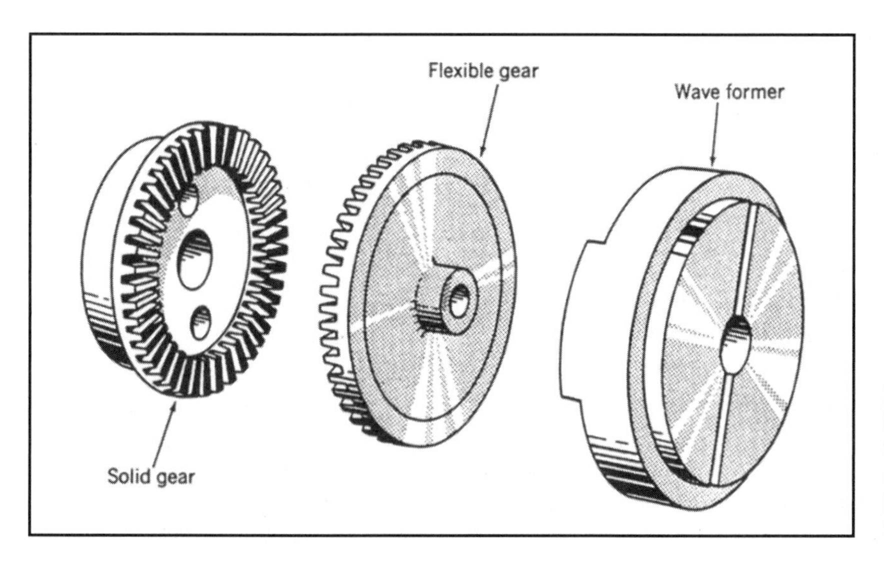

Figure 2-25 A flexible face-gear is flexed by a rotating wave former into contact with a solid gear at point of mesh. The two gears have slightly different numbers of teeth.

Teeth make the difference. Motion between the two gears depends on a slight difference in their number of teeth (usually one or two teeth). But drives with gears that have up to a difference of 10 teeth have been devised.

On each revolution of the wave former, there is a relative motion between the two gears that equals the difference in their numbers of teeth. The reduction ratio equals the number of teeth in the output gear divided by the difference in their numbers of teeth.

Two-stage (Figure 2-26) and four-stage (Figure 2-27) gear reducers are made by combining flexible and solid gears with multiple rows of teeth and driving the flexible gears with a common wave former.

Hermetic sealing is accomplished by making the flexible gear serve as a full seal and by taking output rotation from the solid gear.

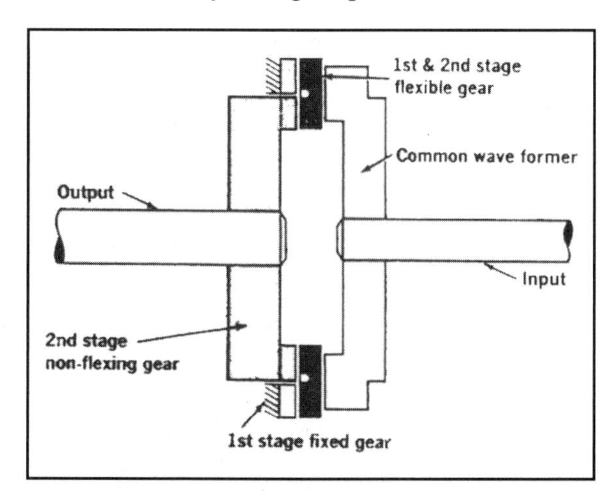

Figure 2-26 A two-stage speed reducer is driven by a common-wave former operating against an integral flexible gear for both stages.

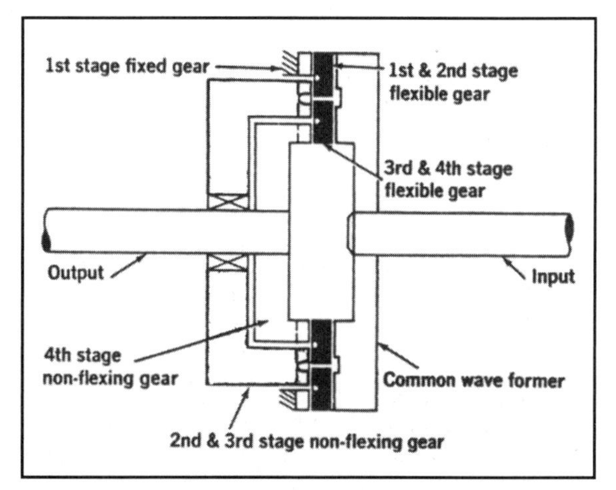

Figure 2-27 A four-stage speed reducer can, theoretically, attain reductions of millions to one. The train is both compact and simple.

HIGH-SPEED GEARHEADS IMPROVE SMALL SERVO PERFORMANCE

The factory-made precision gearheads now available for installation in the latest smaller-sized servosystems can improve their performance while eliminating the external gears, belts, and pulleys commonly used in earlier larger servosystems. The gearheads can be coupled to the smaller, higher-speed servomotors, resulting in simpler systems with lower power consumption and operating costs.

Gearheads, now being made in both in-line and right-angle configurations, can be mounted directly to the drive motor shafts. They can convert high-speed, low-torque rotary motion to a low-speed, high-torque output. The latest models are smaller and more accurate than their predecessors, and they have been designed to be compatible with the smaller, more precise servomotors being offered today.

Gearheads have often been selected for driving long trains of mechanisms in machines that perform such tasks as feeding wire, wood, or metal for further processing. However, the use of an in-line gearhead adds to the space occupied by these machines, and this can be a problem where factory floor space is restricted. One way to avoid this problem is to choose a right-angle gearhead (Figure 2-28). It can be mounted vertically beneath the host machine or even horizontally on the machine bed. Horizontal mounting can save space because the gearheads and motors can be positioned behind the machine, away from the operator.

Bevel gears are commonly used in right-angle drives because they can provide precise motion. Conically shaped bevel gears with straight- or spiral-cut teeth allow mating shafts to intersect at 90° angles. Straight-cut bevel gears typically have contact ratios of about 1.4, but the simultaneous mating of straight teeth along their entire lengths causes more vibration and noise than the mating of spiral-bevel gear teeth. By contrast, spiral-bevel gear teeth engage and disengage gradually and precisely with contact ratios of 2.0 to 3.0, making little noise. The higher contact ratios of spiral-bevel gears permit them to drive loads that are 20 to 30% greater than those possible with straight bevel gears. Moreover, the spiral-bevel teeth mesh with a rolling action that increases their precision and also reduces friction. As a result, operating efficiencies can exceed 90%.

Simplify the Mounting

The smaller servomotors now available force gearheads to operate at higher speeds, making vibrations more likely. Inadvertent misalignment between servomotors and gearboxes, which often occurs during installation, is a common source of vibration. The mounting of conventional

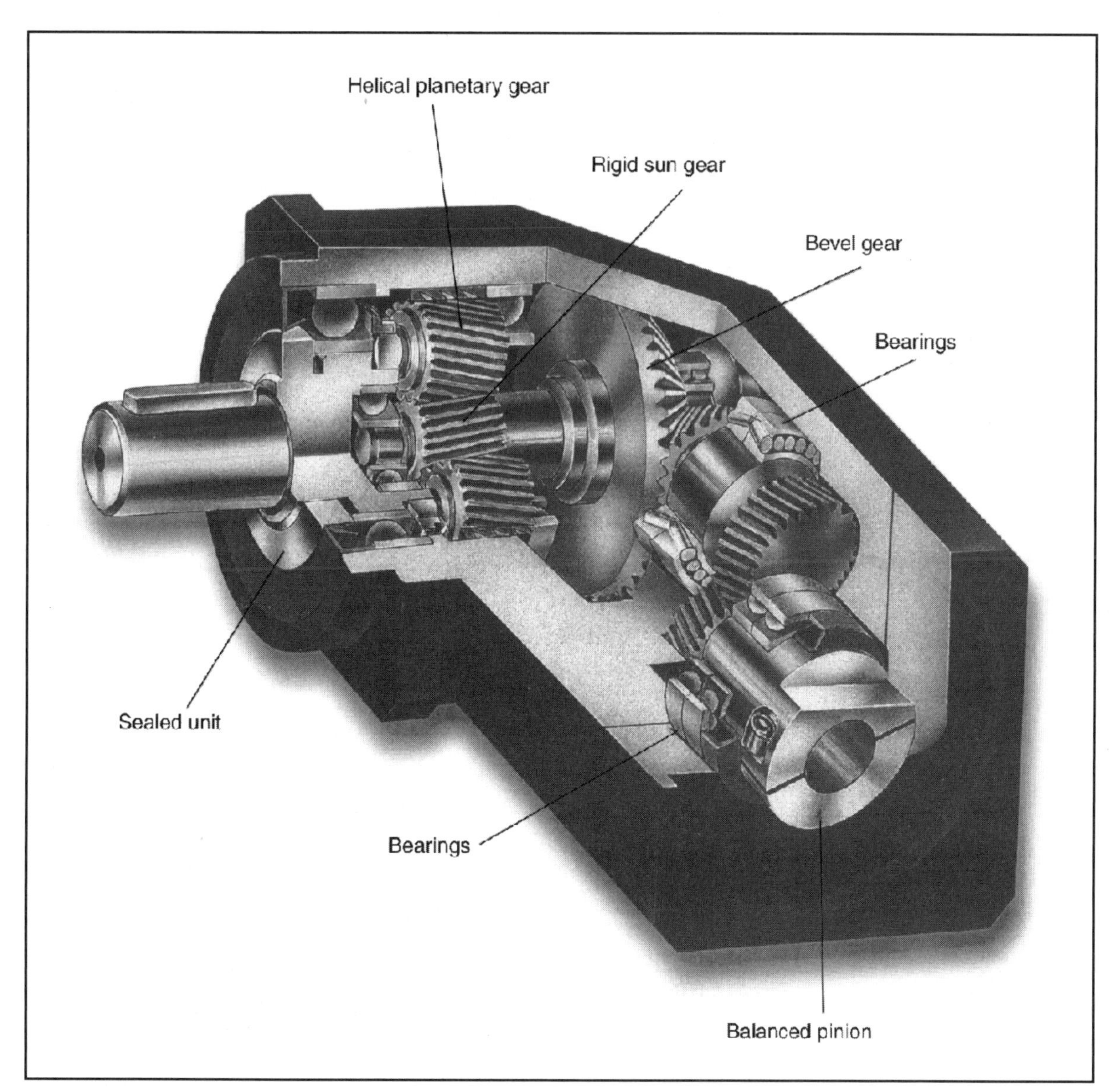

Figure 2-28 This right-angle gearhead is designed for high-performance servo applications. It includes helical planetary output gears, a rigid sun gear, spiral bevel gears, and a balanced input pinion. *Courtesy of Bayside Controls Inc.*

motors with gearboxes requires several precise connections. The output shaft of the motor must be attached to the pinion gear that slips into a set of planetary gears in the end of the gearbox, and an adapter plate must joint the motor to the gearbox. Unfortunately, each of these connections can introduce slight alignment errors that accumulate to cause overall motor/gearbox misalignment.

The pinion is the key to smooth operation because it must be aligned exactly with the motor shaft and gearbox. Until recently it has been standard practice to mount pinions in the field when the motors were connected to the gearboxes. This procedure often caused the assembly to vibrate. Engineers realized that the integration of gearheads into the servomotor package would solve this problem, but the drawback to the integrated unit is that failure of either component would require replacement of the whole unit.

A more practical solution is to make the pinion part of the gearhead assembly because gearheads with built-in pinions are easier to mount to servomotors than gearheads with field-installed pinions. It is only necessary to insert the motor shaft into the collar that extends from the gearhead's rear housing, tighten the clamp with a wrench, and bolt the motor to the gearhead.

Pinions installed at the factory ensure smooth-running gearheads because they are balanced before they are mounted. This procedure permits them to spin at high speed without wobbling. As a result, the balanced pinions minimize friction and thus cause less wear, noise, and vibration than field-installed pinions.

However, the factory-installed pinion requires a floating bearing to support the shaft with a pinion on one end. The Bayside Motion Group of Bayside Controls Inc., Port Washington, New York, developed a self-aligning bearing for this purpose. Bayside gearheads with these pinions are rated for input speeds up to 5000 rpm. A collar on the pinion shaft's other end mounts to the motor shaft. The bearing holds the pinion in place until it is mounted. At that time a pair of bearings in the servomotor support the coupled shaft. The self-aligning feature of the floating bearing lets the motor bearing support the shaft after installation.

The pinion and floating bearing help to seal the unit during its operation. The pinion rests in a blind hole and seals the rear of the gearhead. This seal keeps out dirt while retaining the lubricants within the housing. Consequently, light grease and semifluid lubricants can replace heavy grease.

Cost-Effective Addition

The installation of gearheads can smooth the operation of servosystems as well as reduce system costs. The addition of a gearhead to the system does not necessarily add to overall operating costs because its purchase price can be offset by reductions in operating costs. Smaller servomotors inherently draw less current than larger ones, thus reducing operating costs, but those power savings are greatest in applications calling for low speed and high torque because direct-drive servomotors must be considerably larger than servomotors coupled to gearheads to perform the same work.

Small direct-drive servomotors assigned to high-speed/low-torque applications might be able to perform the work satisfactorily without a gearhead. In those instances servo/gearhead combinations might not be as cost-effective because power consumption will be comparable. Nevertheless, gearheads will still improve efficiency and, over time, even small decreases in power consumption due to the use of smaller-sized servos will result in reduced operating costs.

The decision to purchase a precision gearhead should be evaluated on a case-by-case basis. The first step is to determine speed and torque requirements. Then keep in mind that although in high-speed/low-torque applications a direct-drive system might be satisfactory, low-speed/high-torque applications almost always require gearheads. Then a decision can be made after weighing the purchase price of the gearhead against anticipated servosystem operating expenses in either operating mode to estimate savings.

The planetary gearbox is one of the most efficient and compact gearbox designs. Its internal coaxial layout reduces efficiency robbing side loads on the gear's shafts. Figure 2-29 and the following tables give the formulas required to calculate the input/output ratios. In spite of its higher cost, a planetary gearbox is frequently the best choice for medium ratio power transfer. Because of its even greater precision requirement than spur gears, it is usually better to buy an off-the-shelf gearbox than to design your own.

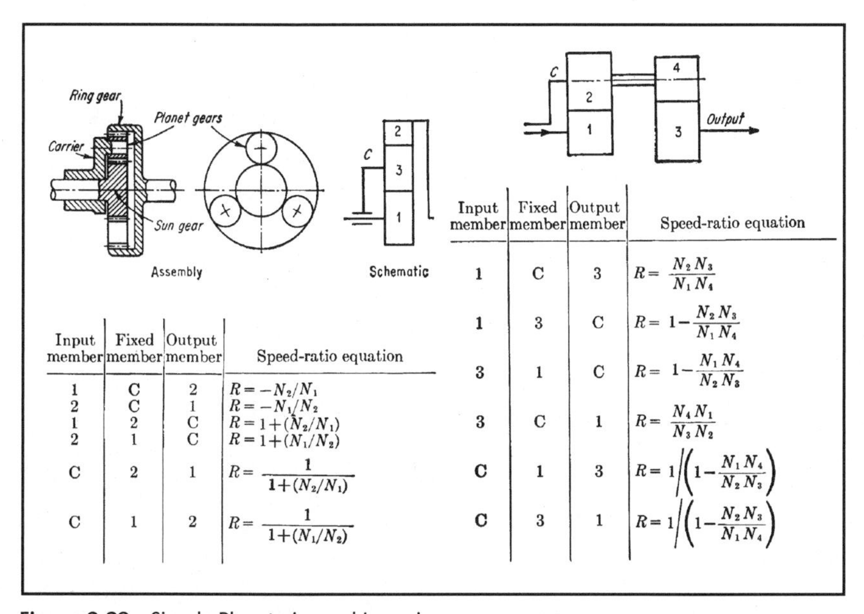

Input member	Fixed member	Output member	Speed-ratio equation
1	C	2	$R=-N_2/N_1$
2	C	1	$R=-N_1/N_2$
1	2	C	$R=1+(N_2/N_1)$
2	1	C	$R=1+(N_1/N_2)$
C	2	1	$R=\frac{1}{1+(N_2/N_1)}$
C	1	2	$R=\frac{1}{1+(N_1/N_2)}$

Input member	Fixed member	Output member	Speed-ratio equation
1	C	3	$R=\frac{N_2 N_3}{N_1 N_4}$
1	3	C	$R=1-\frac{N_2 N_3}{N_1 N_4}$
3	1	C	$R=1-\frac{N_1 N_4}{N_2 N_3}$
3	C	1	$R=\frac{N_4 N_1}{N_3 N_2}$
C	1	3	$R=1\Big/\left(1-\frac{N_1 N_4}{N_2 N_3}\right)$
C	3	1	$R=1\Big/\left(1-\frac{N_2 N_3}{N_1 N_4}\right)$

Figure 2-29 Simple Planetaries and Inversions

Chapter 3 Direct Power Transfer Devices

COUPLINGS

At some point in a mobile robot designer's career there will be a need to couple two shafts together. Fortunately, there are many commercially available couplers to pick from, each with its own strengths and weaknesses. Couplers are available in two major styles: solid and flexible. Solid couplers must be strong enough to hold the shafts' ends together as if they were one shaft. Flexible couplers allow for misalignment and are used where the two shafts are already running in their own bearings, but might be slightly out of alignment. The only other complication is that the shafts may be different diameters, or have different end details like splined, keyed, hex, square, or smooth. The coupler simply has different ends to accept the shafts it is coupling.

Solid couplers are very simple devices. They clamp onto each shaft tight enough to transmit the torque from one shaft onto the other. The shafts styles in each end of the coupler can be the same or different. For shaped shafts, the coupler need only have the same shape and size as the shaft and bolts or other clamping system to hold the coupler to the shaft. For smooth shafts, the coupler must clamp to the shaft tight enough to transmit the torque through friction with the shaft surface. This requires very high clamping forces, but is a common method because it requires no machining of the shafts.

As for online sources of couplers, and more detailed information about torque carrying ability, check out these web sites:

- powertransmission.com
- dodge-pt.com
- flexibleshaftcouplings.com
- wmberg.com
- mcmastercarr.com

METHODS FOR COUPLING ROTATING SHAFTS

Methods for coupling rotating shafts vary from simple bolted flange assembles to complex spring and synthetic rubber assembles. Those including chain belts, splines, bands, and rollers are shown here.

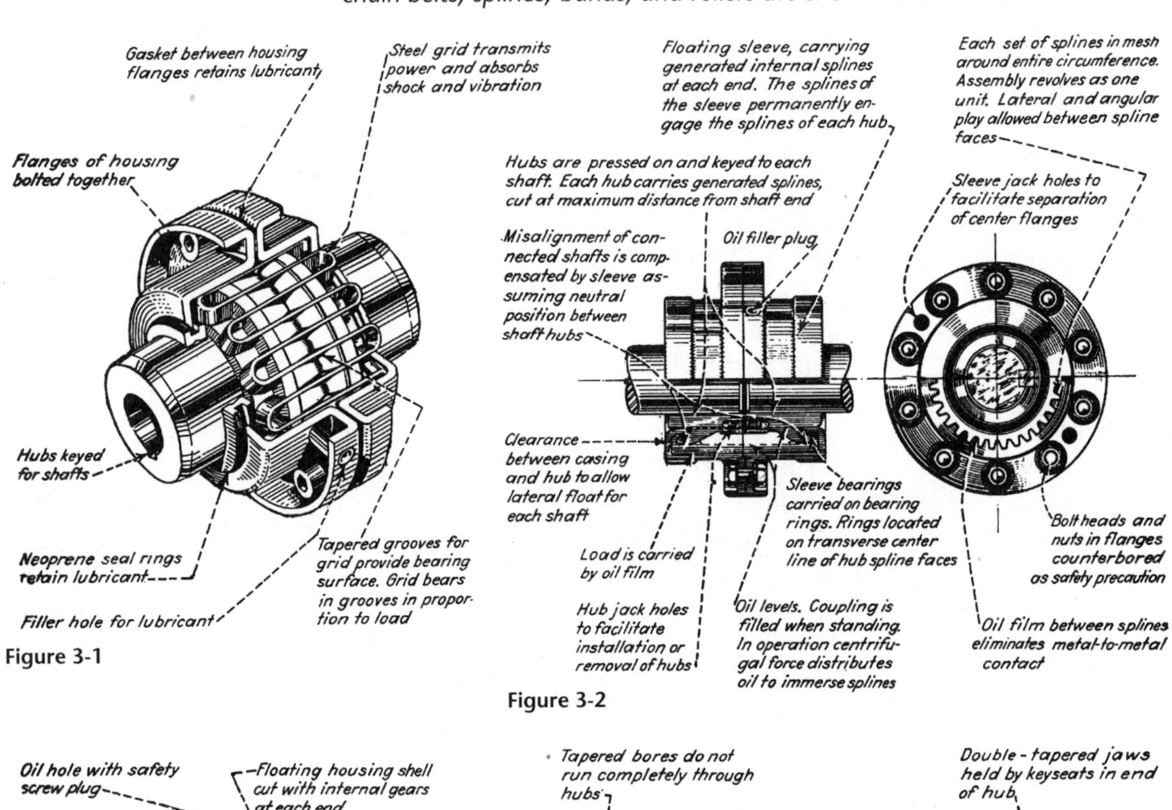

Figure 3-1

Figure 3-2

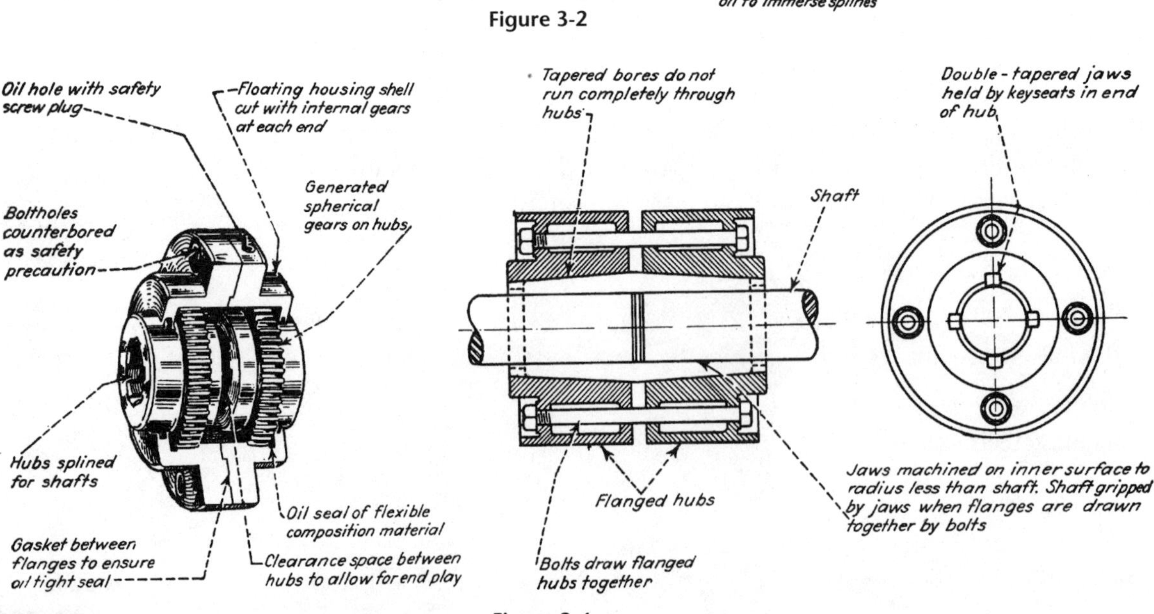

Figure 3-3

Figure 3-4

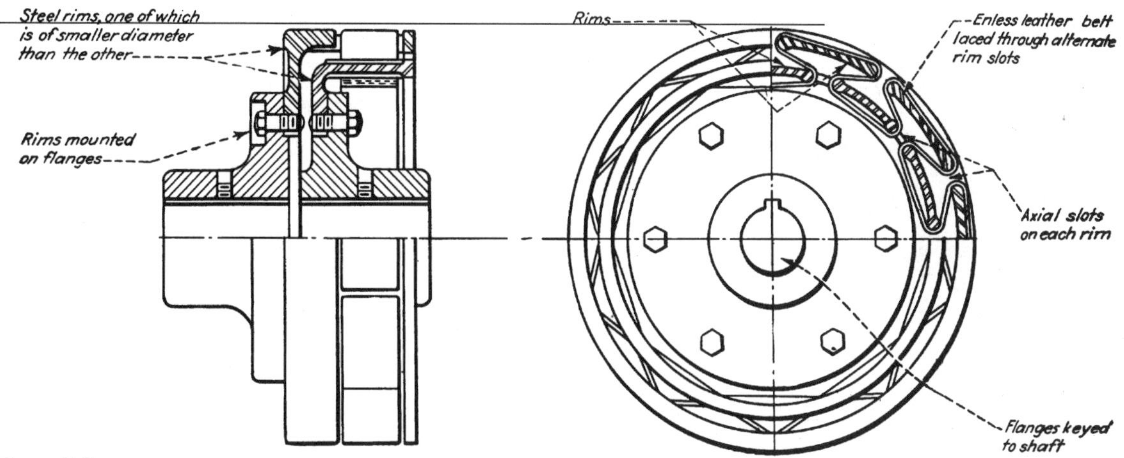

Figure 3-5

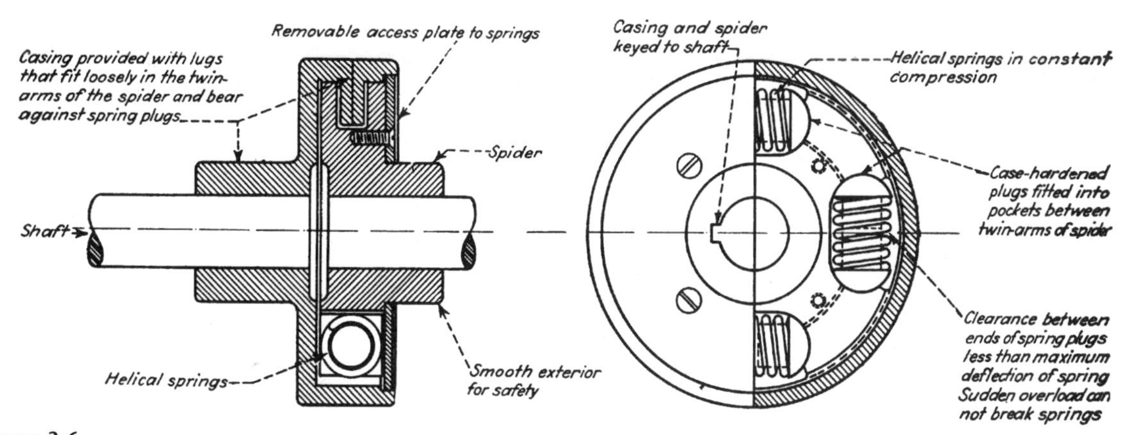

Figure 3-6

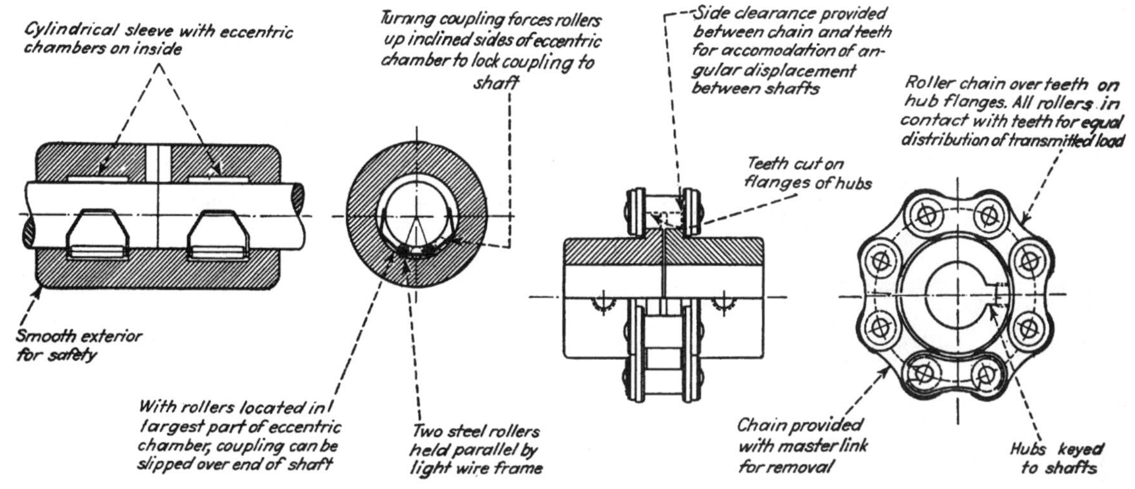

Figure 3-7

Figure 3-8

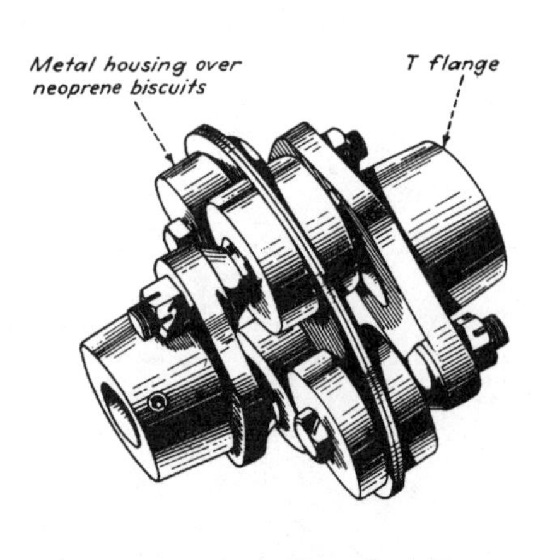

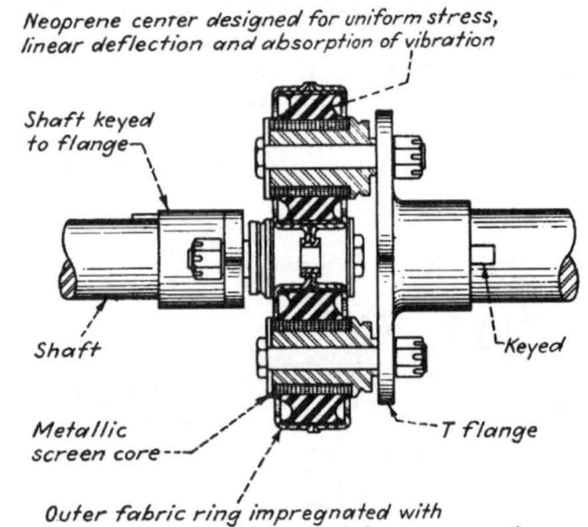

Metal housing over neoprene biscuits
T flange
Neoprene center designed for uniform stress, linear deflection and absorption of vibration
Shaft keyed to flange
Shaft
Keyed
Metallic screen core
T flange
Outer fabric ring impregnated with neoprene, provides support for center section

Figure 3-9

Metal facing plates riveted to flexible disk prevent excessive wear
Flexible disk of vulcanized rubbered fabric
Elongated holes receive boss from mating flange
Flange hubs keyed to shafts
Bosses on metal flanges pass through holes in flexible disk and enter elongated holes in mating flange

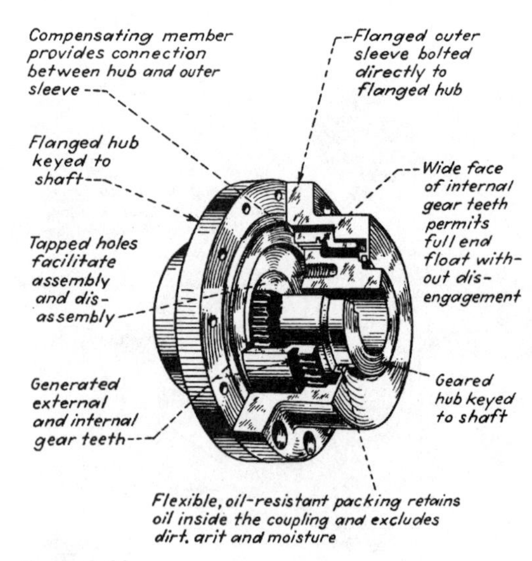

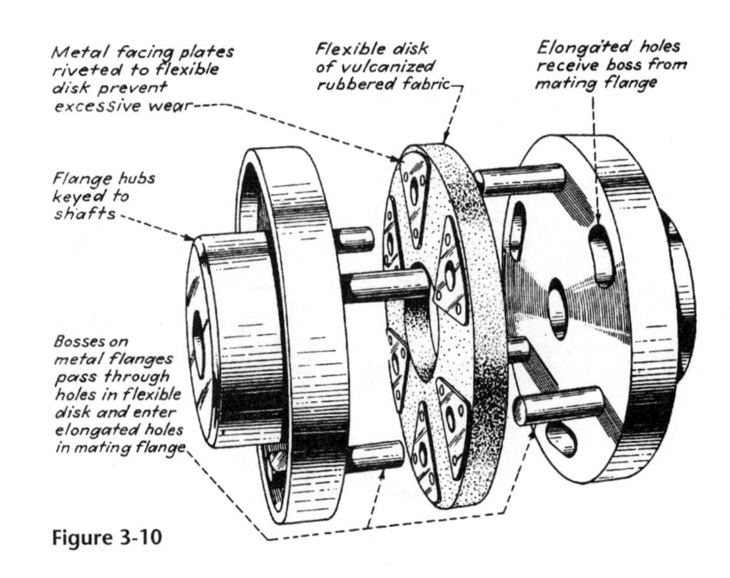

Figure 3-10

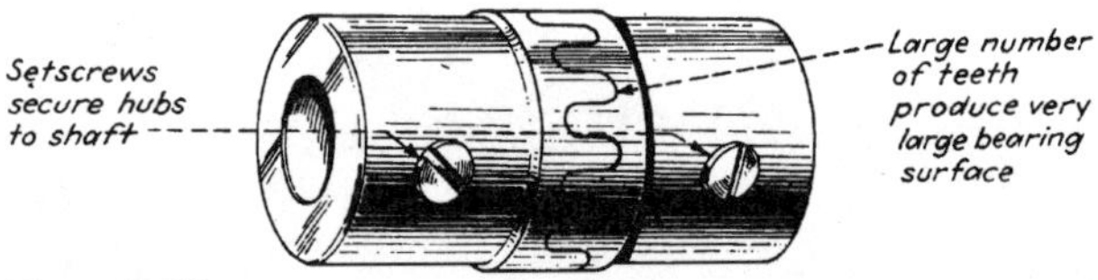

Setscrews secure hubs to shaft
Large number of teeth produce very large bearing surface

Figure 3-11

Compensating member provides connection between hub and outer sleeve
Flanged outer sleeve bolted directly to flanged hub
Flanged hub keyed to shaft
Wide face of internal gear teeth permits full end float without disengagement
Tapped holes facilitate assembly and disassembly
Generated external and internal gear teeth
Geared hub keyed to shaft
Flexible, oil-resistant packing retains oil inside the coupling and excludes dirt, grit and moisture

Figure 3-12

Long gear teeth in sleeve prevent hub from disengaging
Two tapped holes in each hub facilitate assembly and removal
Gasket prevents oil leakage
Clearance between sleeve and hub permits free end float
Load cushioned by oil film between the gear teeth
Solid metal under gear teeth gives added strength and durability
Spherical contour of hub teeth permits free sliding and rocking motion
Flexible, oil resistant packing retains oil inside the coupling and excludes dirt, grit and moisture
Oil chamber
Generated external and internal gear teeth
Machined bands on each hub facilitate accurate alignment
Safety flange with countersunk holes for fitted bolts and self-locking nuts
Two tapped holes in each half of sleeve facilitate assembly and removal
Oil-supply replenished through either of two plugged holes

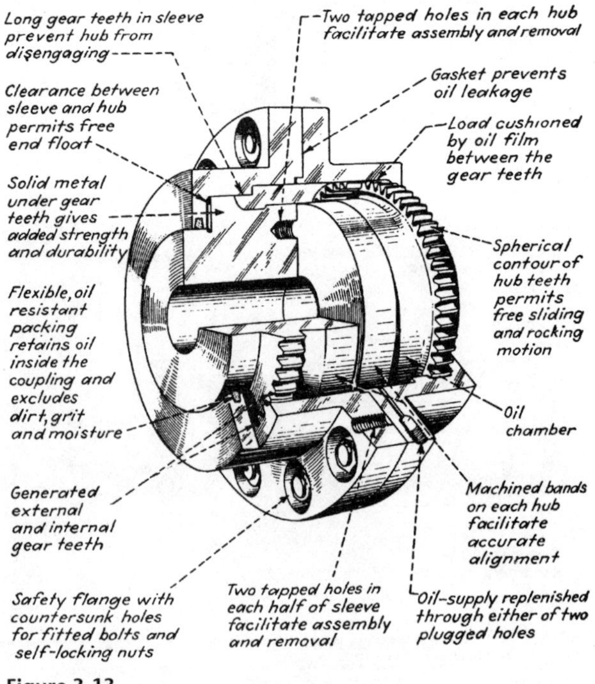

Figure 3-13

Shaft couplings that include internal and external gears, balls, pins, and nonmetallic parts to transmit torque are shown here.

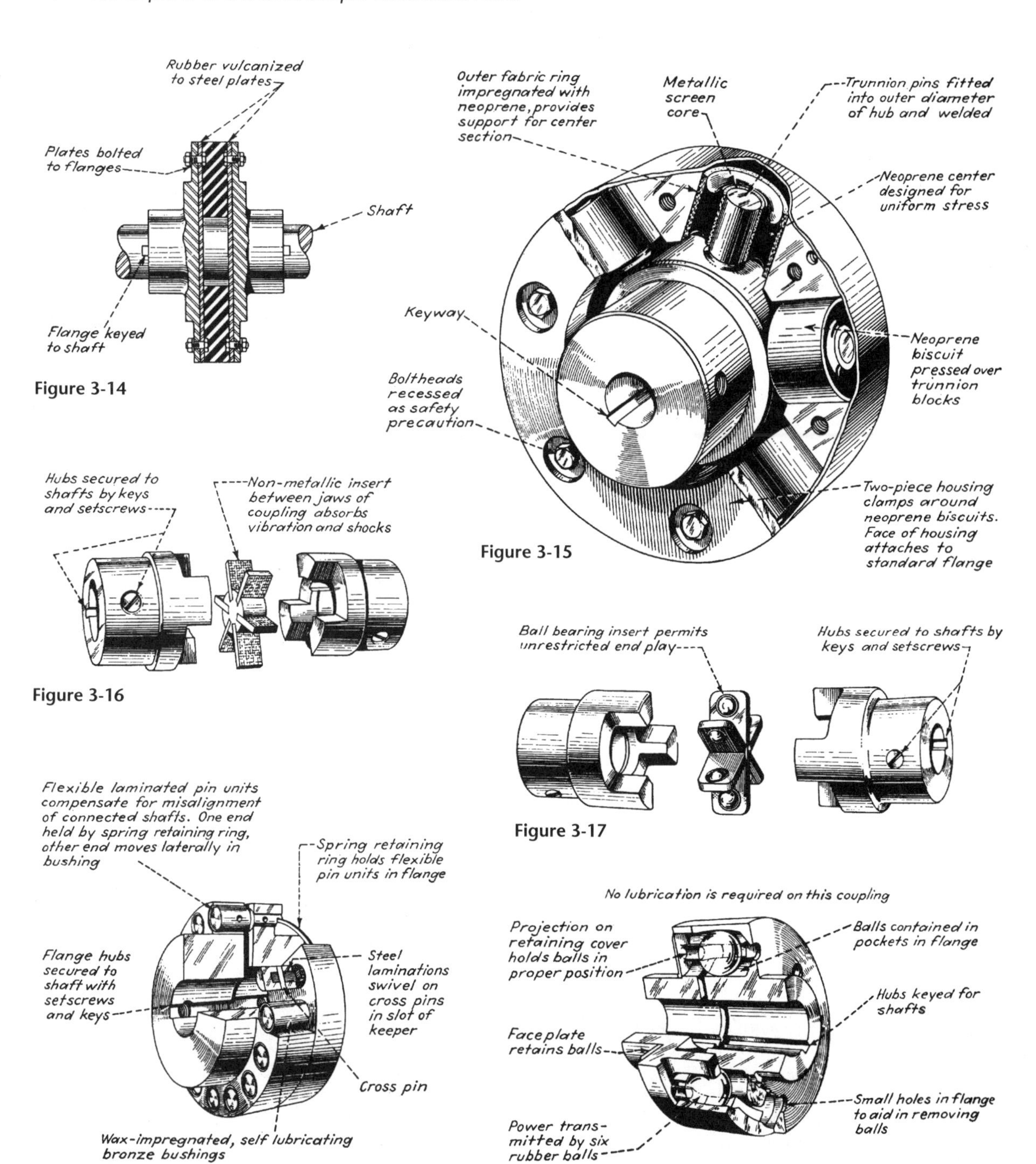

Figure 3-14

Figure 3-15

Figure 3-16

Figure 3-17

Figure 3-18

Figure 3-17

TEN UNIVERSAL SHAFT COUPLINGS

Hooke's Joints

The commonest form of a universal coupling is a *Hooke's joint.* It can transmit torque efficiently up to a maximum shaft alignment angle of about 36°. At slow speeds, on hand-operated mechanisms, the permissible angle can reach 45°. The simplest arrangement for a Hooke's joint is two forked shaft-ends coupled by a cross-shaped piece. There are many variations and a few of them are included here.

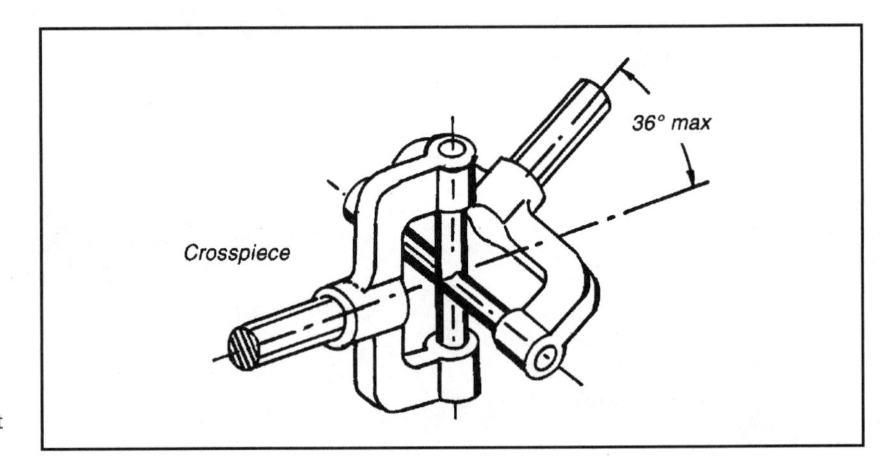

Figure 3-20 The Hooke's joint can transmit heavy loads. Antifriction bearings are a refinement often used.

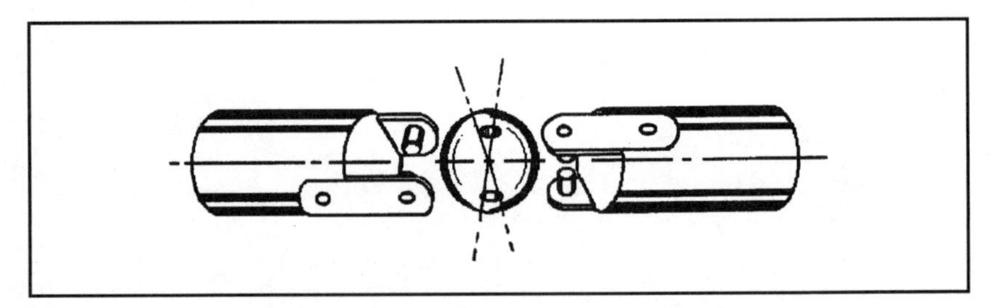

Figure 3-21 A pinned sphere shaft coupling replaces a crosspiece. The result is a more compact joint.

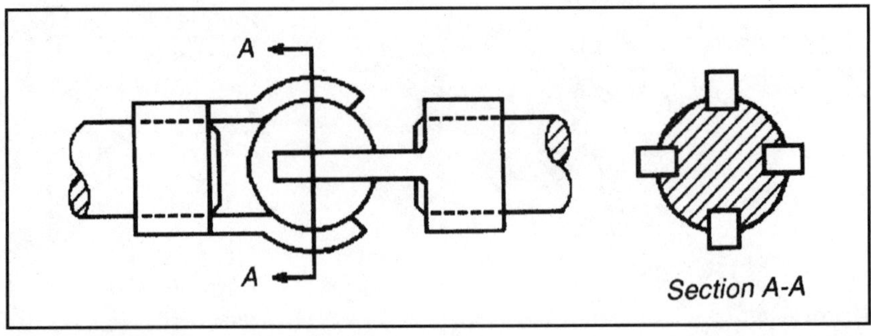

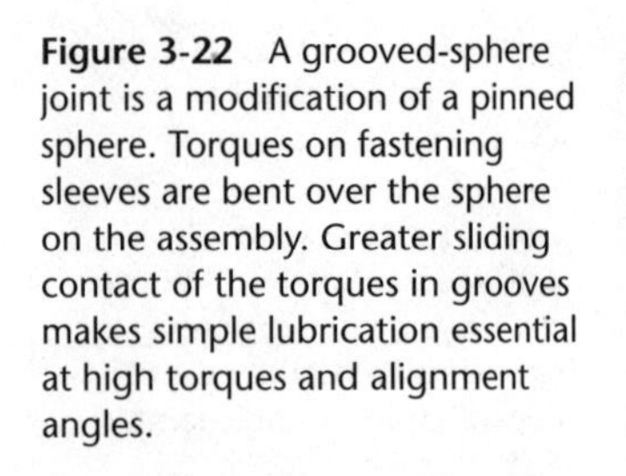

Figure 3-22 A grooved-sphere joint is a modification of a pinned sphere. Torques on fastening sleeves are bent over the sphere on the assembly. Greater sliding contact of the torques in grooves makes simple lubrication essential at high torques and alignment angles.

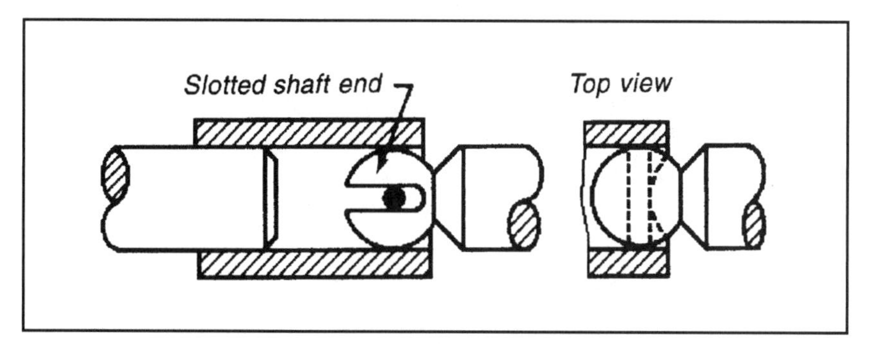

Figure 3-23 A pinned-sleeve shaft-coupling is fastened to one saft that engages the forked, spherical end on the other shaft to provide a joint which also allows for axial shaft movement. In this example, however, the angle between shafts must be small. Also, the joint is only suitable for low torques.

Constant-Velocity Couplings

The disadvantages of a single Hooke's joint is that the velocity of the driven shaft varies. Its maximum velocity can be found by multiplying driving-shaft speed by the secant of the shaft angle; for minimum speed, multiply by the cosine. An example of speed variation: a driving shaft rotates at 100 rpm; the angle between the shafts is 20°. The minimum output is 100 × 0.9397, which equals 93.9 rpm; the maximum output is 1.0642 × 100, or 106.4 rpm. Thus, the difference is 12.43 rpm. When output speed is high, output torque is low, and vice versa. This is an objectionable feature in some mechanisms. However, two universal joints connected by an intermediate shaft solve this speed-torque objection.

This single constant-velocity coupling is based on the principle (Figure 3-25) that the contact point of the two members must always lie on the homokinetic plane. Their rotation speed will then always be equal because the radius to the contact point of each member will always be equal. Such simple couplings are ideal for toys, instruments, and other light-duty mechanisms. For heavy duty, such as the front-wheel drives of

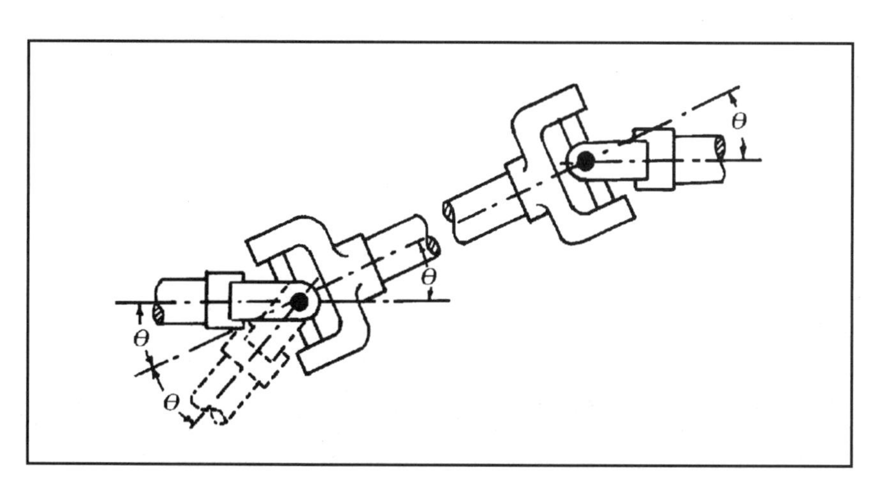

Figure 3-24 A constant-velocity joint is made by coupling two Hooke's joints. They must have equal input and output angles to work correctly. Also, the forks must be assembled so that they will always be in the same plane. The shaft-alignment angle can be double that for a single joint.

military vehicles, a more complex coupling is shown diagrammatically in Figire 3-26A. It has two joints close-coupled with a sliding member between them. The exploded view (Figure 3-26B) shows these members. There are other designs for heavy-duty universal couplings; one, known as the Rzeppa, consists of a cage that keeps six balls in the homokinetic plane at all times. Another constant-velocity joint, the Bendix-Weiss, also incorporates balls.

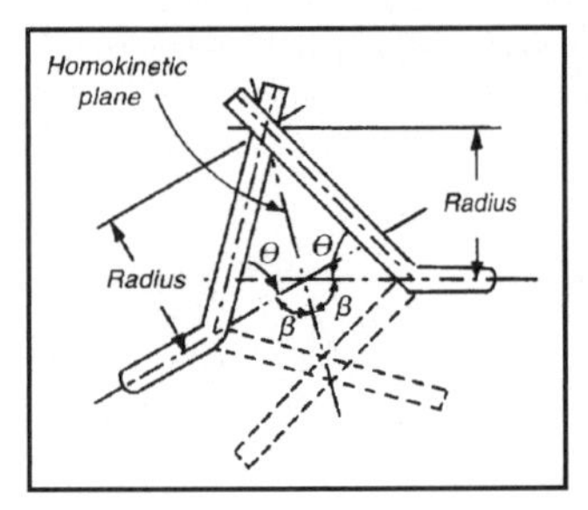

Figure 3-25

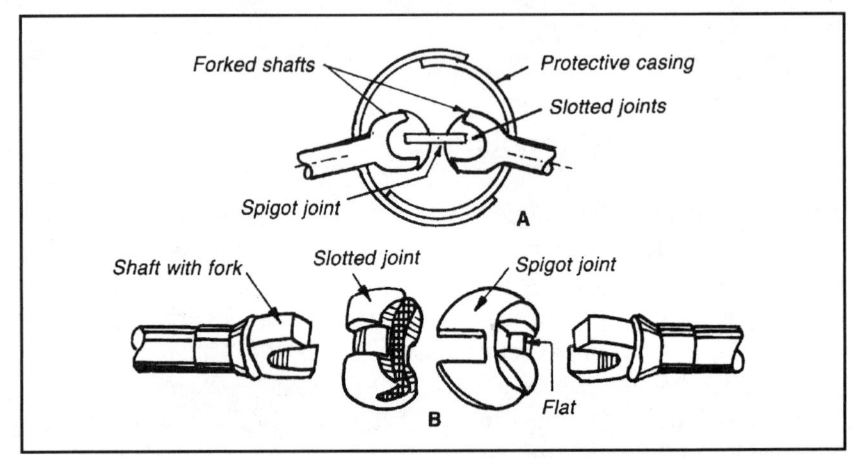

Figure 3-26

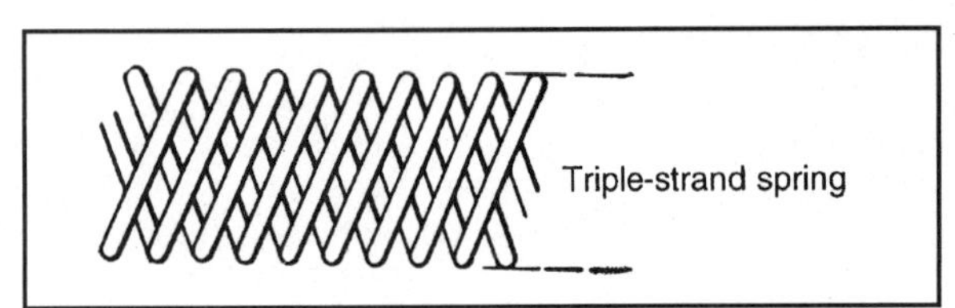

Figure 3-27 This flexible shaft permits any shaft angle. These shafts, if long, should be supported to prevent backlash and coiling.

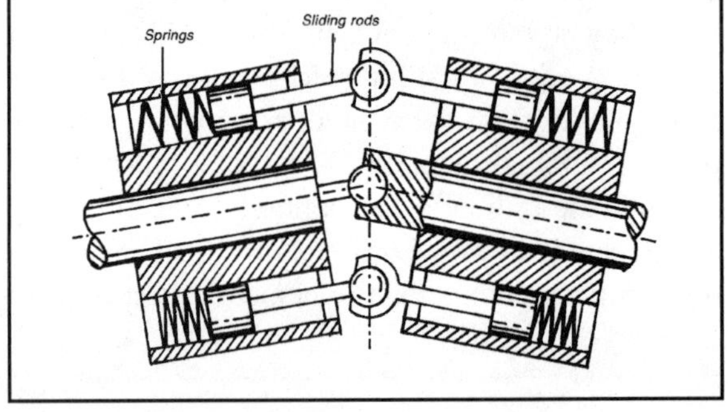

Figure 3-28 This pump-type coupling has the reciprocating action of sliding rods that can drive pistons in cylinders.

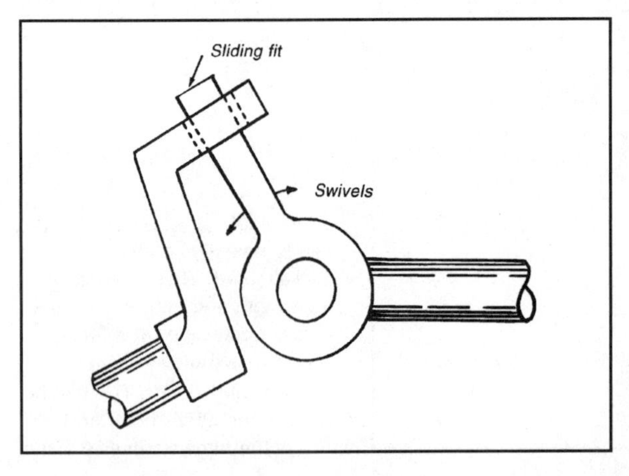

Figure 3-29 This light-duty coupling is ideal for many simple, low-cost mechanisms. The sliding swivel-rod must be kept well lubricated at all times.

COUPLING OF PARALLEL SHAFTS

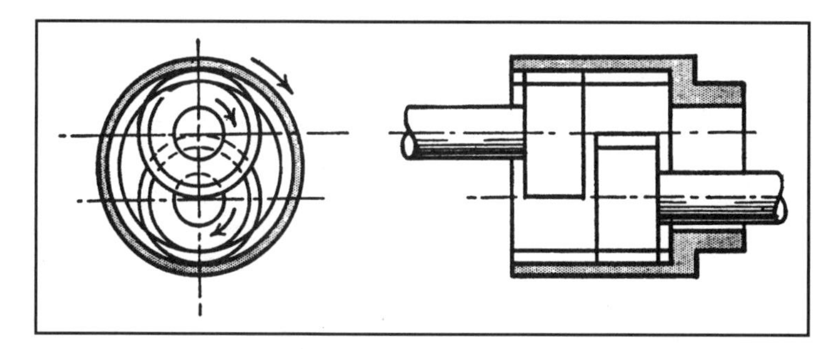

Figure 3-30 One method of coupling shafts makes use of gears that can replace chains, pulleys, and friction drives. Its major limitation is the need for adequate center distance. However, an idler can be used for close centers, as shown. This can be a plain pinion or an internal gear. Transmission is at a constant velocity and there is axial freedom.

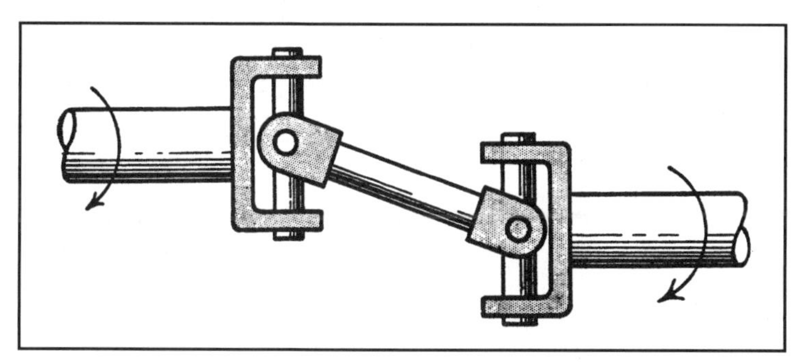

Figure 3-31 This coupling consists of two universal joints and a short shaft. Velocity transmission is constant between the input and output shafts if the shafts remain parallel and if the end yokes are arranged symmetrically. The velocity of the central shaft fluctuates during rotation, but high speed and wide angles can cause vibration. The shaft offset can be varied, but axial freedom requires that one shaft be spline mounted.

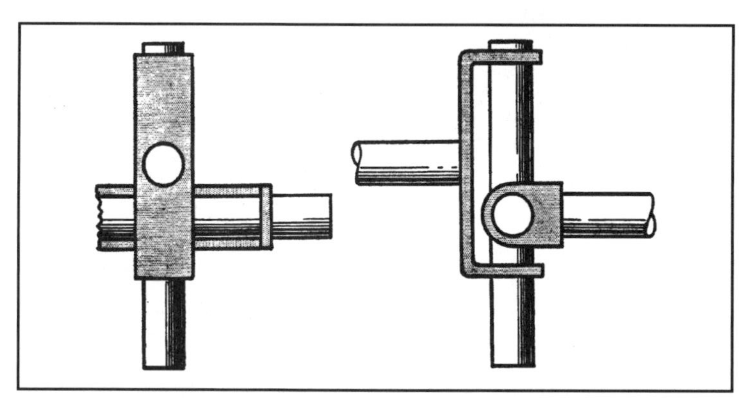

Figure 3-32 This crossed-axis yoke coupling is a variation of the mechanism shown in Fig. 2. Each shaft has a yoke connected so that it can slide along the arms of a rigid cross member. Transmission is at a constant velocity, but the shafts must remain parallel, although the offset can vary. There is no axial freedom. The central cross member describes a circle and is thus subjected to centrifugal loads.

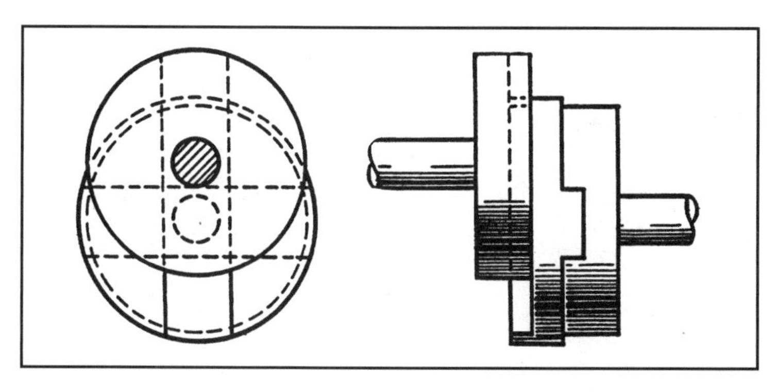

Figure 3-33 This Oldham coupling provides motion at a constant velocity as its central member describes a circle. The shaft offset can vary, but the shafts must remain parallel. A small amount of axial freedom is possible. A tilt in the central member can occur because of the offset of the slots. This can be eliminated by enlarging its diameter and milling the slots in the same transverse plane.

TEN DIFFERENT SPLINED CONNECTIONS

Cylindrical Splines

Figure 3-34 Sqrare Splines make simple connections. They are used mainly for transmitting light loads, where accurate positioning is not critical. This spline is commonly used on machine tools; a cap screw is required to hold the enveloping member.

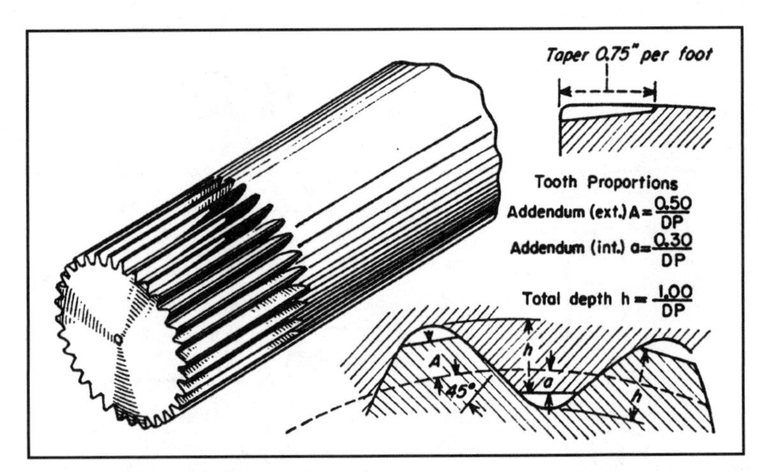

Figure 3-35 Serrations of small size are used mostly for transmitting light loads. This shaft forced into a hole of softer material makes an inexpensive connection. Originally straight-sided and limited to small pitches, 45° serrations have been standardized (SAE) with large pitches up to 10 in. dia. For tight fits, the serrations are tapered.

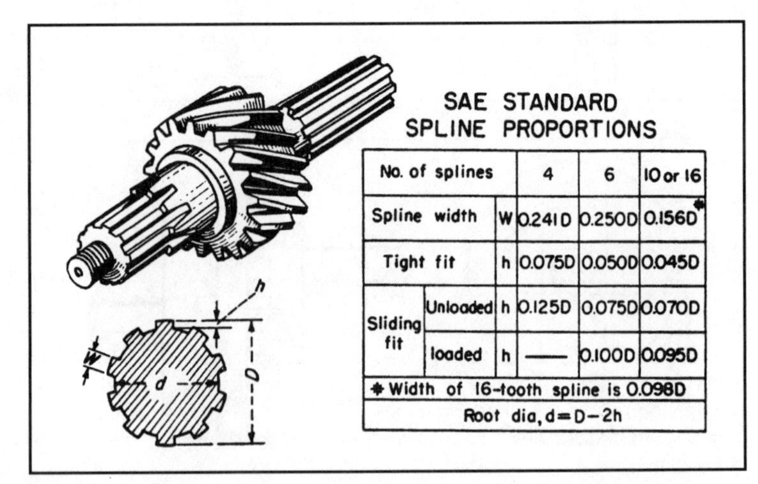

SAE STANDARD SPLINE PROPORTIONS

No. of splines			4	6	10 or 16
Spline width		W	0.241D	0.250D	0.156D*
Tight fit		h	0.075D	0.050D	0.045D
Sliding fit	Unloaded	h	0.125D	0.075D	0.070D
	loaded	h	——	0.100D	0.095D
* Width of 16-tooth spline is 0.098D					
Root dia, d = D − 2h					

Figure 3-36 Straight-Sided splines have been widely used in the automotive field. Such splines are often used for sliding members. The sharp corner at the root limits the torque capacity to pressures of approximately 1,000 psi on the spline projected area. For different applications, tooth height is altered, as shown in the table above.

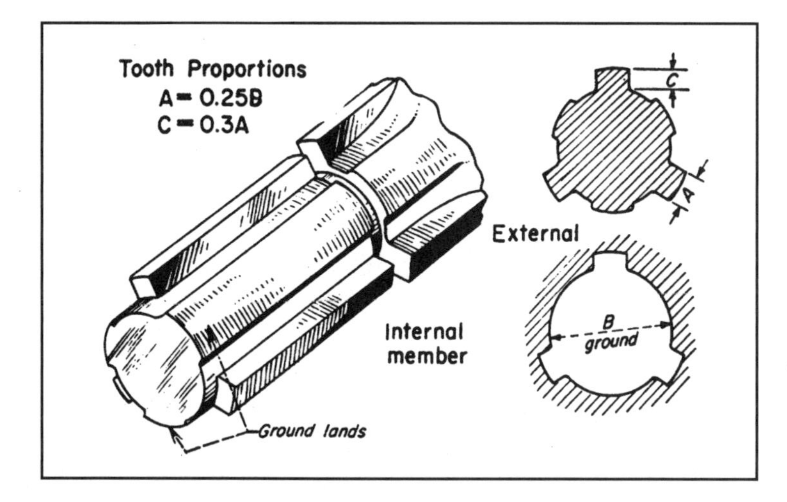

Figure 3-37 Machine-Tool splines have wide gaps between splines to permit accurate cylindrical grinding of the lands—for precise positioning. Internal parts can be ground readily so that they will fit closely with the lands of the external member.

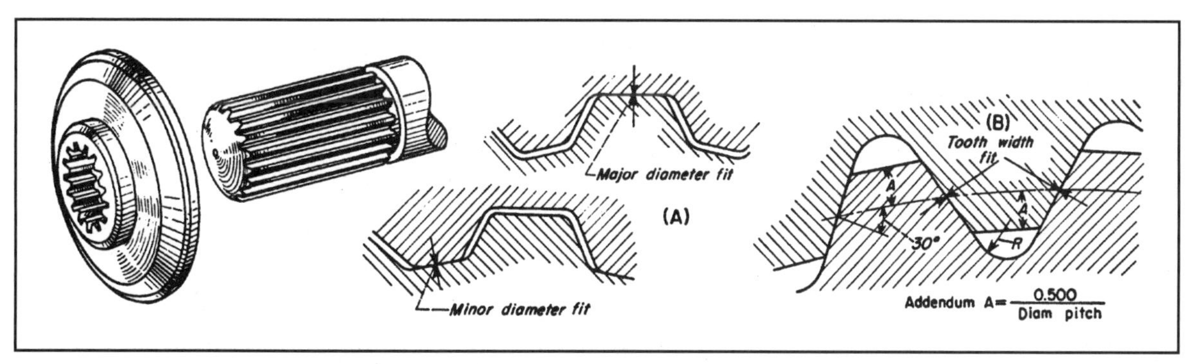

Figure 3-38 Involute-Form splines are used where high loads are to be transmitted. Tooth proportions are based on a 30° stub tooth form. (A) Splined members can be positioned either by close fitting major or minor diameters. (B) Use of the tooth width or side positioning has the advantage of a full fillet radius at the roots. Splines can be parallel or helical. Contact stresses of 4,000 psi are used for accurate, hardened splines. The diametral pitch shown is the ratio of teeth to the pitch diameter.

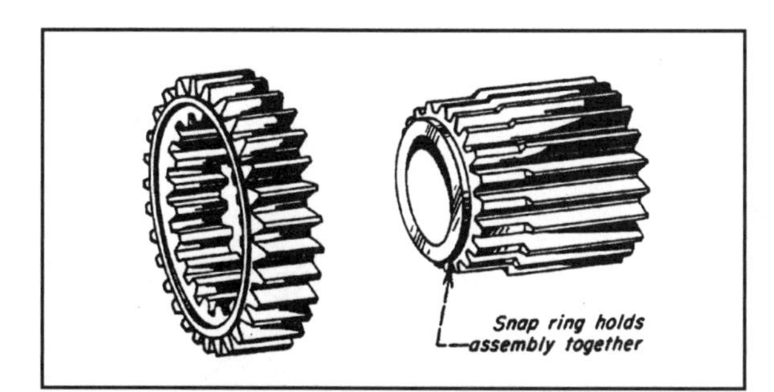

Figure 3-39 Special Involute splines are made by using gear tooth proportions. With full depth teeth, greater contact area is possible. A compound pinion is shown made by cropping the smaller pinion teeth and internally splining the larger pinion.

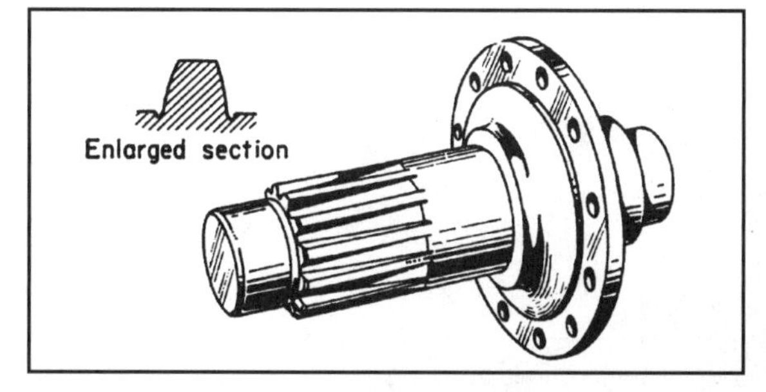

Figure 3-40 Taper-Root splines are for drivers that require positive positioning. This method holds mating parts securely. With a 30° involute stub tooth, this type is stronger than parallel root splines and can be hobbed with a range of tapers.

Face Splines

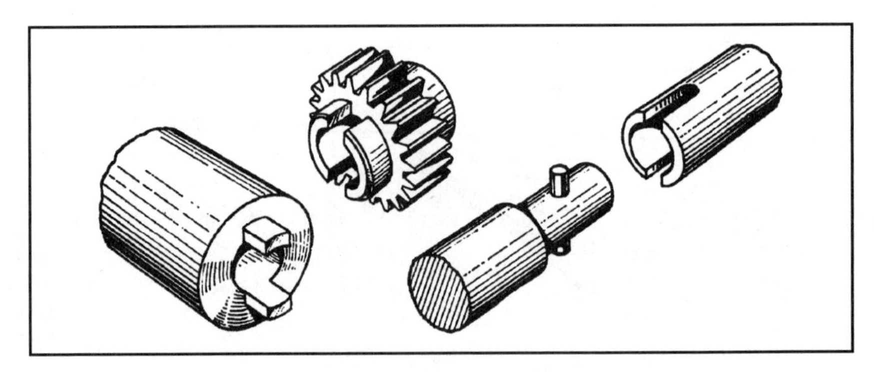

Figure 3-41 Milled Slots in hubs or shafts make inexpensive connections. This spline is limited to moderate loads and requires a locking device to maintain positive engagement. A pin and sleeve method is used for light torques and where accurate positioning is not required.

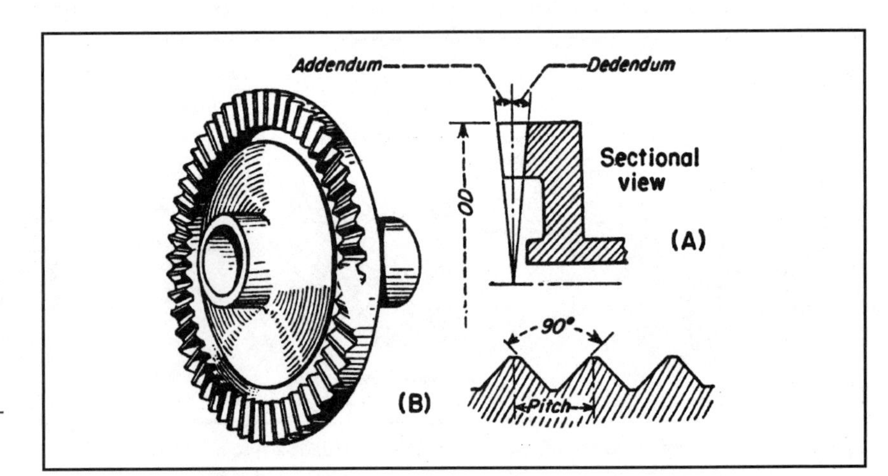

Figure 3-42 Radical Serrations made by milling or shaping the teeth form simple connections. (A) Tooth proportions decrease radially. (B) Teeth can be straight-sided (castellated) or inclined; a 90° angle is common.

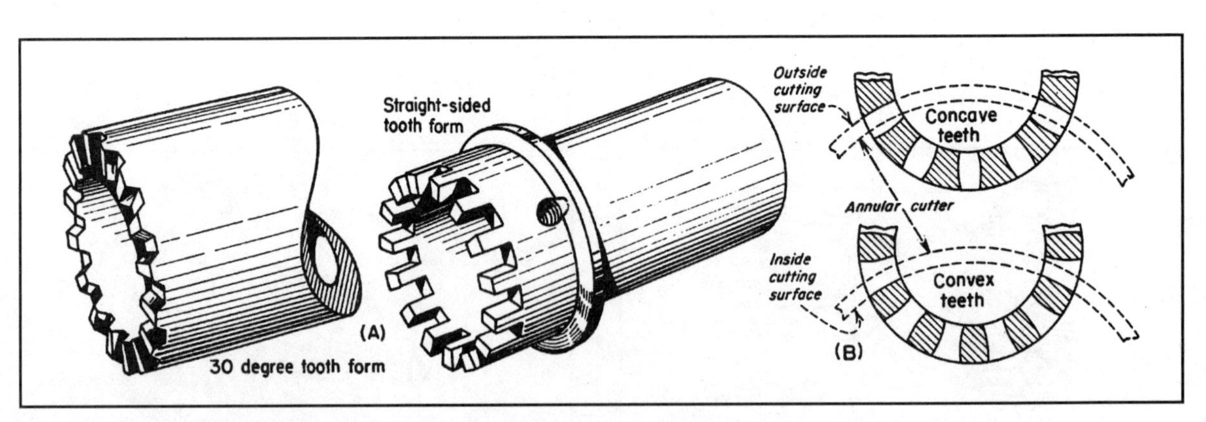

Figure 3-43 Curvic Coupling teeth are machined by a face-mill cutter. When hardened parts are used that require accurate positioning, the teeth can be ground. (A) This process produces teeth with uniform depth. They can be cut at any pressure angle, although 30° is most common. (B) Due to the cutting action, the shape of the teeth will be concave (hour-glass) on one member and convex on the other—the member with which it will be assembled.

TORQUE LIMITERS

Robots powered by electric motors can frequently stop effectively without brakes. This is done by turning the drive motor into a generator, and then placing a load across the motor's terminals. Whenever the wheels turn the motor faster than the speed controller tries to turn the motor, the motor generates electrical power. To make the motor brake the robot, the electrical power is fed through large load resistors, which absorb the power, slowing down the motor. Just like normal brakes, the load resistors get very hot. The energy required to stop the robot is given off in this heat. This method works very well for robots that travel at slow speeds.

In a case where the rotating shaft suddenly jams or becomes overloaded for some unexpected reason, the torque in the shaft could break the shaft, the gearbox, or some other part of the rotating system. Installing a device that brakes first, particularly one that isn't damaged when it is overloaded, is sometimes required. This mechanical device is called a torque limiter.

There are many ways to limit torque. Magnets, rubber bands, friction clutches, ball detents, and springs can all be used in one way or another, and all have certain advantages and disadvantages. It must be remembered that they all rely on giving off heat to absorb the energy of stopping the rotating part, usually the output shaft. Figures 3-44 through 3-53 show several torque limiters, which are good examples of the wide variety of methods available.

TEN TORQUE-LIMITERS

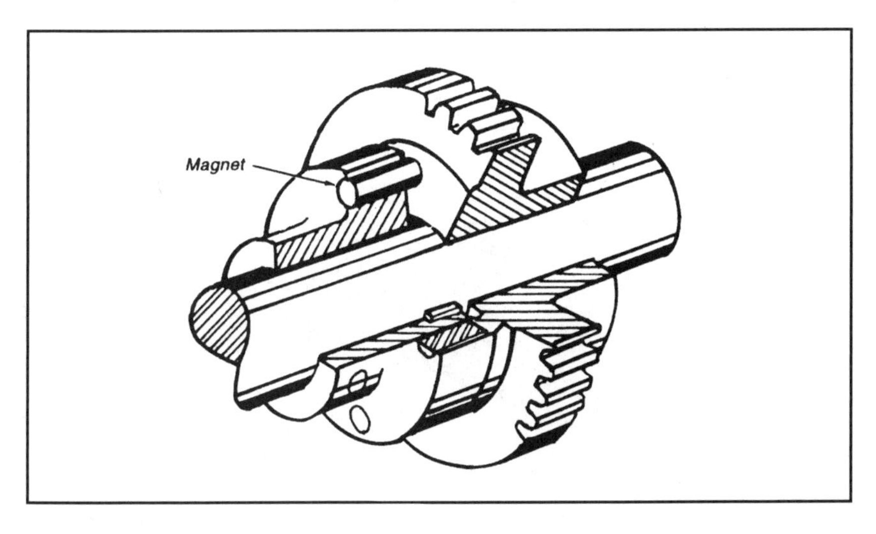

Figure 3-44 Permanent magnets transmit torque in accordance with their numbers and size around the circumference of the clutch plate. Control of the drive in place is limited to removing magnets to reduce the drive's torque capacity.

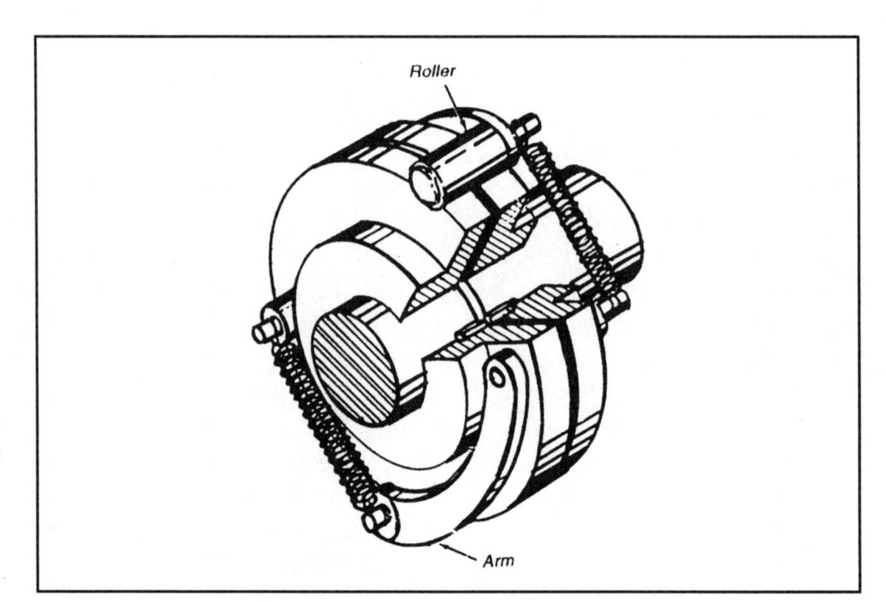

Figure 3-45 Arms hold rollers in the slots that are cut across the disks mounted on the ends of butting shafts. Springs keep the roller in the slots, but excessive torque forces them out.

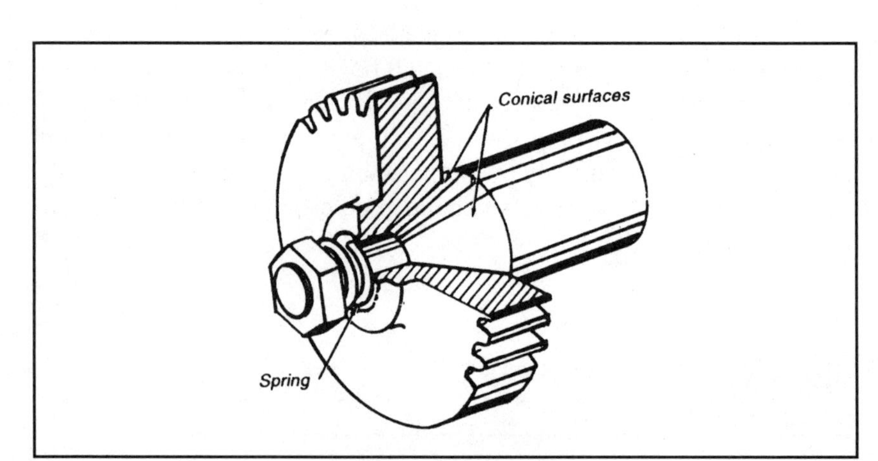

Figure 3-46 A cone clutch is formed by mating a taper on the shaft to a beveled central hole in the gear. Increasing compression on the spring by tightening the nut increases the drive's torque capacity.

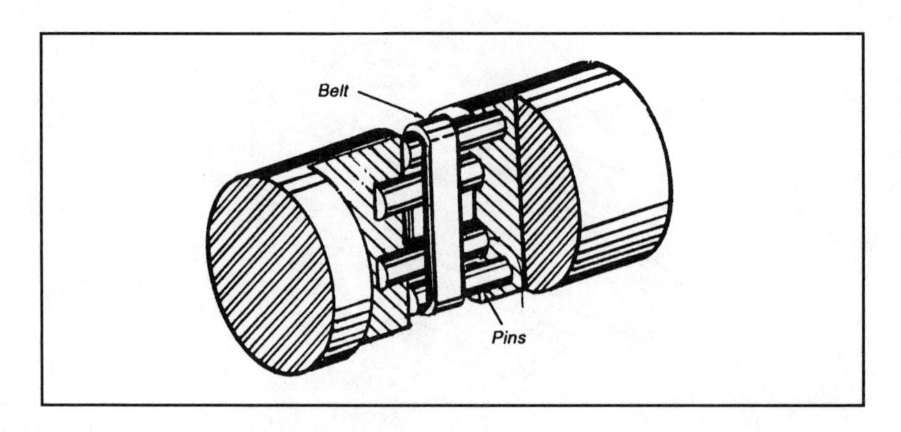

Figure 3-47 A flexible belt wrapped around four pins transmits only the lightest loads. The outer pins are smaller than the inner pins to ensure contact.

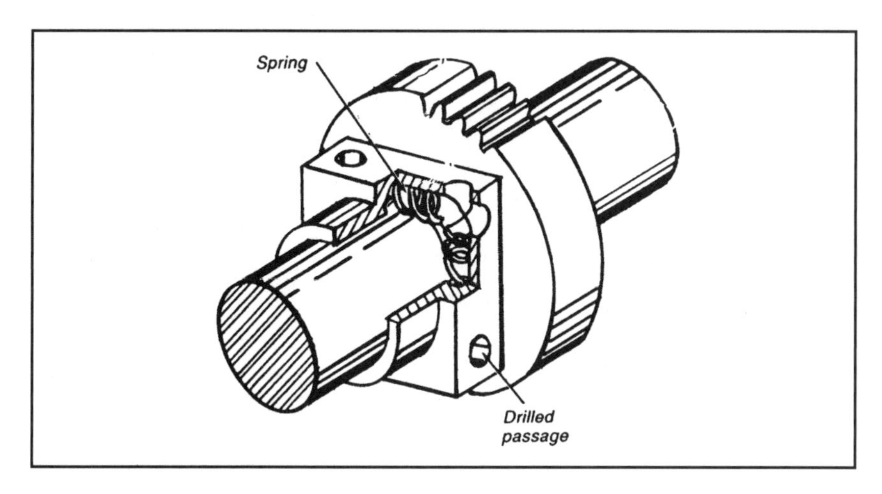

Figure 3-48 Springs inside the block grip the shaft because they are distorted when the gear is mounted to the box on the shaft.

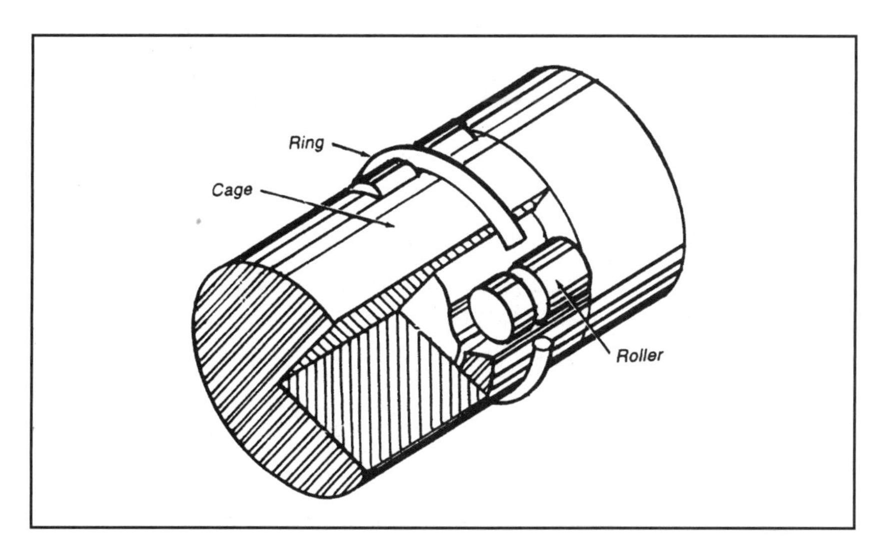

Figure 3-49 The ring resists the natural tendency of the rollers to jump out of the grooves in the reduced end of one shaft. The slotted end of the hollow shaft acts as a cage.

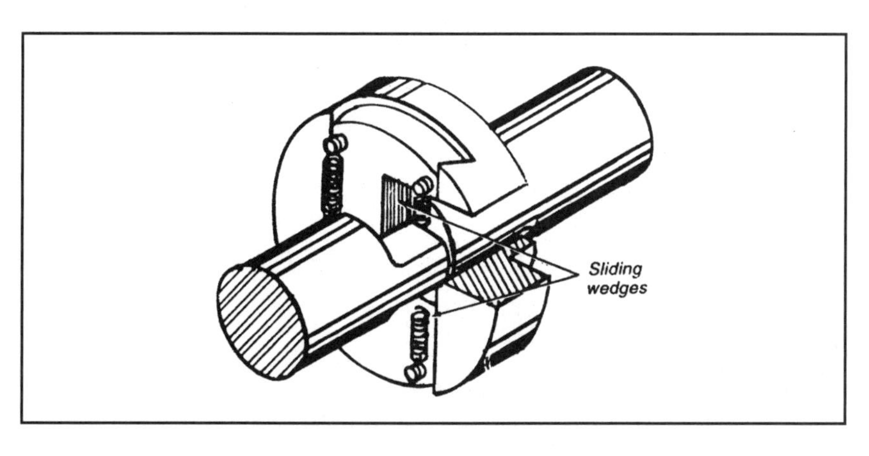

Figure 3-50 Sliding wedges clamp down on the flattened end of the shaft. They spread apart when torque becomes excessive. The strength of the springs in tension that hold the wedges together sets the torque limit.

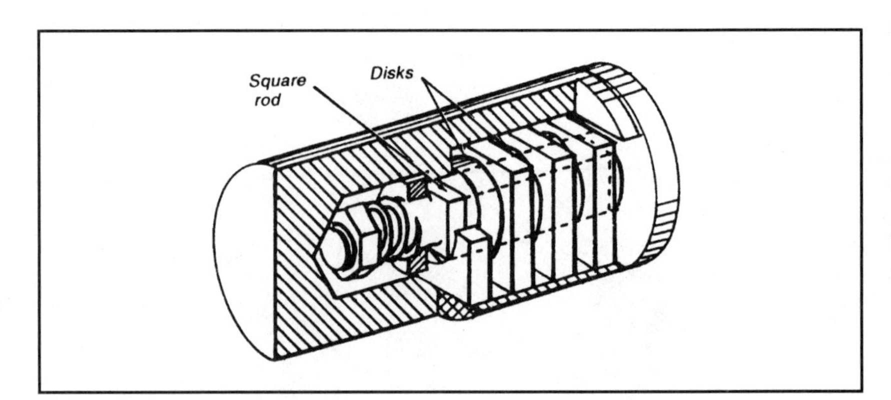

Figure 3-51 Friction disks are compressed by an adjustable spring. Square disks lock into the square hole in the left shaft, and round disks lock onto the square rod on the right shaft.

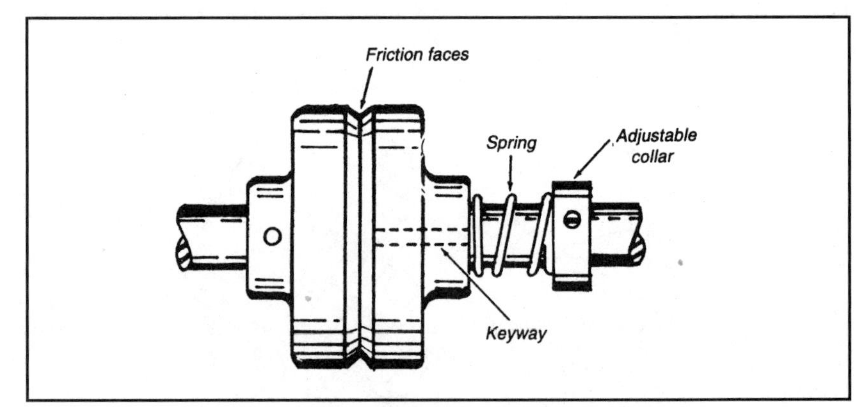

Figure 3-52 Friction clutch torque limiter. Adjustable spring tension holds the two friction surfaces together to set the overload limit. As soon as an overload is removed, the clutch reengages. A drawback to this design is that a slipping clutch can destroy itself if it goes undetected.

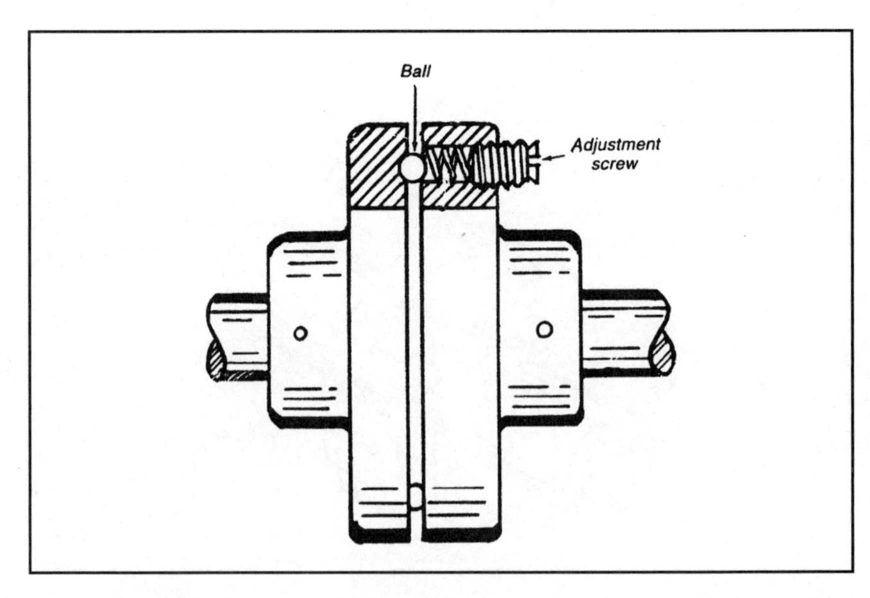

Figure 3-53 Mechanical keys. A spring holds a ball in a dimple in the opposite face of this torque limiter until an overload forces it out. Once a slip begins, clutch face wear can be rapid. Thus, this limiter is not recommended for machines where overload is common.

ONE TIME USE TORQUE LIMITING

In some cases, the torque limit can be set very high, beyond the practical limit of a torque limiter, or the device that is being protected needs only a one-time protection from damage. In this case, a device called a shear pin is used. In mobile robots, particularly in autonomous robots, it will be found that a torque limiter is the better choice, even if a large one is required to handle the torque. With careful control of motor power, both accelerating and braking, even torque limiters can be left out of most designs.

Torque limiters should be considered as protective devices for motors and gearboxes and are not designed to fail very often. They don't often turn up in the drive system of mobile robots, because the slow moving robot rarely generates an overload condition. They do find a place in manipulators to prevent damage to joints if the manipulator gets overloaded. If a torque limiter is used in the joint of a manipulator, the joint must have a proprioceptive sensor that senses the angle or extension of the joint so that the microprocessor has that information after the joint has slipped. Figure 3-54 shows a basic shear pin torque limiter.

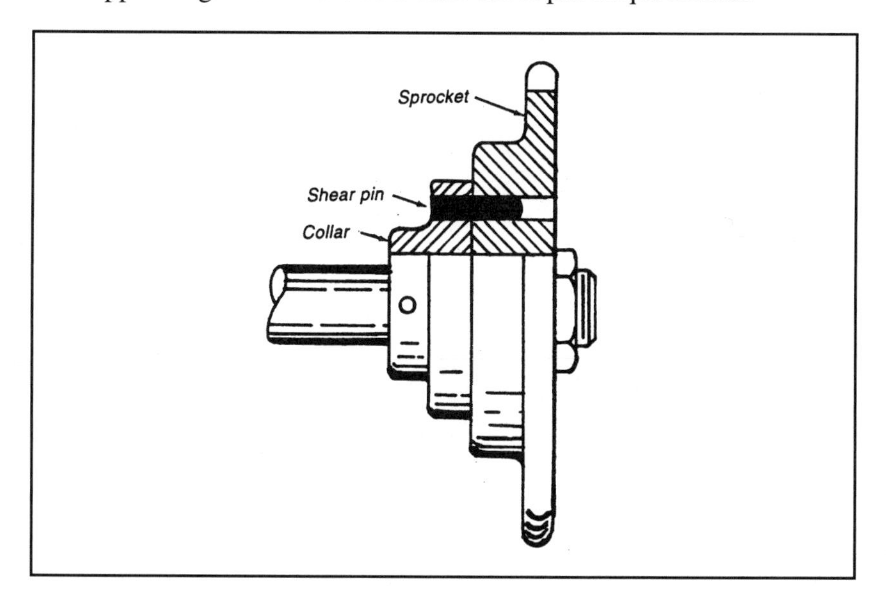

Figure 3-54 A shear pin is a simple and reliable torque limiter. However, after an overload, removing the sheared pin stubs and replacing them with a new pin can be time consuming. Be sure that spare shear pins are available in a convenient location.

Chapter 4 Wheeled Vehicle Suspensions and Drivetrains

Given the definition of robot in the introduction to this book, the most vital mechanical part of a robot must be its mobility system, including the suspension and drivetrain, and/or legs and feet. The ability of the these systems to effectively traverse what ever terrain is required is paramount to the success of the robot, but to my knowledge, there has never been an apples to apples comparison of mobility systems.

First, just what is a mobility system? A mobility system is all parts of a vehicle, a land-based robot for the purposes of this book, that aid in locomoting from one place to another. This means all motors, gearboxes, suspension pieces, transmissions, wheels, tires, tracks, springs, legs, foot pads, linkages, mechanisms for moving the center of gravity, mechanisms for changing the shape or geometry of the vehicle, mechanisms for changing the shape or geometry of the drivetrain, mechanisms and linkages for steering, etc., are parts of mobility systems.

The systems and mechanisms described in this book are divided into four general categories: wheeled, tracked, walkers, and special cases. Each gets its own chapter, and following the chapter on special cases is a separate chapter devoted to comparing the effectiveness of many of the systems.

There are some that are described in the text that are not discussed in Chapter Nine. These are mostly very interesting designs that are worth describing, but their mobility or some other trait precludes comparing them to the other designs. Most of the systems discussed in Chapter Eight fall into this category because they are designed to move through very specific environments and are not general enough to be comparable. Some wheeled designs are discussed simply because they are very simple even though their mobility is limited. This chapter deals with wheeled systems, everything from one-wheeled vehicles to eight-wheeled vehicles. It is divided into four sections: vehicles with one to three wheels and four-wheeled diamond layouts, four- and five-wheeled layouts, six-wheeled layouts, and eight-wheeled layouts.

WHEELED MOBILITY SYSTEMS

By far the most common form of vehicle layout is the four-wheeled, front-steer vehicle. It is a descendant of the horse-drawn wagon, but has undergone some subtle and some major changes in the many decades since a motor was added to replace the horses. The most important changes (other than the internal combustion engine) were to the suspension and steering systems. The steering was changed from a solid center-pivot axle to independently pivoting front wheels, which took up less space under the carriage. Eventually the suspension was developed into the nearly ubiquitous independently suspended wheels on all four corners of the vehicle.

Although the details of the suspensions used today are widely varied, they all use some form of spring and shock combination to provide good control and a relatively comfortable ride to the driver. Most suspensions are designed for high-speed control over mostly smooth surfaces, but more importantly, they are designed to be controlled by a human. In spite of their popularity and sometimes truly fantastic performance in racecars and off-road vehicles, there are very few sprung suspension systems discussed in this book. The exception is sprung bogies in some of the tracked vehicle layouts and a sprung fourth wheel in a couple four-wheel designs.

WHY NOT SPRINGS?

Springs are so common on people-controlled vehicles, why not include them in the list of suspension systems being discussed?

Springs do seem to be important to mobility, but what they are really addressing is rider comfort and control in vehicles that travel more than about 8m/s. Below that speed, they are actually a hindrance to mobility because they change the force each wheel exerts on the ground as bumps are negotiated. A four-wheeled conventional independent suspension vehicle appears to keep all wheels equally on the ground, but the wheels that are on the bumps, being lifted, are carrying more weight than the other wheels. This reduces the traction of the lightly loaded wheels. The better solution, at low speeds, is to allow some of the wheels to rise, relative to the chassis, over bumps without changing the weight distribution or changing it as little as possible. This is precisely what happens in rocker and rocker/bogie suspensions.

Ground pressures across all vehicles range from twenty to eighty kilopascals (the average human foot exerts a pressure on the ground of about

35 kilo-pascals) for the majority of vehicles of all types. Everything from the largest military tank to the smallest motor cycle falls within that range, though some specialized vehicles designed for travel on loose powder snow have pressures of as low as five kilo-pascals. This narrow range of pressures is due to the relatively small range of densities and materials of which the ground is made. Vehicles with relatively low ground pressure will perform better on softer materials like loose sand, snow, and thick mud. Those with high pressures mostly perform better on harder packed materials like packed snow, dirt, gravel, and common road surfaces. The best example of this fact are vehicles designed to travel on both hard roads and sand. The operator must stop and deflate the tires, reducing ground pressure, as the vehicle is driven off a road and onto a stretch of sand. Several military vehicles like the WWII amphibious DUKS were designed so tire pressure could be adjusted from inside the cab, without stopping. This is now also possible on some modified Hummers to extend their mobility, and might be a practical trick for a wheeled robot that will be working on both hard and soft surfaces.

This also points to the advantage of maintaining as even a ground pressure as possible on all tires, even when some of them may be lifted up onto a rock or fallen tree. Suspension systems that do this well will theoretically work better on a wider range of ground materials. Suspension systems that can change their ground pressure in response to changes in ground materials, either by tire inflation pressure, variable geometry tires, or a method of changing the number of tires in contact with the ground, will also theoretically work well on a wider range of ground materials.

This chapter focuses on suspension systems that are designed to work on a wide range of ground materials, but it also covers many layouts that are excellent for indoor or relatively benign outdoor environments. The latter are shown because they are simple and easy to implement, allowing a basic mobile platform to be quickly built to ease the process of getting started building an autonomous robot. Vehicles intended for use in any arbitrary outdoor environment tend to be more complicated, but some, with acceptably high mobility, are surprisingly simple.

SHIFTING THE CENTER OF GRAVITY

A trick that can be applied to mobile robots that extends the robot's mobility, independent of the mobility system, is to move the center of gravity (cg) of the robot, thereby changing which wheels, tracks, or legs are carrying the most weight. A discussion of this concept and some layouts are included in this chapter, but the basic concept can be applied to almost any mobile robot.

Shifting the center of gravity can be accomplished by moving a dedicated weight, shifting the cargo, or reorienting the manipulator. Moving the cg can allow the robot to move across wider gaps, climb steeper slopes, and get over or onto higher steps. If it is planned to move the manipulator, then the manipulator must make up a significant fraction of the total weight of the vehicle for the concept to work effectively. While moving the cg seems very useful, all but the manipulator technique require extra space in the robot for the weight and/or mechanism that moves the weight.

The figures show the basic concept and several variations of cg shifting that might be tried if no other mobility system can be designed to negotiate a required obstacle, or if the concept is being applied as a retrofit to extend an existing robot's mobility. Functionally, as a gap in the terrain approaches, the cg is shifted aft, allowing the mobility system's front ground contact point to reach across the gap without the robot tipping forward. When those parts reach the far side of the gap, the robot is driven forward until it is almost across, then the cg is shifted forward, lifting the rear ground contact points off the ground. The vehicle is then driven across the gap the rest of the way.

For stair climbing or steep slopes, the cg is shifted forward so it remains over the center of area of the mobility system. For climbing up a single bump or step, it is shifted back just as the vehicle climbs onto the step. This reduces the tendency of the robot to slam down on the front parts of the mobility system. It must be noted that cg shifting can be controlled autonomously fairly easily if there is an inclinometer or accelerometer onboard the robot that can give inclination. The control loop would be set to move the cg in relation to the fore and aft tilt of the robot. In fact, it might be possible to make the cg shifting system completely automatic and independent of all other systems on the robot, but no known example of this has been tested. Figures 4-1 and 4-2 show two basic techniques for moving the cg.

The various figures in this chapter show wheel layouts without showing drive mechanisms. The location of the drive motor(s) is left to the designer, but there are a few unusual techniques for connecting the drive motor to the wheels that affect mobility that should be discussed. Some of the figures show the chassis located in line with the axles of the wheels, and some show it completely above the wheels, which increases ground clearance at the possible expense of increased complexity of the coupling mechanism. In many cases, the layouts that show the chassis down low can be altered to have it up high, and vise-versa.

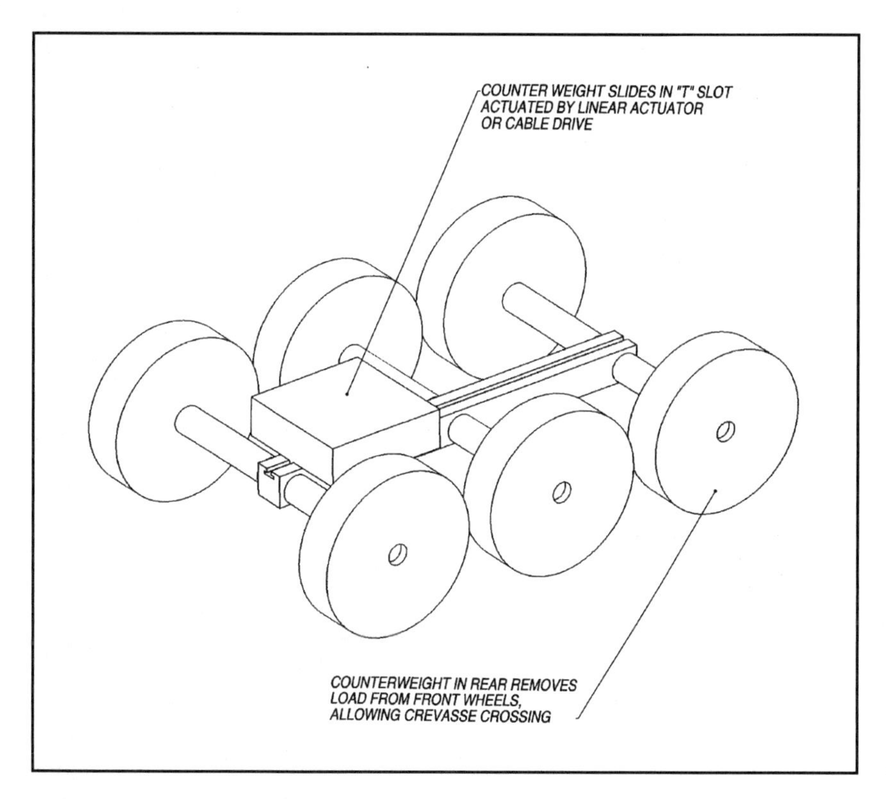

Figure 4-1 Method for shifting the center of gravity on a linear slide

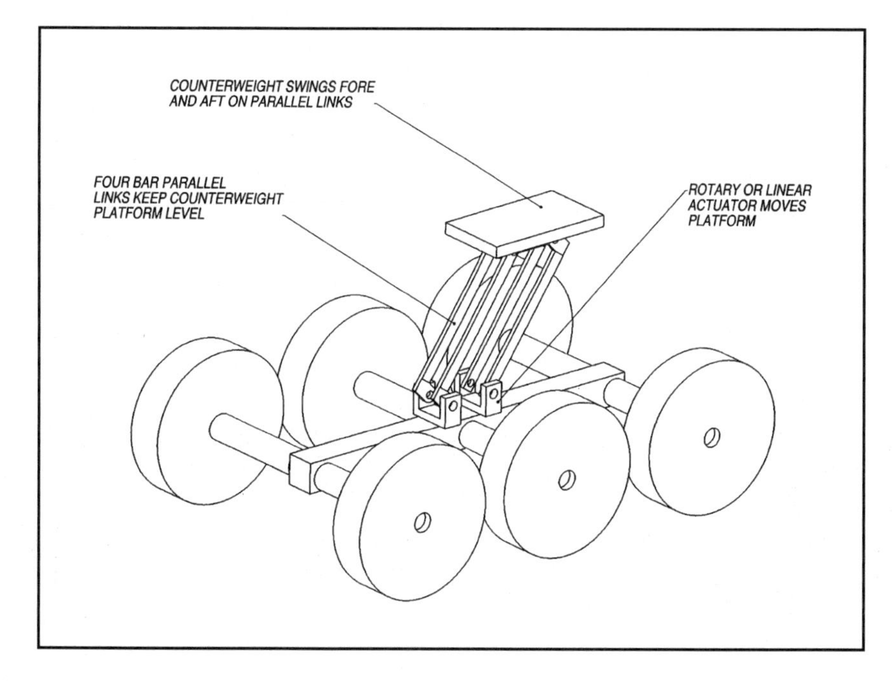

Figure 4-2 Shifting the cg on a swinging arm

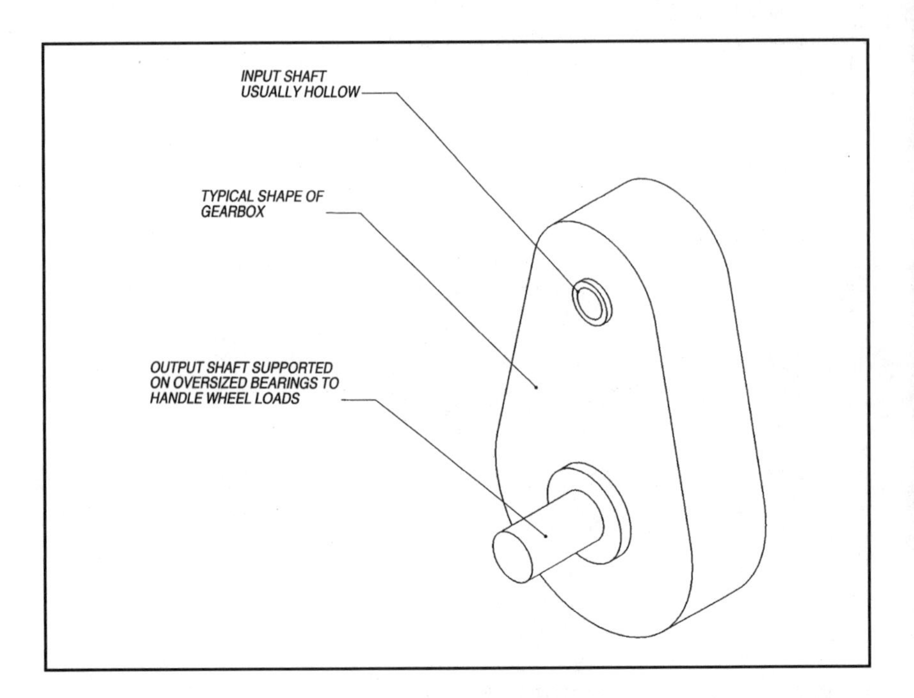

Figure 4-3 Geared offset wheel hub

For the raised layouts, the drive axle is coupled to the wheel through a chain, belt drive, or gearbox. The US Army's High Mobility Multipurpose Wheeled Vehicle (HMMWV, HumVee, or Hummer), uses geared offset hubs (Figure 4-3) resulting in a ground clearance of 16" with tires that are 37" in diameter. This shows how effective the raised chassis layout can be.

WHEEL SIZE

In general, the larger the wheel, the larger the obstacle a given vehicle can get over. In most simple suspension and drivetrain systems, a wheel will be able to roll itself over a step-like bump that is about one-third the diameter of the wheel. In a well-designed four-wheel drive off-road truck, this can be increased a little, but the limit in most suspensions is something less than half the diameter of the wheel. There are ways around this though. If a driven wheel is pushed against a wall that is taller than the wheel diameter with sufficient forward force relative to the vertical load on it, it will roll up the wall. This is the basis for the design of rocker bogie systems.

Three wheels are the minimum required for static stability, and three-wheeled robots are very common. They come in many varieties, from very simple two-actuator differential steer with fixed third wheel types, to relatively complex roller-walkers with wheels at the end of two or even three DOF legs. Mobility and complexity are increased by adding even more wheels. Let's take a look at wheeled vehicles in rough order of complexity.

The most basic vehicle would have the least number of wheels. Believe it or not, it is possible to make a one-wheeled vehicle! This vehicle has limited mobility, but can get around relatively benign environments. Its wheel is actually a ball with an internal movable counterweight that, when not over the point of contact of the ball and the ground, causes the ball to roll. With some appropriate control on the counterweight and how it is attached and moved within the ball, the vehicle can be steered around clumsily. Its step-climbing ability is limited and depends on what the actual tire is made of, and the weight ratio between the tire and the counterweight.

There are two obvious two wheeled layouts, wheels side by side, and wheels fore and aft. The common bicycle is perhaps one of the most recognized two-wheeled vehicles in the world. For robots, though, it is quite difficult to use because it is not inherently stable. The side by side layout is also not inherently stable, but is easier to control, at low speeds, than a bike. Dean Kamen developed the Segway two-wheeled balancing vehicle, proving it is possible, and is actually fairly mobile. It suffers from

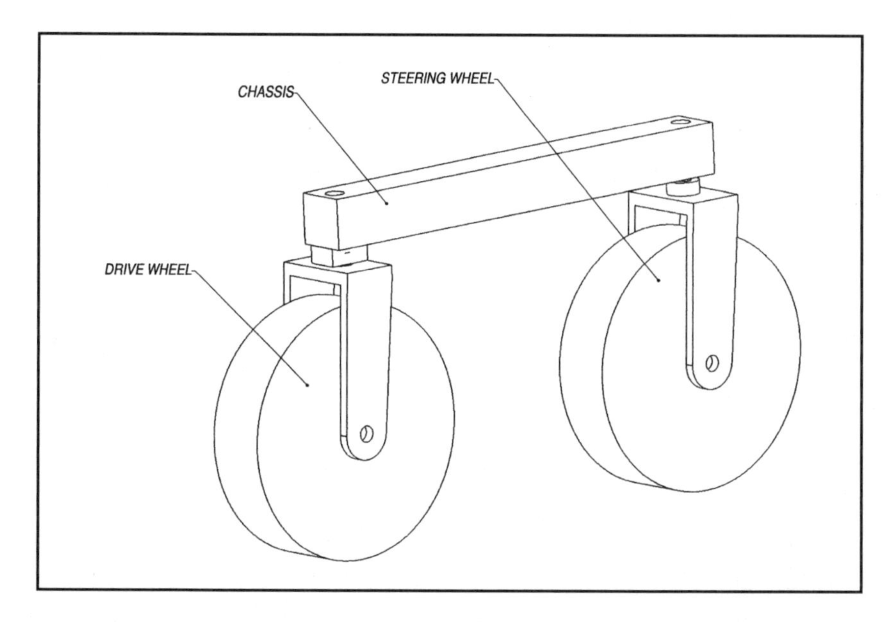

Figure 4-4 Bicycle

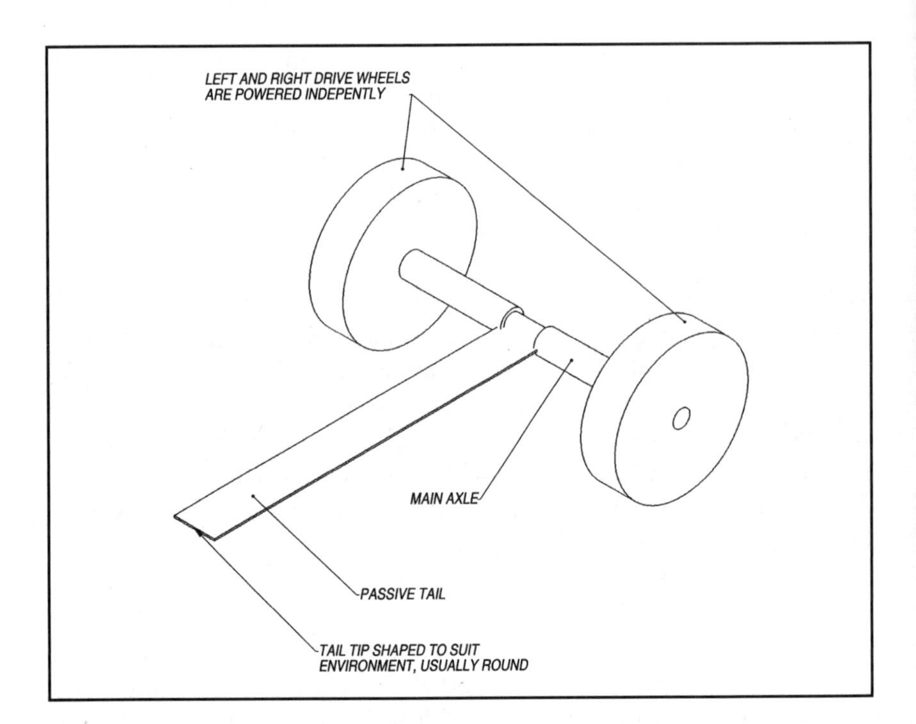

Figure 4-5 Tail dragger

the same limitation the single wheeled ball suffers from and cannot get over bumps much higher than one quarter a wheel height.

The third, less obvious layout is to drag a passive leg or tail behind the vehicle. This tail counteracts the torque produced by the wheels, makes the vehicle statically stable, and increases, somewhat, the height of obstacle the robot can get over. The tail dragger is ultra-simple to control by independently varying the speed of the wheels. This serves to control both velocity and steering. The tail on robots using this layout must be light, strong, and just long enough to gain the mobility needed. Too long and it gets in the way when turning, too short and it doesn't increase mobility much at all. It can be either slightly flexible, or completely stiff. The tail end slides both fore and aft and side to side, requiring it to be of a shape that does not hang up on things. A ball shape, or a shape very similar, made of a low friction material like Teflon or polyethylene, usually works out best.

THREE-WHEELED LAYOUTS

The tail dragger demonstrates the simplest statically stable wheeled vehicle, but, unfortunately, it has limited mobility. Powering that third contact point improves mobility greatly. Three wheels can be laid out in several ways. Five varieties are pictured in the following figures. The most common and easiest to implement, but with, perhaps, the least mobility of the five three-wheeled types, is represented by a child's tricycle. On the kid-powered version, the front wheel provides both propulsion and steers. Robots destined to be used indoors, in a test lab or other controllable space, can use this simple layout with ease, but it has extremely poor mobility. Just watch any child struggling to ride their tricycle on anything but a flat smooth road or sidewalk. Powering only one of the three wheels is the major cause of this problem. Nevertheless, there have been many successful indoor test platforms that use this layout precisely because of its simplicity.

In order to improve the mobility and stability of motorcycles, the three wheeled All Terrain Cycle (ATC) was developed. This vehicle demonstrates the next step up in the mobility of three wheeled vehicles. The rear two wheels are powered through a differential, and the front steers. This design is still simple, but although ATCs seemed to have high mobility, they did not do well in forest environments filled with rocks and logs, etc. The ATC was eventually outlawed because of its major flaw, very poor stability. Putting the single wheel in front lead to reduced resistance to tipping over the front wheel. This is also the most common form of accident with a child's tricycle.

Increasing the stability of a tricycle can be easily accomplished by reversing the layout, putting the two wheels in front. This layout works fine for relatively low speeds, but the geometry is difficult to control when turning at higher speeds as the forces on the rear steering wheel tend to make the vehicle turn more sharply until eventually it is out of control. This can be minimized by careful placement of the vehicle's center of gravity, moving it forward just the right amount without going so far that a hard stop flips the vehicle end over end. A clever version of this tail dragger-like layout gets around the problem of flipping over by virtue of its ability to flip itself back upright simply by accelerating rapidly. The vehicle flips over because there is no lever arm to resist the torque in the wheels. Theoretically, this could be done with a tricycle also. At low speeds, this layout has similar mobility to a tail dragger and, in fact, they are very similar vehicles.

Steering with the front wheels on a reversed tricycle removes the steering problem, but adds the complexity of steering and driving both wheels. This layout does allow placing more weight on the passive rear wheel, significantly reducing the flipping over tendencies, and mobility is moderately good. The layout is still dragging around a passive wheel, however, and mobility is further enhanced if this wheel is powered.

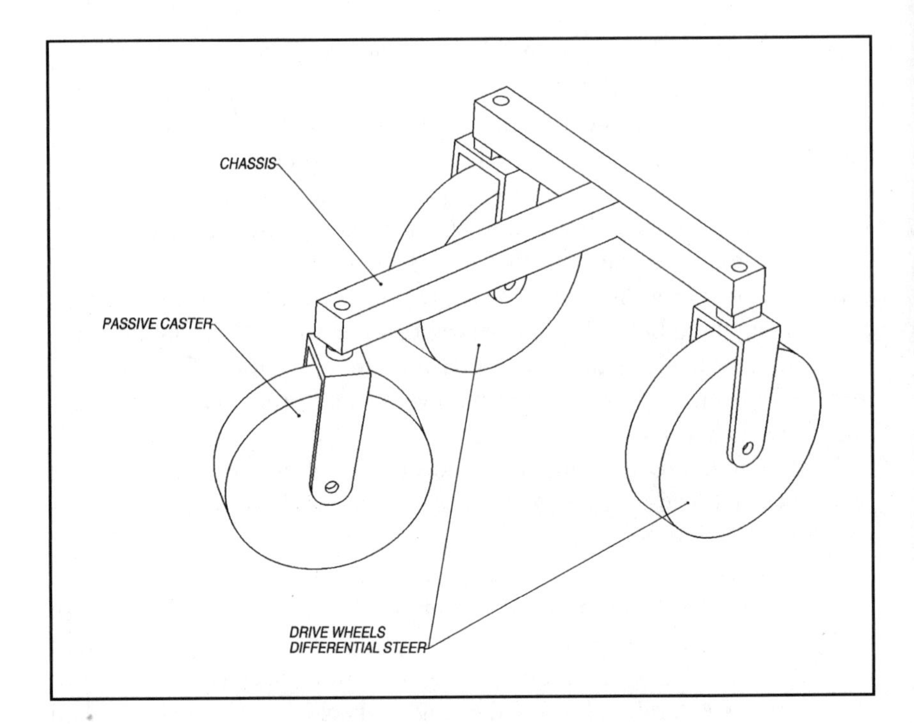

Figure 4-6 Reversed tricycle, differential steer

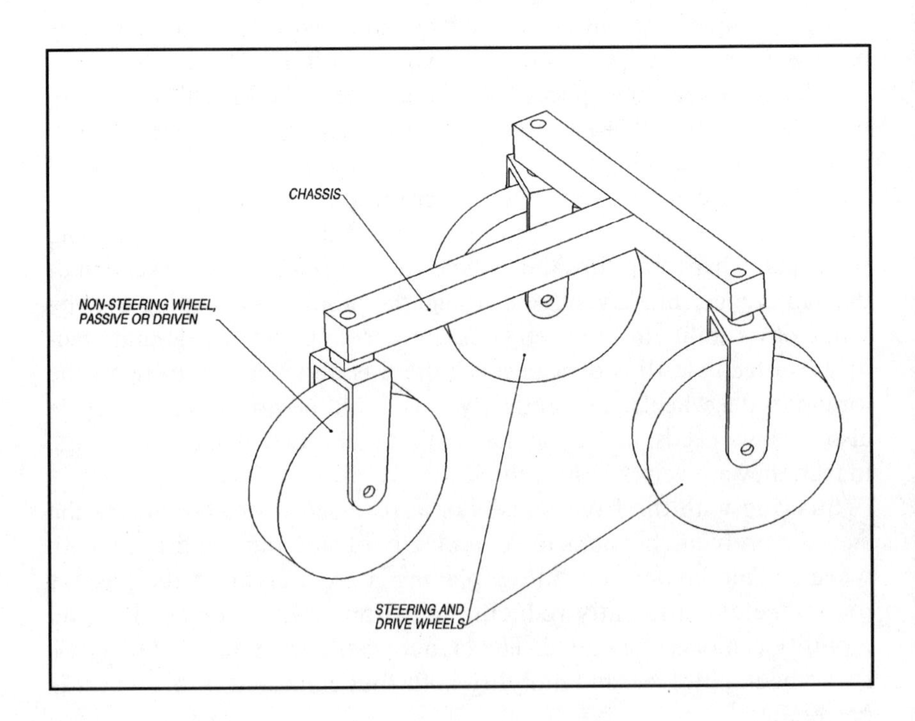

Figure 4-7 Reversed tricycle, front steer

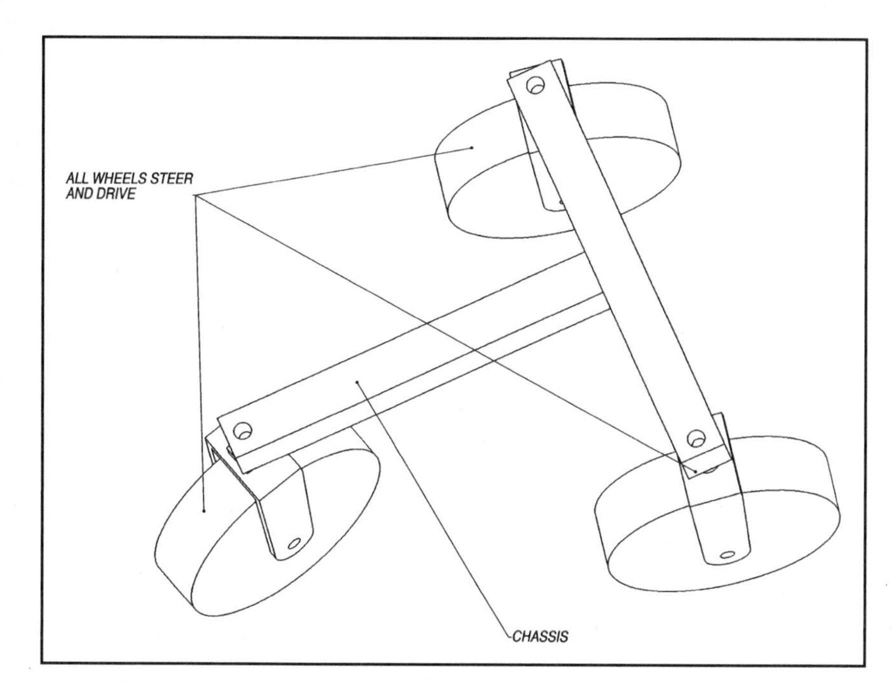

Figure 4-8 Reversed tricycle, all drive, all steer

The most complicated and highest mobility three-wheeled layout is one where all wheels are powered and steered. This layout is extremely versatile, providing motion in any direction without the need to be moving; it can turn in place. This ability is called holonomic motion and is very useful for mobile robots because it can significantly improve mobility in cluttered terrain. Of the vehicles discussed so far, all, except the front steer reversed tricycle, can be made holonomic if the third wheel lies on the circumference of the circle whose center is midway between the two opposing wheels, and the steering or passive wheel can swing through 180 degrees. To be truly holonomic, even in situations where the vehicle is enclosed on three sides, like in a dead-end hallway, the vehicle itself must be round. This enables it to turn at any time to find a path out of its trap. See Figure 4-8.

Before we investigate four-wheeled vehicles, there is a mechanism that must be, at least basically, understood—the differential. The differential (Figure 4-9) gets its name from the fact that it differentiates the rotational velocity of two wheels driven from one drive shaft. The most basic differential uses a set of gears mounted inside a larger gear, but on an axis that lies along a radius of the larger gear. These gears rotate with the large gear, and are coupled to the axles through crown gears on the ends of the axles. When both wheels are rolling on relatively high fric-

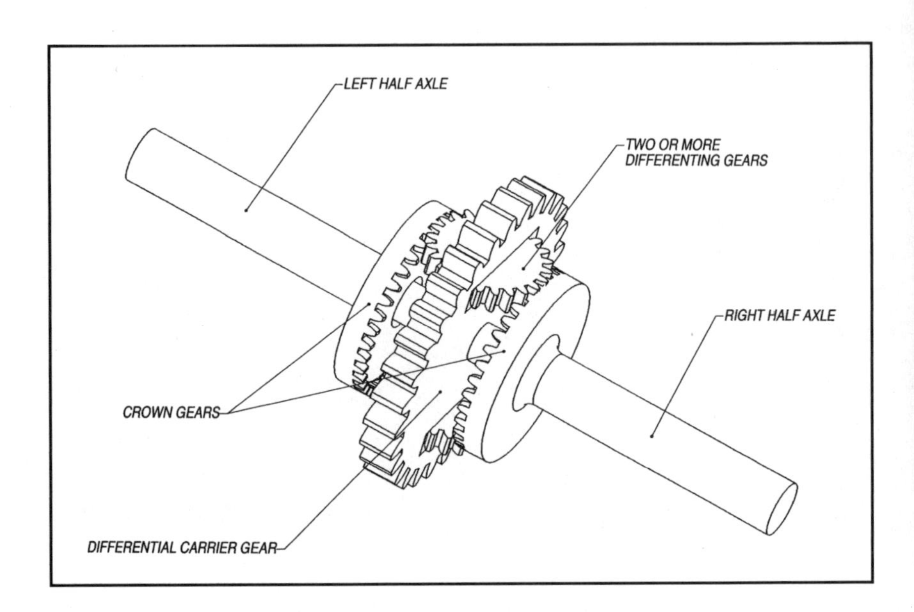

Figure 4-9 The common and unpredictable differential

tion surfaces, and the vehicle is going straight, the wheels rotate at the same rpm. If the vehicle turns a corner, the outside wheel is traversing a longer path and therefore must be turning faster than the inside wheel. The differential facilitates this through the internal gears, which rotate inside the large gear, allowing one axle to rotate relative to the other. This system, or something very much like it, is what is inside virtually every car and truck on the road today. It obviously works well.

The simple differential has one drawback. If one wheel is rolling on a surface with significantly less friction, it can slip and spin much faster than the other wheel. As soon as it starts to slip, the friction goes down further, exacerbating the problem. This is almost never noticed by a human operator, but can cause mobility problems for vehicles that frequently drive on slippery surfaces like mud, ice, and snow.

There are a couple of solutions. One is to add clutches between the axles that slide on each other when one wheel rotates faster than the other. This works well, but is inefficient because the clutches absorb power whenever the vehicle goes around a corner. The other solution is the wonderfully complicated Torsen differential, manufactured by Zexel.

The Torsen differential uses specially shaped worm gears to tie the two axles together. These gears allow the required differentiation between the two wheels when turning, but do not allow one wheel to spin as it looses traction. A vehicle equipped with a Torsen differential can effectively drive with one wheel on ice and the other on hard dry pavement! This differential uses very complex gear geometries. The best explanation of how it works can be found on Zexel's web site: www.torsen.com.

FOUR-WHEELED LAYOUTS

The most basic four-wheeled vehicle actually doesn't even use a differential. It has two wheels on each side that are coupled together and is steered just like differential steered tricycles. Since the wheels are in line on each side and do not turn when a corner is commanded, they slide as the vehicle turns. This sliding action gives this steering method its name—Skid Steer. Notice that this layout does not use differentials, even though it is also called differential steering.

Skid steered vehicles are a robust, simple design with good mobility, in spite of the inefficiency of the sliding wheels. Because the wheels don't turn, it is easy to attach them to the chassis, and they don't take up the space required to turn. There are many industrial off-road skid steered vehicles in use, popularly called Bobcats. Figure 4-10 shows that a skid steered vehicle is indeed very simple.

The problem with skid steered, non-suspended drivetrains is that as the vehicle goes over bumps, one wheel necessarily comes off the ground. This problem doesn't exist in two or three wheeled vehicles, but is a major thing to deal with on vehicles with more than three wheels. Though not a requirement for good mobility, it is better to use some mechanism that keeps all the wheels on the ground. There are many ways to accomplish this, starting with a design that splits the chassis in two.

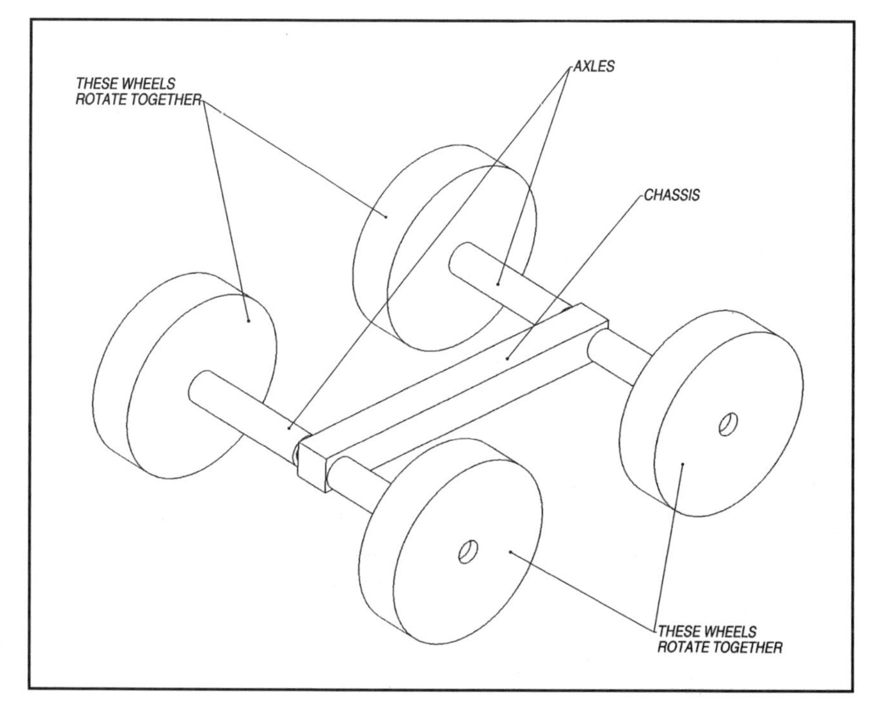

Figure 4-10 All four fixed, skid steered

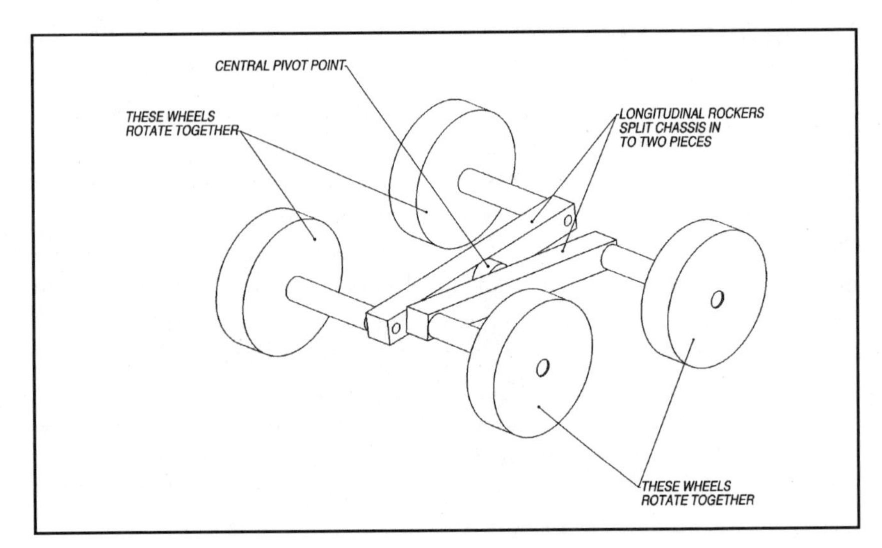

Figure 4-11 Simple longitudinal rocker

The longitudinal rocker design divides the entire vehicle right down the middle and places a passive pivot joint in between the two halves. This joint is connected on each end to a rocker arm, which in turn carry a wheel at each of their ends. This layout allows the rocker arms to pivot when any wheel tries to go higher or lower than the rest. This passive pivoting action keeps the load on all four wheels almost equal, increasing mobility simply by maintaining driving and braking action on all wheels at all times. Longitudinal rocker designs are skid steered, with the wheels on each side usually mechanically tied together like a simple skid steer, but sometimes, to increase mobility even further, the wheels are independently powered. Figure 4-11 shows the basic layout, developed by Sandia Labs for a vehicle named Ratler.

The well-known forklift industrial truck uses a sideways version of the rocker system. Since its front wheels carry most of all loads lifted by the vehicle, structurally tying the wheels together is a more robust layout. These vehicles have four wheels without any suspension, and, therefore, require some method of keeping all the wheels on the ground. The most common layout has the front wheels tied together and a rocker installed transversely and coupled to the rear wheels, which are usually the steering wheels. Figure 4-12 shows this layout.

The weakness of the forklift is that it is usually only two-wheel drive. This works well for its application, and because so much of the weight of the vehicle is over the front wheels. In general, though, powering all four wheels provides much higher mobility. In a two-wheel drive vehicle, the driven wheels must provide traction not only for whatever they are trying to get over, but also must push or pull the non-driven wheels. Many of the wheeled layouts are complex enough that they require a motor for

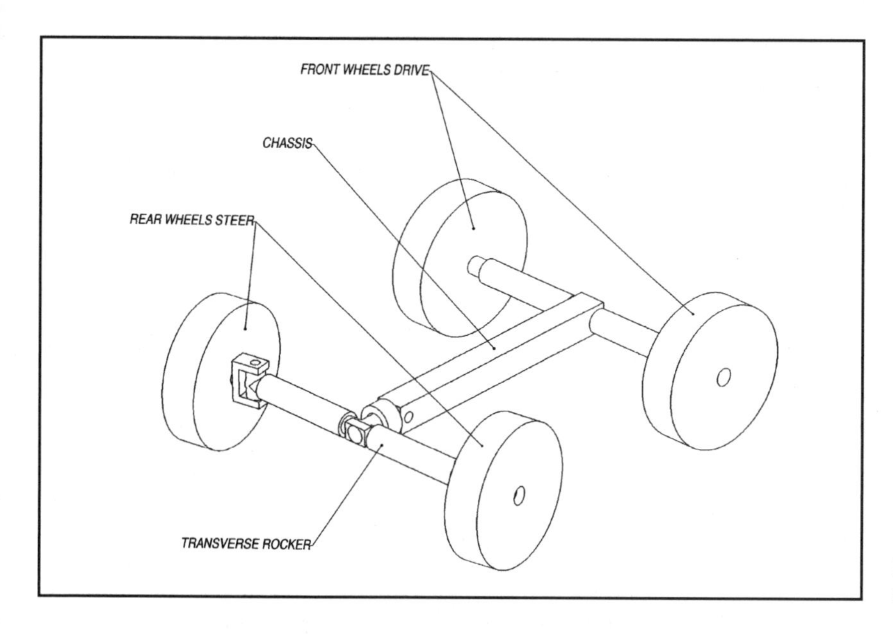

Figure 4-12 Rear transverse rocker, rear steer

every wheel. Although this seems like a complicated solution from an electrical and control standpoint, it is simpler mechanically.

Steering with the rear wheels is effective for a human controlled vehicle, especially in an environment with few obstacles that must be driven around. The transverse rocker layout can also be used with a front steered layout (Figure 4-13) which makes it very much like an automobile. Couple this layout with all wheel drive, and this is a good performer.

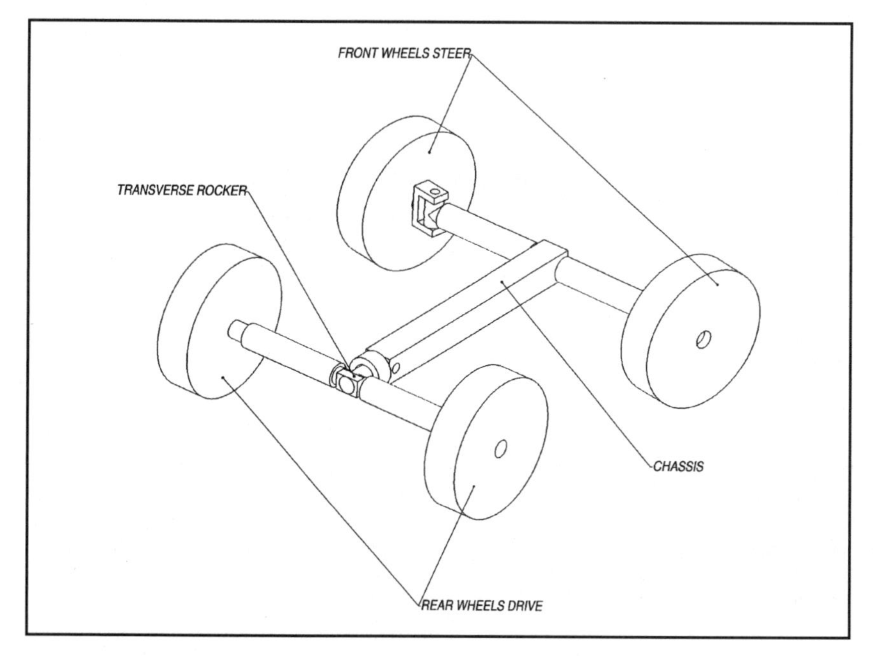

Figure 4-13 Rear transverse rocker, front steer

ALL-TERRAIN VEHICLE WITH SELF-RIGHTING AND POSE CONTROL

Wheels dr iven by gearmotors are mounted on pivoting struts.

NASA's Jet Propulsion Laboratory, Pasadena, California

A small prototype robotic all-terrain vehicle features a unique drive and suspension system that affords capabilities for self righting, pose control, and enhanced maneuverability for passing over obstacles. The vehicle is designed for exploration of planets and asteroids, and could just as well be used on Earth to carry scientific instruments to remote, hostile, or otherwise inaccessible locations on the ground. The drive and suspension system enable the vehicle to perform such diverse maneuvers as flipping itself over, traveling normal side up or upside down, orienting the main vehicle body in a specified direction in all three dimensions, or setting the main vehicle body down onto the ground, to name a few. Another maneuver enables the vehicle to overcome a common weakness of traditional all-terrain vehicles—a limitation on traction and drive force that makes it difficult or impossible to push wheels over some obstacles: This vehicle can simply lift a wheel onto the top of an obstacle.

The basic mode of operation of the vehicle can be characterized as four-wheel drive with skid steering. Each wheel is driven individually by a dedicated gearmotor. Each wheel and its gearmotor are mounted at the free end of a strut that pivots about a lateral axis through the center of

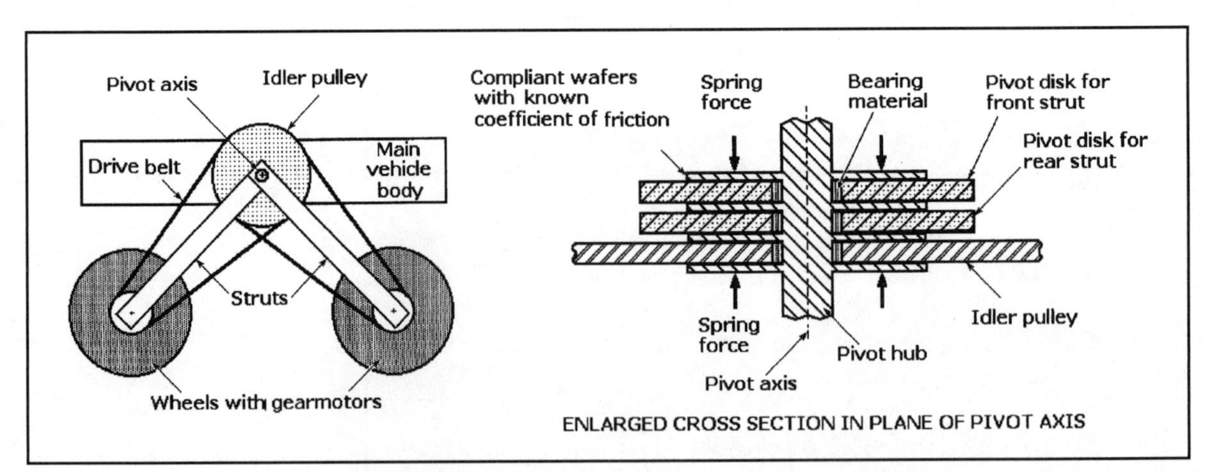

Figure 4-14 Each wheel Is driven by a dedicated gearmotor and is coupled to the idler pulley. The pivot assembly imposes a constant frictional torque T, so that it is possible to (a) turn both wheels in unison while both struts remain locked, (b) pivot one strut, or (c) pivot both struts in opposite directions by energizing the gearmotors to apply various combinations of torques T/2 or T.

gravity of the vehicle (see figure). Through pulleys or other mechanism attached to their wheels, both gearmotors on each side of the vehicle drive a single idler disk or pulley that turns about the pivot axis.

The design of the pivot assembly is crucial to the unique capabilities of this system. The idler pulley and the pivot disks of the struts are made of suitably chosen materials and spring-loaded together along the pivot axis in such a way as to resist turning with a static frictional torque T; in other words, it is necessary to apply a torque of T to rotate the idler pulley or either strut with respect to each other or the vehicle body.

During ordinary backward or forward motion along the ground, both wheels are turned in unison by their gearmotors, and the belt couplings make the idler pulley turn along with the wheels. In this operational mode, each gearmotor contributes a torque T/2 so that together, both gearmotors provide torque T to overcome the locking friction on the idler pulley. Each strut remains locked at its preset angle because the torque T/2 supplied by its motor is not sufficient to overcome its locking friction T.

If it is desired to change the angle between one strut and the main vehicle body, then the gearmotor on that strut only is energized. In general, a gearmotor acts as a brake when not energized. Since the gearmotor on the other strut is not energized and since it is coupled to the idler pulley, a torque greater than T would be needed to turn the idler pulley. However, as soon as the gearmotor on the strut that one desires to turn is energized, it develops enough torque (T) to begin pivoting the strut with respect to the vehicle body.

It is also possible to pivot both struts simultaneously in opposite directions to change the angle between them. To accomplish this, one energizes the gearmotors to apply equal and opposite torques of magnitude T: The net torque on the idler pulley balances out to zero, so that the idler pulley and body remain locked, while the applied torques are just sufficient to turn the struts against locking friction. If it is desired to pivot the struts through unequal angles, then the gearmotor speeds are adjusted accordingly.

The prototype vehicle has performed successfully in tests. Current and future work is focused on designing a simple hub mechanism, which is not sensitive to dust or other contamination, and on active control techniques to allow autonomous planetary rovers to take advantage of the flexibility of the mechanism.

This work was done by Brian H. Wilcox and Annette K. Nasif of Caltech for **NASA's Jet Propulsion Laboratory**.

If a differential is installed between the halves of a longitudinal rocker layout, with the axles of the differential attached to each longitudinal rocker, and interesting effect happens to the differential input gear as the

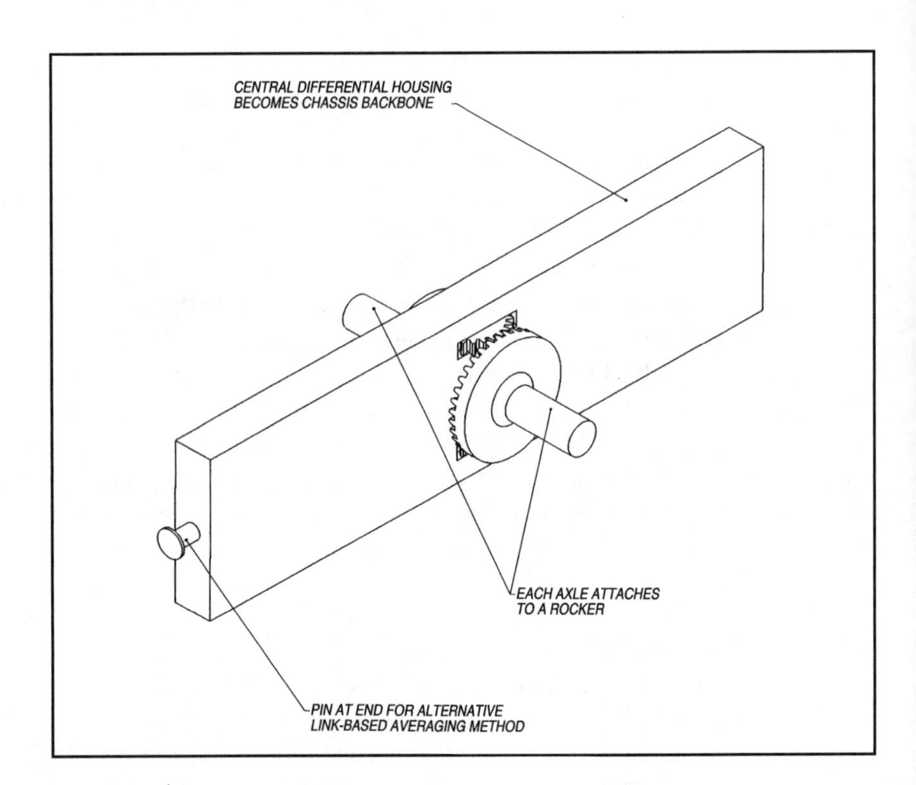

Figure 4-15 Pitch averaging mechanism

vehicle traverses bumpy terrain. If you attach the chassis to this gear, the pitch angle of the chassis is half the pitch angle of either side rocker. This pitch averaging effectively reduces the pitching motion of the chassis, maintaining it at a more level pose as either side of the suspension system travels over bumps. This can be advantageous in vehicles under camera control, and even a fully autonomous sensor driven robot can benefit from less rocking motion of the main chassis. This mechanism also tends to distribute the weight more evenly on all four wheels, increasing traction, and, therefore, mobility. Figure 4-15 shows the basic mechanism and Figure 4-16 shows it installed in a vehicle.

Another mechanical linkage gives the same result as the differential-based chassis pitch-averaging system. See figure 4-17. This design uses a third rocker tied at each end to a point on the side rockers. The middle of the third rocker is then tied to the middle of the rear (or front) of the chassis, and, therefore, travels up and down only half the distance each end of a given rocker travels. The third rocker design can be more volumetrically efficient and perhaps lighter than the differential layout.

Another layout that can commonly be found in large industrial vehicles is one where the vehicle is divided into two sections, front and rear, each with its own pair of wheels. The two sections are connected through

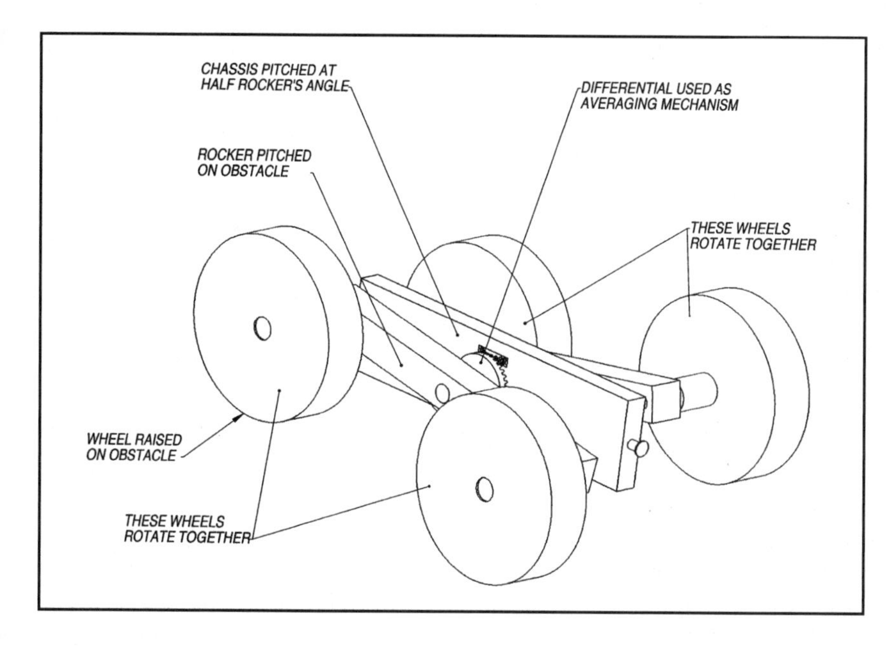

Figure 4-16 Chassis pitch averaging mechanism using differential

an articulated (powered) vertical-axis joint. In the industrial truck version, the front and rear sections' wheels are driven through differentials, but higher traction would be obtained if the differentials were limited slip, or lockable. Even better would be to have each wheel driven with its

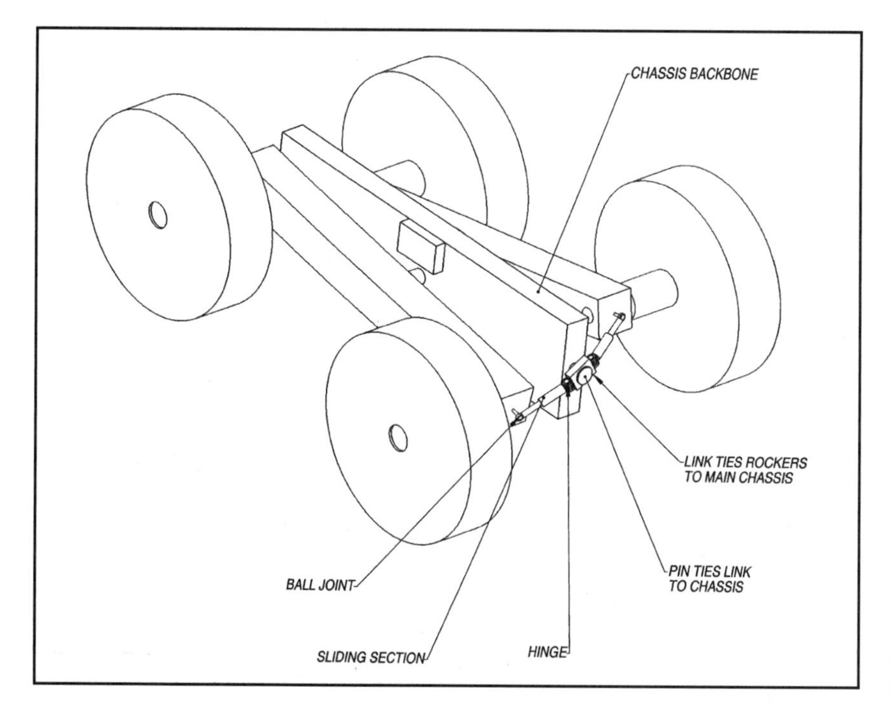

Figure 4-17 Chassis link-based pitch averaging mechanism

own motor. This design cannot turn in place, but careful layout can produce a vehicle that can turn in little more than twice its width.

Greater mobility is achieved if the center joint also allows a rolling motion between the two sections. This degree of freedom keeps all four wheels on the ground while traversing uneven terrain or obstacles. It also improves traction while turning on bumps. Highest mobility for this layout would come from powering both the pivot and roll joints with their own motors and each wheel individually powered for a total of six motors. Alternatively, the wheels could be powered through limited slip differentials and the roll axis left passive for less mobility, but only three motors. Figures 4-18 and 4-19 show these two closely related layouts.

An unusual and unintuitive layout is the five-wheeled drivetrain. This is basically the tricycle layout, but with an extra pair of wheels in the back to increase traction and ground contact area. The front wheel is not normally powered and is only for steering. Figure 4-20 shows this is a fairly simple layout relative to its mobility, especially if the side wheel pairs are driven together through a simple chain or belt drive. Although the front wheels must be pushed over obstacles, there is ample traction from all that rubber on the four rear wheels.

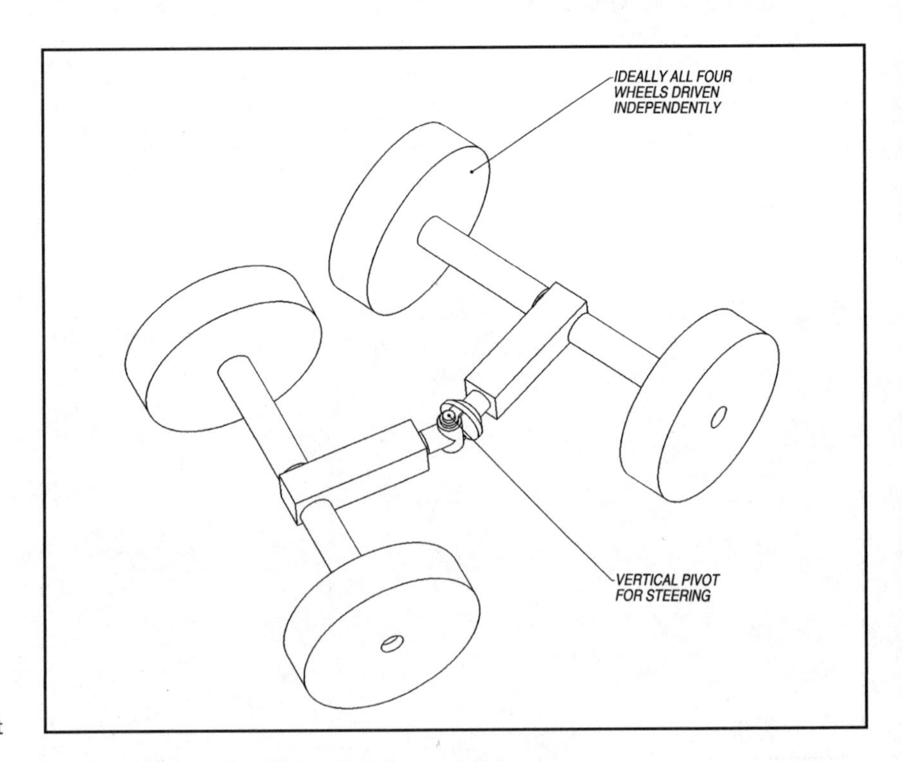

Figure 4-18 Two-sections connected through vertical axis joint

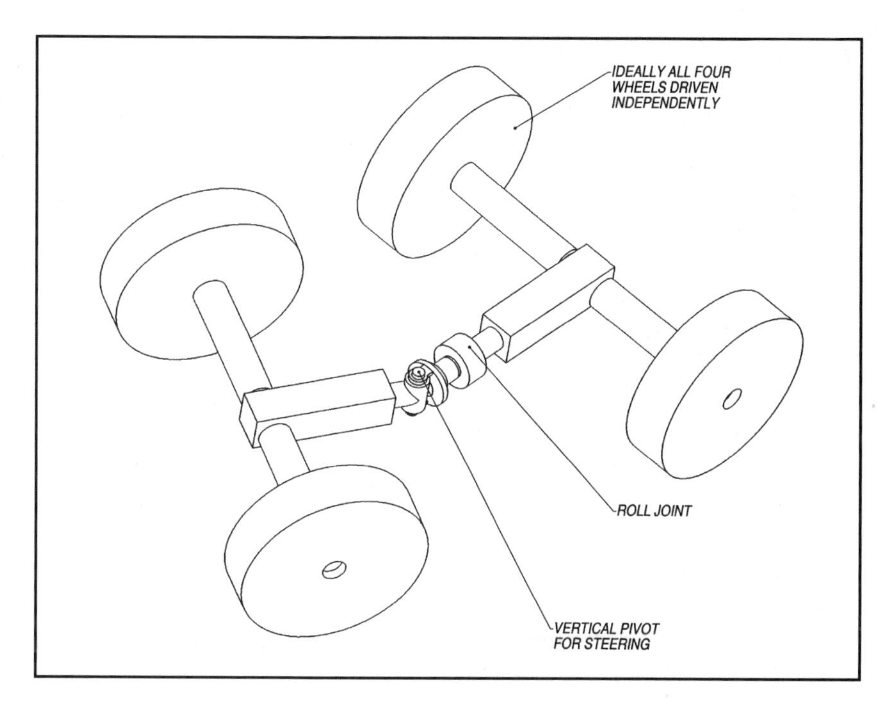

Figure 4-19 Two sections connected through both a vertical axis and a longitudinal axis joint

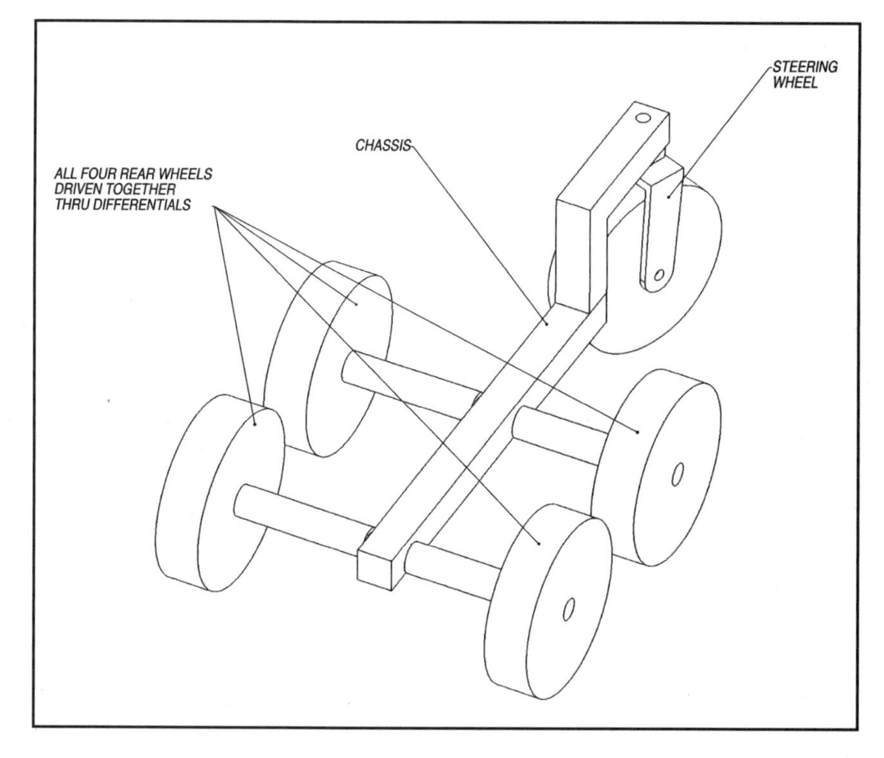

Figure 4-20 Five wheels

SIX-WHEELED LAYOUTS

Beyond four- and five-wheeled vehicles is the large class of six-wheeled layouts. There are many layouts, suspensions, and drivetrains based on six wheels. Six wheels are generally the best compromise for high-mobility wheeled vehicles. Six wheels put enough ground pressure, traction, steering mobility, and obstacle-negotiating ability on a vehicle without, in most cases, very much complexity. Let's take a look at the more practical variations of six-wheeled layouts.

The most basic six-wheeled vehicle, shown in Figure 4-21, is the skid-steered non-suspended design. This is very much like the four-wheeled design with improved mobility simply because there is more traction and less ground pressure because of the third wheel on each side. The wheels can be driven with chains, belts, or bevel gearboxes in a simple way, making for a robust system.

An advantage of the third wheel in the skid-steer layout is that the middle wheel on each side can be mounted slightly lower than the other two, reducing the weight the front and rear wheel pairs carry. The lower weight reduces the forces needed to skid them around when turning, reducing turning power. The offset center axle can make the vehicle wobble a bit. Careful planning of the location of the center of gravity is required to minimize this problem. Figure 4-22 shows the basic concept.

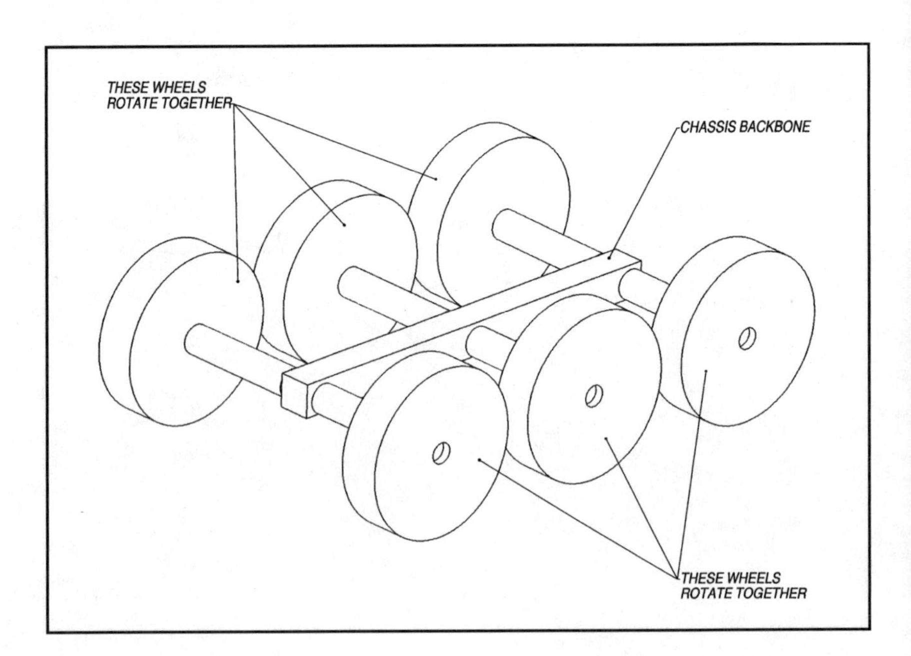

Figure 4-21 Six wheels, all fixed, skid steer

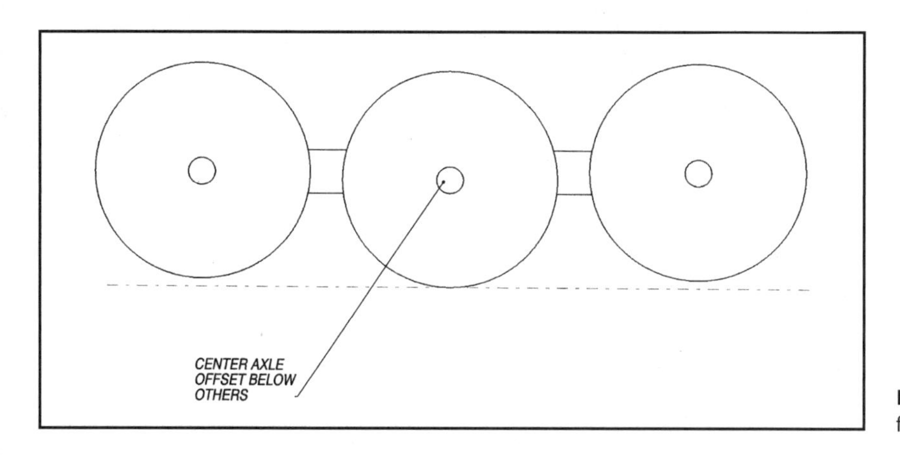

Figure 4-22 Six wheels, all fixed, skid steer, offset center axle

An even trickier layout adds two pairs of four-bar mechanisms supporting the front and rear wheel pairs (Figure 4-23). These mechanisms are moved by linear actuators, which raise and lower the wheels at each corner independently. This semi-walking mechanism allows the vehicle to negotiate obstacles that are taller than the wheels, and can aid in traversing other difficult terrain by actively controlling the weight on each wheel. This added mobility comes at the expense of many more moving parts and four more actuators.

Skid steering can be improved by adding a steering mechanism to the front pair of wheels, and grouping the rear pair more closely together.

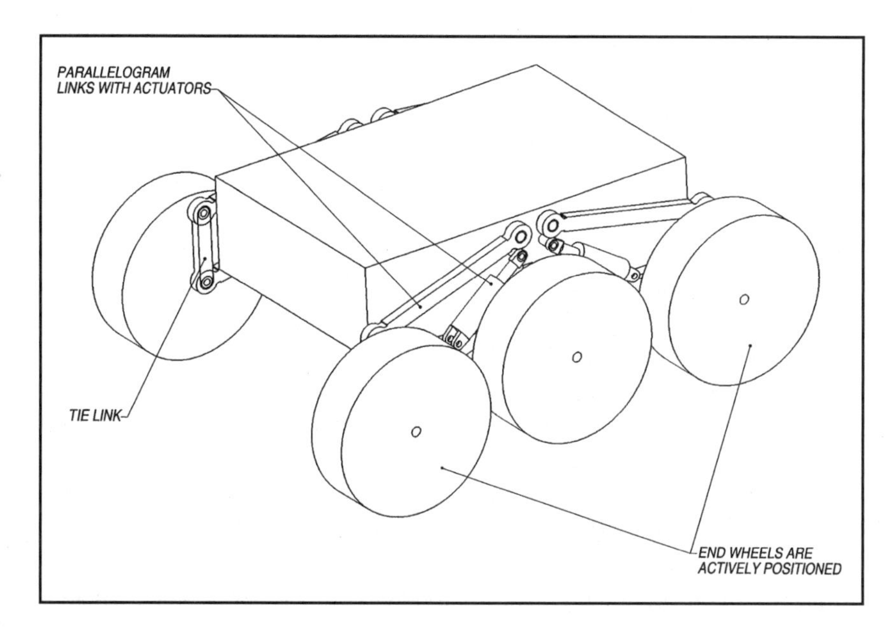

Figure 4-23 Six wheels, all corner wheels have adjustable height, skid steer

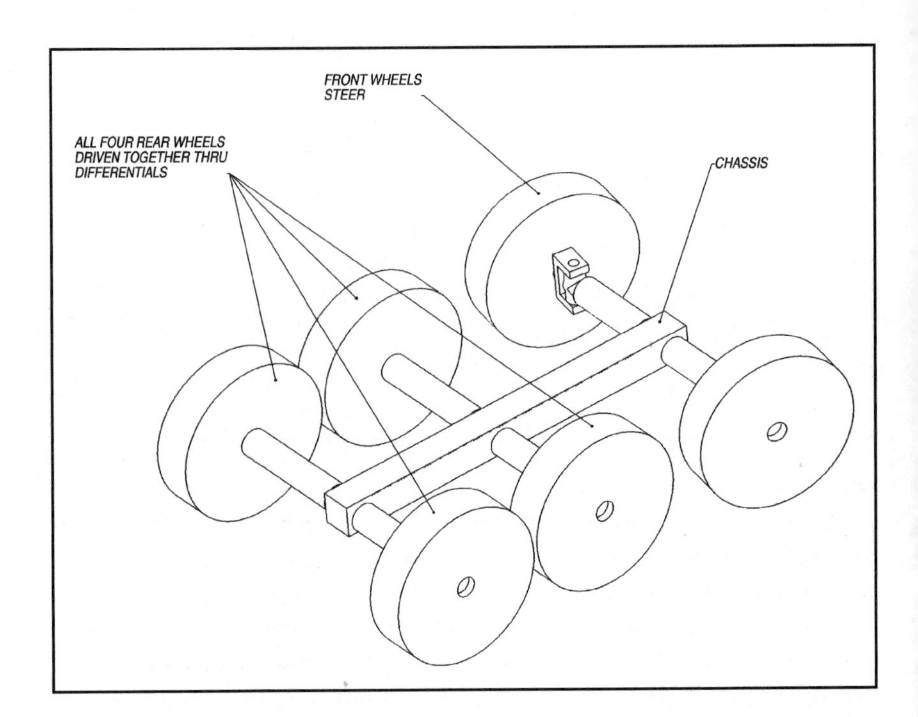

Figure 4-24 Six wheels, front pair steer

This has better steering efficiency, but, surprisingly, not much better mobility. Incorporating the Ackerman steering layout removes the ability of the robot to turn in place. This can be a real handicap in tight places. Figure 4-24 shows the basic layout. Remember that the relative sizes of wheels and the spacing between them can be varied to produce different mobility characteristics.

The epitome of complexity in a once commercially available six-wheeled vehicle, not recommended to be copied for autonomous robot use, is the Alvis Stalwart. This vehicle was designed with the goal of going anywhere in any conditions. It was a six-wheeled (all independently suspended on parallel links with torsion arms) vehicle whose front four wheels steered. Each bank of three wheels was driven together through bevel gears off half-shafts. It had offset wheel hub reduction gear boxes, a lockable central differential power transfer box with integral reversing gears, and twin water jet drives for amphibious propulsion. All six wheels could be locked together for ultimate straight-ahead traction. No sketch is included for obvious reasons, but a website with good information and pictures of this fantastically complicated machine is *www.4wdonline.com/Mil/alvis/stalwart.html*.

The main problem with these simple layouts is that when one wheel is up on a bump, the lack of suspension lifts the other wheels up, drastically reducing traction and mobility. The ideal suspension would keep the load

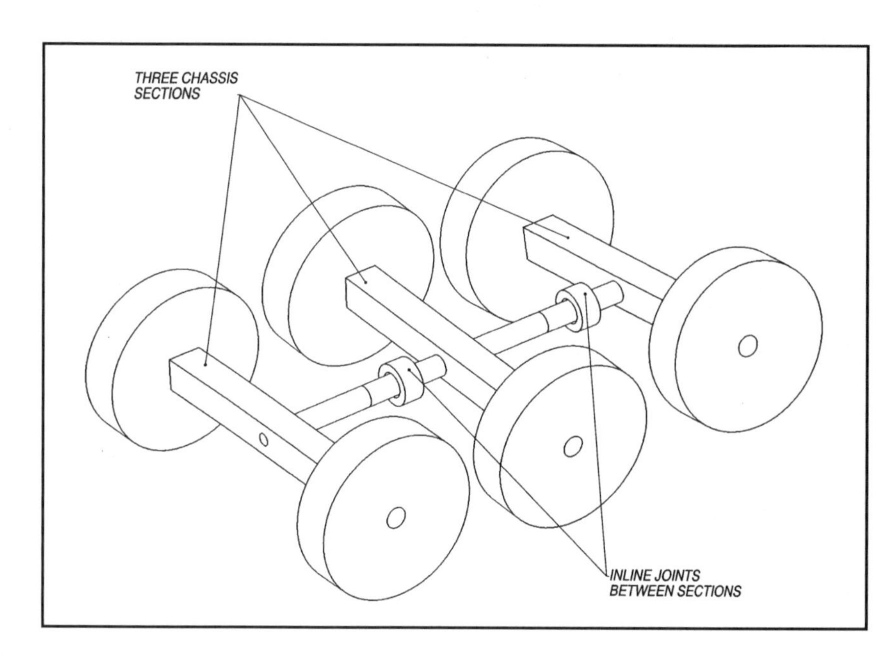

Figure 4-25 Six wheels, three sections, one DOF between each section, skid steer

on each wheel the same no matter at what height any one wheel is. The following suspension systems even out the load on each wheel—some more than others.

An interesting layout that does a good job of maintaining an even load distribution divides the robot into three sections connected by a single degree of freedom joint between each section (Figure 4-25). The center section has longitudinal joints on its front and back that attach to the cross pieces of the front and rear sections. These joints allow each section to roll independently. This movement keeps all six wheels on the ground. The roll axes are passive, requiring no actuators, but the separation of the wheeled sections usually forces putting a motor at each wheel, and the vehicle is skid steered. This layout has been experimented with by researchers and has very high mobility. The only drawback is that the roll joints must be sized to handle the large forces generated when skid steering.

The rocker bogie suspension system shown in Figure 4-26 uses an extension of the basic four-wheel rocker layout. By adding a bogie to one end of the rocker arm, two wheels can be suspended from one end and one from the other end. Although this layout looks like it would produce asymmetrical loads on the wheels, if the length of the bogie is half that of the rocker, and the rocker is attached to the chassis one third of its length from the bogie end, the load on each wheel is actually identical. The proportions can be varied to produce uneven loads, which can improve mobility incrementally for one travel direction, but the basic layout has very good mobility. The rocker bogie's big advantage is that it

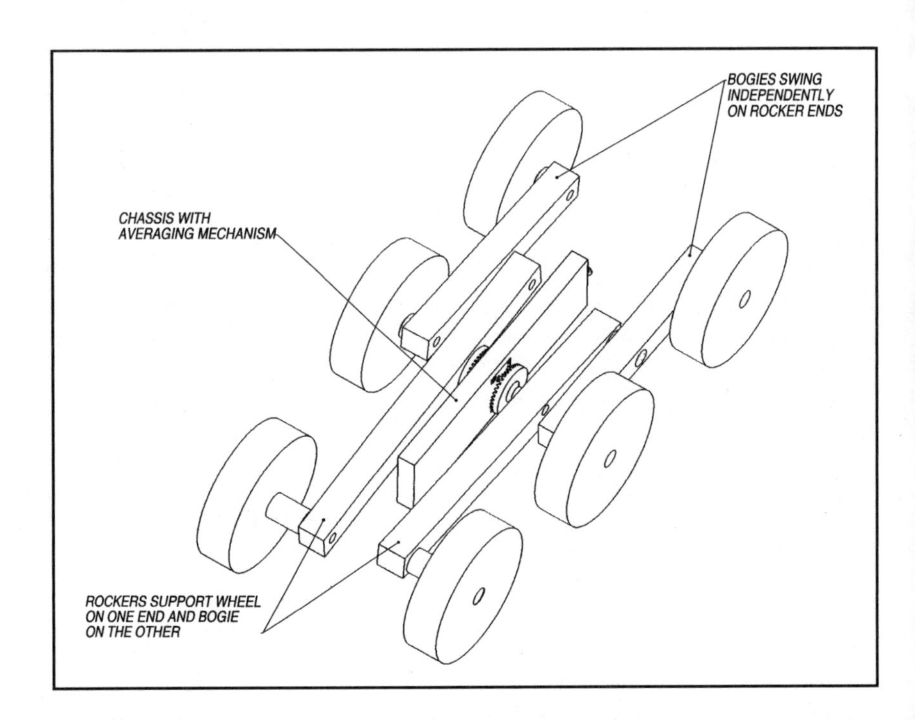

Figure 4-26 Rocker bogie

can negotiate obstacles that are twice the wheel height. This figure shows only the basic parts of the mobility layout. The part labeled "chassis" is the backbone or main support piece for the main body, which is not shown.

The very fact that each wheel is passively loaded by the rocker bogie suspension reduces its negotiable chasm width. Lockable pivots on the bogie can extend the negotiable chasm width by making the center wheels able to support the weight of the entire vehicle. This adds yet another actuator to this already complicated layout. This actuator can be a simple band or disc brake.

The rocker bogie suspension can be skid steered, but the side forces on the wheels produce moments in the rockers for which the rockers must be designed. Since the wheels are at the end of arms that move relative to each other, the most common layout puts a motor in each wheel. Steering is done by turning both the front and the rear wheels with their own steering motors. This means that this layout uses 10 motors to achieve its very high mobility. In this design, the large number of actuators reduces the number of moving parts and over all complexity.

The steering geometry allows turning in place with no skidding at all. This is the layout used on Sojourner, the robot that is now sitting on Mars after completing an entirely successful exploration mission on the Red Planet. Mobility experts claim this layout has the highest mobility possible in a wheeled vehicle, but this high mobility comes at the cost of those ten actuators and all their associated control electronics and debug time.

There is a layout that is basically six-wheeled, but with an extra pair of wheels mounted on flippers at the front. These wheels are powered with the three on each side and the vehicle is skid steered, but the front set of wheels are only placed on the ground for extra traction and stair climbing. This layout is in the same category as several layouts of tracked vehicles, as are several of the eight-wheeled layouts.

The next logical progression, already commercially available from Remotec in a slight variation, is to put the four center wheels on the ground, and put both end pairs on flippers. The center pair, instead of wheels, could be tracks, as it is on Remotec's Andros. The flippers carry either wheels or short tracks. This vehicle is rather complicated, but has great mobility since it can reconfigure itself into a long stair-climbing or crevasse-crossing layout, or fold up into a short vehicle about half as long.

EIGHT-WHEELED LAYOUTS

If six wheels are good then eight wheels are better, right? For a certain set of requirements, eight wheels can be better than six. There is, theoretically, more surface area simply because there are more wheels, but this is true mostly if there is a height limitation on the robot. If the robot needs to be particularly low for its size, then eight wheels may be the answer.

The most common layout for eight wheels, since inherently there are more moving parts already, is to skid-steer with fixed wheels. Lowering the center two pairs aids in skid steering just like on a six wheeled skid steer, but the four wheels on the ground means there is less wobbling when stopping and starting. Figure 4-27 shows this basic layout with the center wheels lowered slightly.

With all the wheels fixed there are many times when several of the wheels will be lifted off the ground, reducing traction greatly. A simple step to reduce this problem is to put the wheels on rockers, in pairs on each side. A set of wheels may still leave the ground in some terrains, but the other six wheels should remain mostly in contact with the ground to give some traction. Adding steering motors at the attachment point of each rocker would produce four-corner steering with minimal skidding. Since the bogie is a fairly simple arm connecting only a pair of wheels, a single motor could potentially be mounted near the center of the bogie and through a power transfer system, drive both wheels. This would

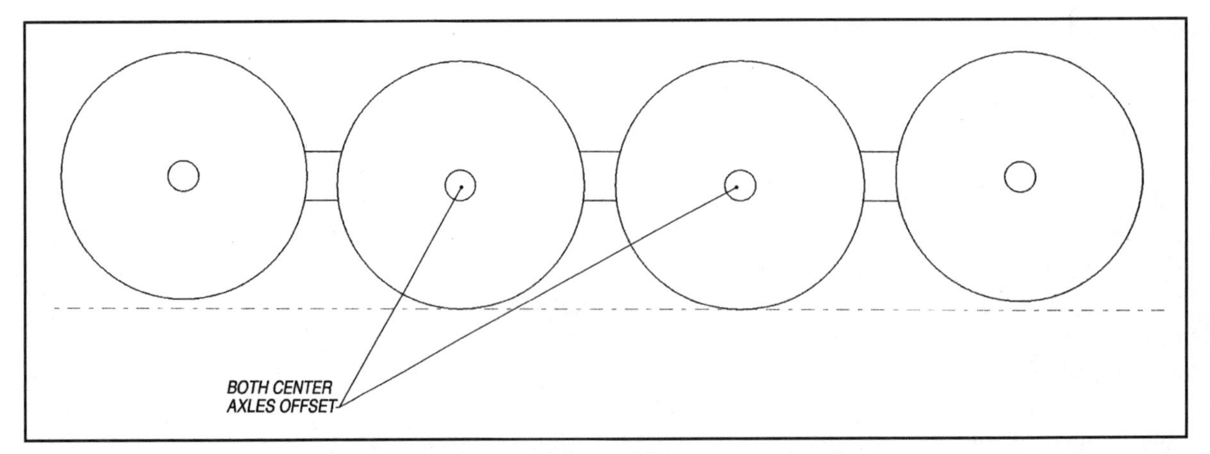

Figure 4-27 Eight wheels, all fixed, center axles offset

reduce the number of actuators, even with four corner steering, to eight. No known instances of this layout, shown in Figure 4-28, have been built for testing, though it seems like an effective layout.

With eight wheels, there is the possibility of dividing the vehicle into two sections, each with four wheels. The two parts are then either connected through a passive joint and individually skid steered, or the joint is articulated and steering is done by bending the vehicle in the middle. This is identical to the four tracked layouts discussed and shown in chapter five. This can be a very effective layout for obstacle negotiation and crevasse crossing, but cannot turn in place. Figure 4-29 shows an example of a two-part passive joint eight-wheeled layout. Figure 4-30 adds a roll joint to aid in keeping more wheels on the ground.

Another eight-wheeled layout, also applicable to a four-tracked vehicle, uses a transverse pivot, which allows the two halves to pitch up and down. It is skid steered, and is suited for bumpy terrain, but which has few obstacles it must go around. Imagine the vehicle in Figure 4-30, but with the pivot axis on its side. This layout is similar to the double rocker layout, with similar mobility and fewer moving parts.

The two halves of an eight-wheeled layout can also be coupled together with a ball joint. The ball joint allows pitch, roll, and yaw between the two parts which facilitates keeping all eight wheels on the ground most of the time. The ball joint is a simple joint and can be made robust. It has a limited range of motion around two of the axis, but the third axis can rotate three hundred sixty degrees. Aligning this axis vertically aligns it in the steering axis. This allows the vehicle to have a tighter steering radius, but it cannot turn in place. Figure 4-31 shows the four-wheeled sections connected through a vertical axis ball joint. The ball joint is difficult to use with a four-wheeled vehicle because the

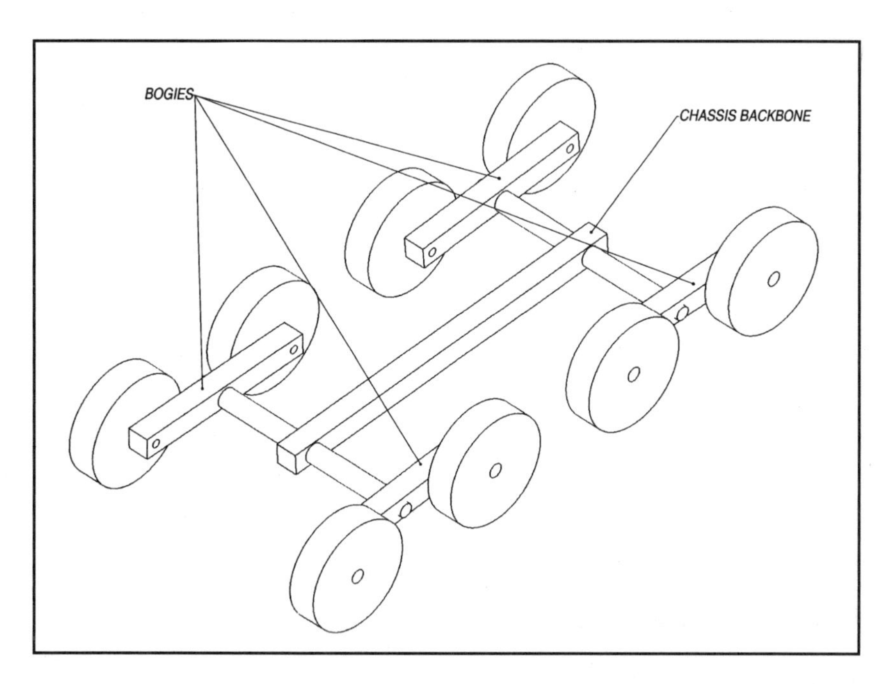

Figure 4-28 Eight wheels, double bogie

wheel torque would try to spin the section around the wheels. This problem can be reduced if the wheels are coupled together so their torques are always nearly indentical.

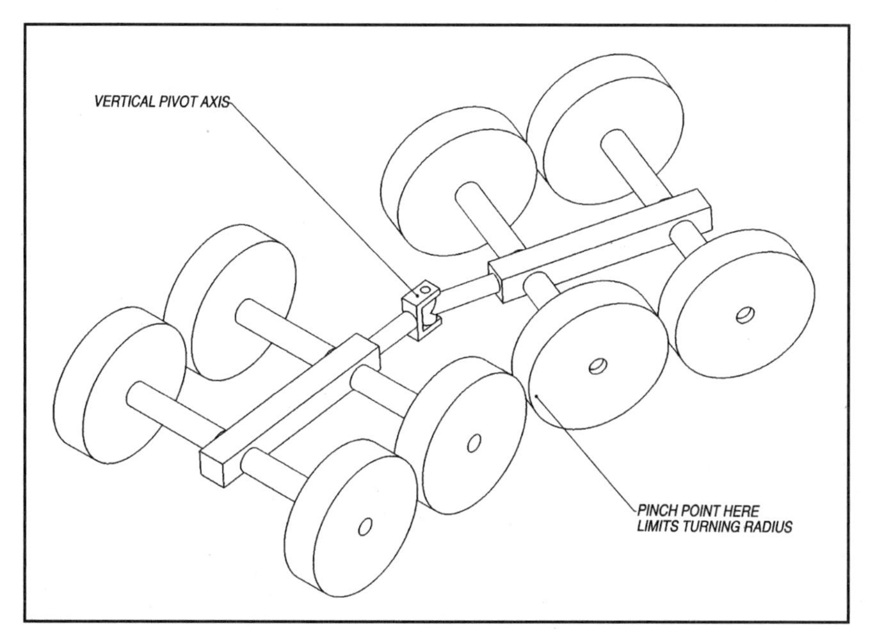

Figure 4-29 Two part, eight wheeled, vertical center pivot

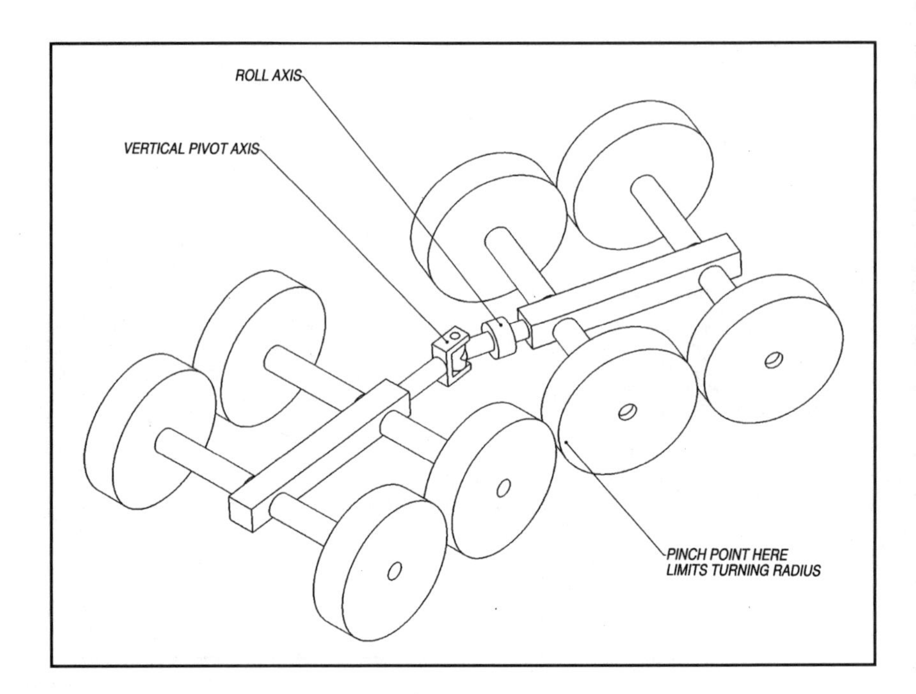

Figure 4-30 Two part, eight wheeled, vertical and roll joints

For a truly complicated wheeled drive mechanism, the Tri-star Land-Master from the movie *Damnation Alley* is probably the most impressive. This vehicle, of which only one was built, is a two-section, center pivot steered layout with a Tri-star wheel at each corner. The Tri-star wheels consist of three wheels, all driven together, arranged in a three-

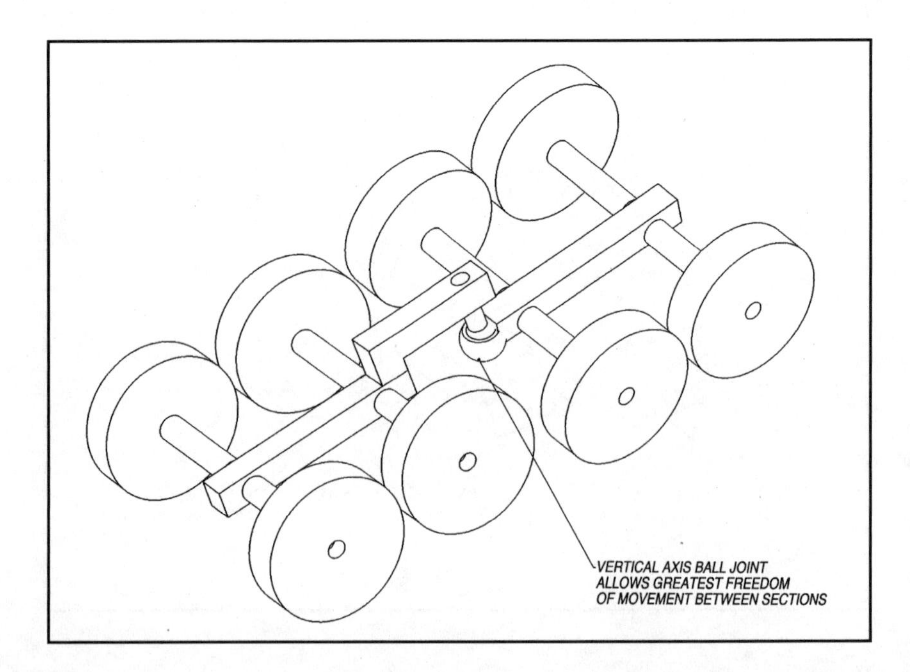

Figure 4-31 Two part, eight wheeled, vertical ball joint

pointed star on a shared hub that is also driven by the same shaft that drives the wheels. When a bump or ditch is encountered that the wheels alone cannot traverse, the whole three-wheeled system rotates around the center hub and the wheels essentially become very large cleats. The Tri-star wheels are driven through differentials on the Land-Master, but powering each with its own motor would increase mobility even further.

Chapter 5 Tracked Vehicle Suspensions and Drivetrains

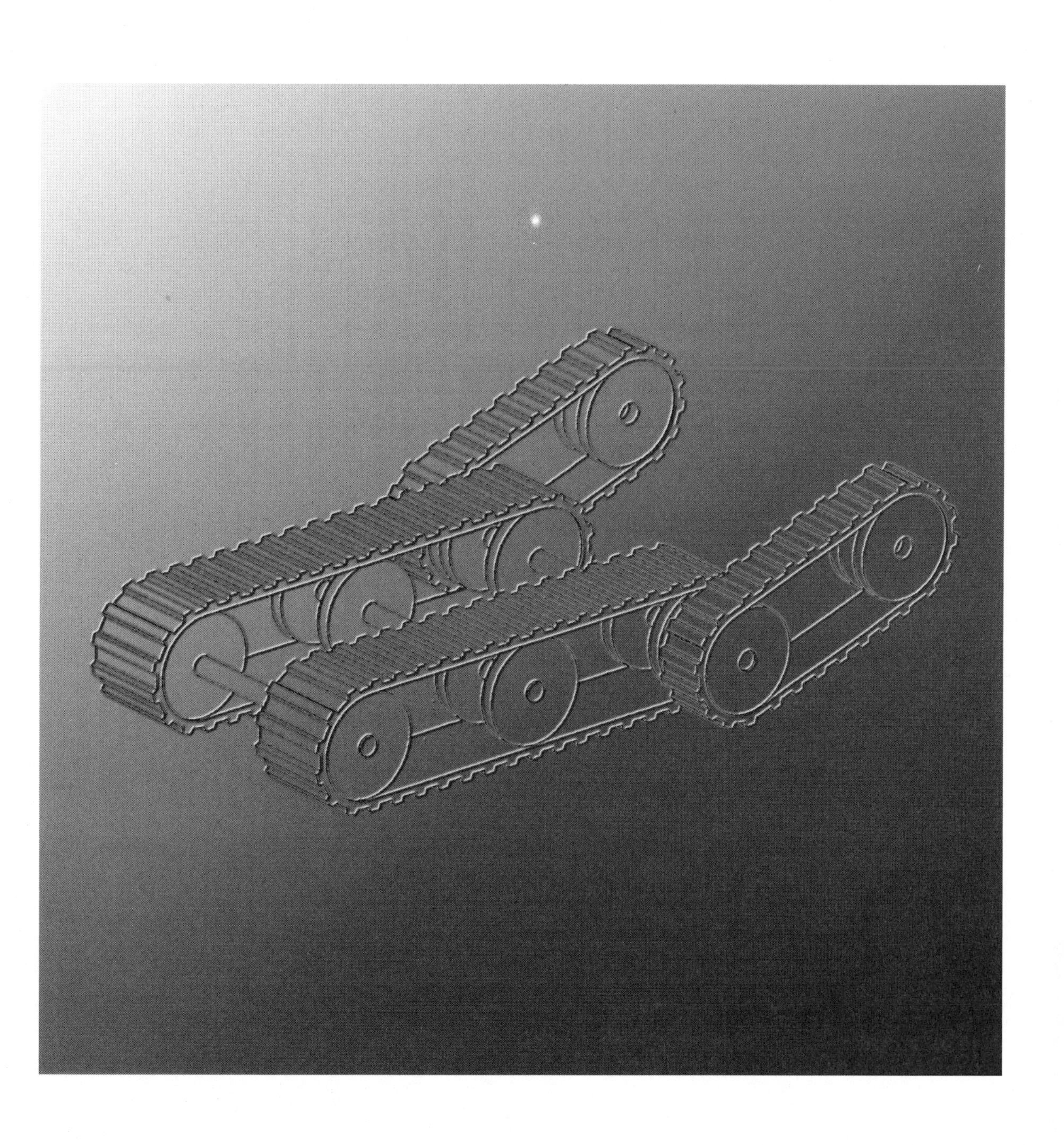

There has long been a belief that tracks have inherently better mobility than wheels and anyone intending to design a high mobility vehicle should use tracks. While tracks can breeze through situations where wheels would struggle, there are only a few obstacles and terrains which would stop a six wheeled rocker bogie vehicle, but not stop a similar sized tracked vehicle. They are

- very soft terrain: loose sand, deep mud, and soft powder snow
- obstacles of a size that can get jammed between wheels
- crevasses

They get this higher mobility at a cost of greater complexity and lower drive efficiency, so tracks are *better* for these situations, but not inherently better overall.

Tracked vehicles first started to appear in the early 1900s and were used extensively in WWI. The basic layout used then is still in use today on heavy construction equipment; a drive sprocket at one end, an idler wheel at the other that usually serves as a tensioner, and something in between to support the tracks on the ground. This basic, simple layout is robust and easy to control. Even in its most simple form, this track layout has all of the improvements over wheels previously listed.

The continuous surface in contact with the ground is what produces the benefits of tracks. The long surface combined with widths similar to wheels puts a large surface on the ground. This lowers ground pressure, allowing traveling on softer surfaces. It also provides more area for treads, increasing the number of teeth on the ground.

The continuous surface eliminates a wheeled vehicle's problem of becoming high centered between the wheels on one side. A correctly sized obstacle can get caught between the wheels on one side, but the track stays on top. The wheeled vehicle can get stuck in these situations, where the track would simply roll over the obstacle.

Perhaps the most important capability the continuous surface facilitates that a wheeled suspension cannot match (without undue complexity) is the ability to cross crevasses. Clever suspension components can be added to a six-wheeled or eight-wheeled vehicle to increase its negotiable crevasse width, but these add complexity to the wheeled vehicle's inherent simplicity. Tracks, however, have the ability to cross crevasses built in to their design. Add a mechanism for shifting the center of gravity, and a tracked vehicle can cross crevasses that are wider than half the length of the vehicle.

Most types have many more moving parts than a wheeled layout, all of which tend to increase rolling friction, but a well-designed track can actually be more efficient than a wheeled vehicle on very soft surfaces. The greater number of moving parts also increase complexity, and one of the major problems of track design is preventing the track from being thrown off the suspension system. Loosing a track stops the vehicle completely.

Track systems are made up of track, drive sprocket, idler/tension wheel, suspension system, and, sometimes, support rollers. There are several variations of the track system, each with its own set of both mobility and robustness pros and cons.

- The design of the track itself (steel links with hinges, continuous rubber, tread shapes)
- Method of keeping the tracks on the vehicle (pin-in-hole, guide knives, V-groove)
- Suspension system that supports the track on the ground (sprung and unsprung road wheels, fixed guides)
- Shape of the one end or both ends of the track system (round or ramped)
- Relative size of the idler and/or drive sprocket

Variations of most of these system layouts have already been tried, some with great success, others with apparently no improvements in mobility.

There are also many varieties of track layouts and layouts with different numbers of tracks. These various layouts have certain advantages and disadvantages over each other.

- One track with a separate method for steering
- The basic two track side-by-side
- Two tracks and a separate method for steering
- Two track fore-and-aft
- Several designs that use four tracks

- A six-tracked layout consisting of two main tracks and two sets of flipper tracks and each end

The six-track layout may be overkill because there is a patented track layout that has truly impressive mobility that has four tracks and uses only three actuators.

Robots are slowly coming into common use in the home and one tough requirement in the otherwise benign indoor environment is climbing stairs. It is just plain difficult to climb stairs with any rolling drive system, even one with tracks. Tracks simplify the problem somewhat and can climb stairs more smoothly than wheeled drivetrains, allowing higher speeds, but they have difficulty staying aligned with the stairs. They can quickly become tilted over, requiring steering corrections that are tricky even for a human operator. At the time of this writing there is no known autonomous vehicle that can climb a full flight of stairs without human input.

This chapter covers all known track layouts that have been or are being used on production vehicles ranging in size from thirty centimeters long (about a foot) to over forty five meters (a city block). Tracks can be used with good effectiveness on small vehicles, but problems can develop due to the stiffness of the track material. Toys only ten centimeters long have used tracks, and at least one robotics researcher has constructed tiny robots with tracks about twenty-five millimeters long. These fully autonomous robots were about the size of a quarter. Inuktun (www.inuktun.com) makes track units for use in pipe crawling robots that are about twenty-five centimeters long

The opposite extreme is large construction equipment and military tanks like the M1A2 Abrams. The M1A2's tracks are .635 meters wide (the width of a comfortable chair) and 4.75 meters long (longer than most cars) and together, including the suspension components, make up nearly a quarter of the total weight of the tank. A much larger tracked vehicle is NASA's Crawler Transporter used to move the Mobile Launch Pad of the Space Shuttle program. A single link of the Crawler Transporter's tracks is nearly 2m long and weighs nearly eight thousand newtons (about the same as a mid-sized car). There are 57 links per track and eight tracks mounted in pairs at each corner of what is the largest vehicle in the world. Although mobility of this behemoth is limited, it is designed to climb the five-percent grade up to the launch site while holding the Space Shuttle exactly vertical on a controllable pitch platform. It blazes along at a slow walk for the whole trip. Most large vehicles like these use metal link tracks because of the very large forces on the track.

On a more practical scale for mobile robots, urethane belts with molded-in steel bars for the drive sprocket and molded-in steel teeth for traction are increasingly replacing all-metal tracks. The smaller sizes can use solid urethane belts with no steel at all. Urethane belts are lighter and surprisingly more durable if sized correctly. They also cause far less damage to hard surface roads in larger sizes. If properly designed and sized, they can be quite efficient, though not like the mechanical efficiency of a wheeled vehicle. They do not stretch, rust, or require any maintenance like a metal-link track.

The much larger surface area in contact with the ground allows a heavier vehicle of the same size without increasing ground pressure, which facilitates a heavier payload or more batteries. Even the very heavy M1A2 has a ground pressure of about eighty-two kilo pascals (roughly the same pressure as a large person standing on one foot). At the opposite end of the scale the Bv206 four-tracked vehicle has a ground pressure of only ten kilo pascals. This low ground pressure allows the Bv206 to drive over and through swamps, bogs, or soft snow that even humans would have trouble getting through. Nevertheless, the Bv206 does not have the lowest pressure. That is reserved for vehicles designed specifically for use on powdery snow. These vehicles have pressures as low as five kilo-pascals. This is a little more than the pressure exerted on a table by a one-liter bottle of Coke.

When compared to wheeled drivetrains, the track drive unit can appear to be a relatively large part of the vehicle. The sprockets, idlers, and road wheels inside the track leave little volume for anything else. This is a little misleading, though, because a wheeled vehicle with a drivetrain scaled to negotiate the same size obstacles as a tracked unit would have suspension components that take up nearly the same volume. In fact, the volume of a six wheeled rocker bogie suspension is about the same as that of a track unit when the negotiable obstacle height is the baseline parameter.

The last advantage of tracks over wheels is negotiable crevasse width. In this situation, tracks are clearly better. The long contact surface allows the vehicle to extend out over the edge of a crevasse until the front of the track touches the opposite side. A wheeled vehicle, even with eight-wheels, would simply fall into the crevasse as the gap between the wheels cannot support the middle of the vehicle at the crevasse's edge. The clever mechanism incorporated into a six-wheeled rocker bogie suspension shown in Chapter Four is one solution to this problem, but requires more moving parts and another actuator.

To simplify building a tracked robot, there are companies that manufacture the undercarriages of construction equipment. These all-in-one drive units require only power and control systems to be added. They are extremely robust and come in a large variety of styles and are made for both steel and rubber tracks. Nearly all are hydraulic powered, but a few have inputs for a rotating shaft that could be powered by an electric motor. They are not manufactured in sizes smaller than about 1m long, but for larger robots, they should be given consideration in a design because they are designed by companies that understand tracks and undercarriages, they are robust, and they constitute a bolt-on solution to one of the more complex systems of a tracked mobile robot.

STEERING TRACKED VEHICLES

Steering of tracked vehicles is basically a simple concept, drive one track faster than the other and the vehicle turns. This is exactly the same as a skid-steer wheeled vehicle. It is also called differential steering. The skidding power requirements on a tracked vehicle are about the same, or perhaps a little higher, as on a four-wheel skid steer layout. Since brakes were required on early versions of tracked vehicles, the simplest way to steer by slowing one track was to apply the brake on that side.

Several novel layouts improve on this drive-and-brake steering system. Controlling the speed of each track directly adds a second major drive source, but gives fine steering and speed control. A second improvement to drive-and-brake steering uses a fantastically complicated second differential powered by its own motor. One output of this differential is directly connected to one output of the main differential; the other is cross connected to the other output axle of the main differential. Varying the speed of the steering motor varies the relative speed of the two tracks. This also gives fine steering control, but is quite complex.

Another method for steering tracked vehicles is to use some external steerable device. The most familiar vehicle that uses this type of system is the common snow mobile. This is a one-tracked separately steered layout. For use on surfaces other than snow, the skis can be replaced with wheels.

A steering method that can improve mobility is one called articulated steering. This layout has two major sections, both with tracks, which are connected through a joint that allows controlled motion in at least one direction. This joint bends the vehicle in the middle, making it turn a corner. This is the same system as used on wheeled front-end loaders. These systems can aid mobility further if a second degree of freedom is added which allows controlled or passive motion about a transverse pivot joint at nearly the same location as the steering joint.

The same trick that reduces steering power on skid steered wheeled vehicles can be applied to tracks, i.e., lowering the suspension a little at the middle of the track. This has the effect of raising the ends, reducing the power required to skid them around when turning. Since this reduces the main benefit of tracks, having more ground contact surface area, it is not incorporated into tracked vehicles very often.

VARIOUS TRACK CONSTRUCTION METHODS

Tracks are constructed in many different ways. Early tracks were nearly all steel because that was all that was available that was strong enough. Since the advent of Urethane and other very tough rubbers, tracks have moved away from steel. All-steel tracks are very heavy and on smaller vehicles, this can be a substantial problem. On larger vehicles or vehicles designed to carry high loads, steel linked tracks may be the best solution. There are at least six different general construction techniques for tracks.

- All steel hinged links
- Hinged steel links with removable urethane road pads
- Solid urethane
- Urethane with embedded steel tension members
- Urethane with embedded steel tension members and external steel shoes (sometimes called cleats)
- Urethane with embedded steel tension members and embedded steel transverse drive rungs with integral guide teeth

All-steel hinged linked track (Figure 5-1) would seem to be the toughest design for something that gets beat on as much as tracks do, but there are several drawbacks to this design. Debris can get caught in the spaces between the moving links and can jamb the track. A solution to this problem is to mount the hinge point as far out on the track as possible. This reduces the amount that the external surface of the track opens and closes, reducing the size of the pinch volume. This is a subtle but important part of steel track design. This lowered pivot is shown in Figure 5-2.

Tracked vehicles, even autonomous robots, will drive on finished roads at some point in their life, and all-steel tracks tear up macadam. The solution to this problem has been to install urethane pads in the links of the track. These pads are designed to be easily replaceable. The pads are bolted or attached with adhesive to pockets in special links on the track. This allows them to be removed and replaced as they wear out.

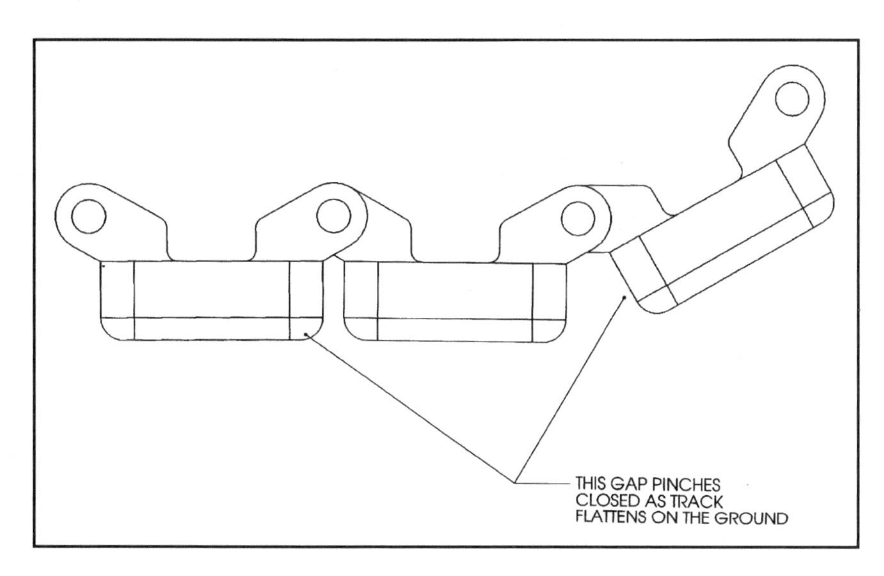

Figure 5-1 Basic steel link layout showing pinch point

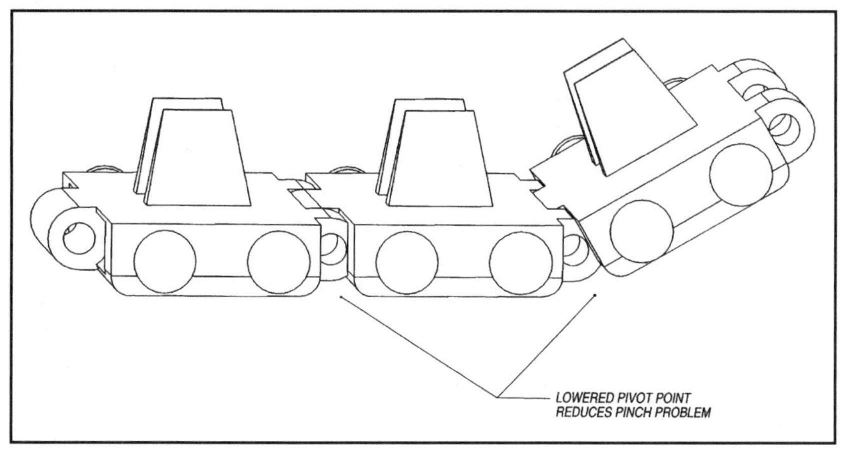

Figure 5-2 Effective hinge location of all-steel track

Figure 5-3 shows the lowered pivot link with an added pocket for the urethane road wheel.

The way to completely remove the pinch point is to make the track all one piece. This is what a urethane track does. There are no pinch points at all; the track is a continuous loop with or without treads. Molding the treads into the urethane works for most surface types. It is very tough, relatively high friction compared to steel, and inexpensive. It also does not damage prepared roads. Ironically, if higher traction is needed, steel cleats can be bolted to the urethane. Just like urethane road pads on steel tracks, the steel links are usually designed to be removable.

Urethane by itself is too stretchy for most track applications. This weakness is overcome by molding the urethane over steel cables. The steel is completely covered by the urethane so there is no corrosion prob-

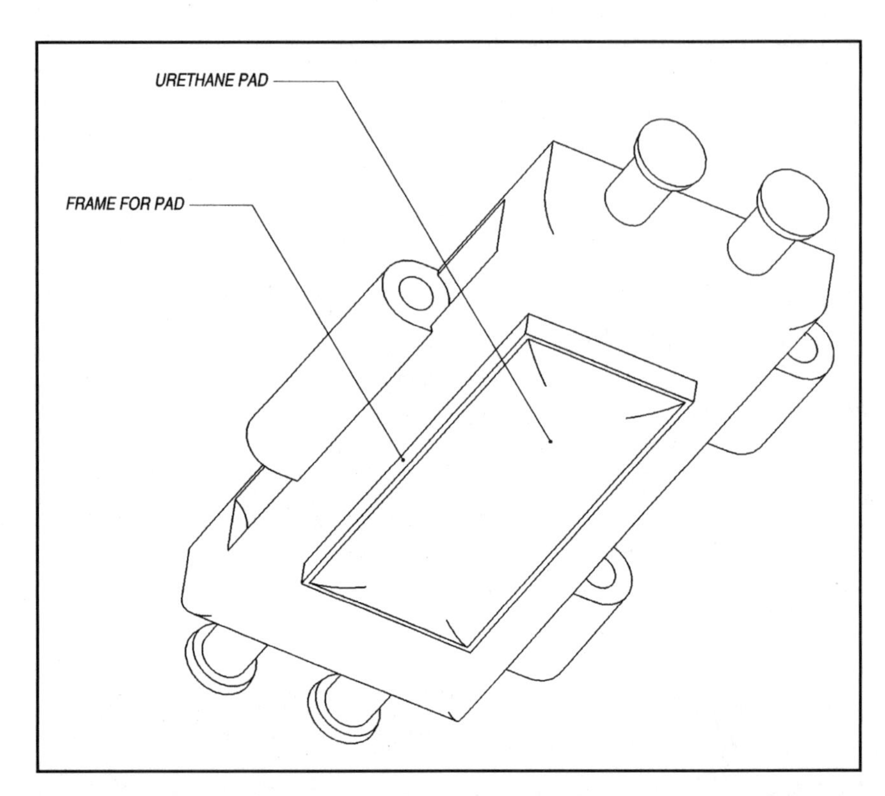

Figure 5-3 Urethane pads for hard surface roads

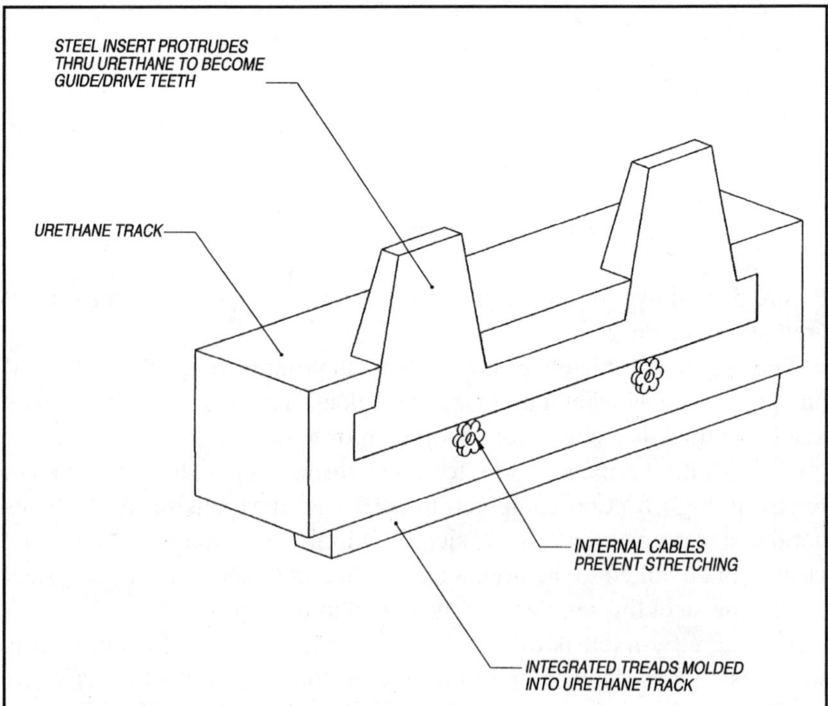

Figure 5-4 Cross section of urethane molded track with strengthening bars and internal cables

lem. The steel eliminates stretching, and adds little weight to the system. For even greater strength, hardened steel crossbars are molded into the track. These bars are shaped and located so that the teeth on the drive sprocket can push directly on them. This gives the urethane track much greater tension strength, and extends its life. Yet another modification to this system is to extend these bars towards the outer side of the track, where they reinforce the treads. This is the most common layout for urethane tracks on industrial vehicles. Figure 5-4 shows a cross section of this layout.

TRACK SHAPES

The basic track formed by a drive sprocket, idler, and road wheels works well in many applications, but there are simple things that can be done to modify this oblong shape to increase its mobility and robustness. Mobility can be increased by raising the front of the track, which aids in getting over taller obstacles. Robustness can be augmented by moving vulnerable components, like the drive sprocket, away from possibly harmful locations. These improvements can be applied to any track design, but are unnecessary on variable or reconfigurable tracks.

The simplest way to increase negotiable obstacle height is to make the front wheel of the system larger. This method does not increase the complexity of the system at all, and in fact can simplify it by eliminating the need for support rollers along the return path of the track. This layout, when combined with locating the drive sprocket on the front axle, also raises up the drive system. This reduces the chance of damaging the drive sprocket and related parts. Many early tanks of WWI used this track shape.

Another way to raise the ends of the track is to make them into ramps. Adding ramps can increase the number of road wheels and therefore the number of moving parts, but they can greatly increase mobility. Ramping the front is common and has obvious advantages, but ramping the back can aid mobility when running in tight spaces that require backing up over obstacles. As shown in Figure 5-5 (a–d), ramps are created by raising the drive and/or idler sprocket higher than the road wheels. Some of these designs increase the volume inside the track system, but this volume can potentially be used by other components of the robot.

More than one company has designed and built track systems that can change shape. These variable geometry track systems use a track that is more flexible than most, which allows it to bend around smaller sprockets and idler wheels, and to bend in both directions. The road wheels are

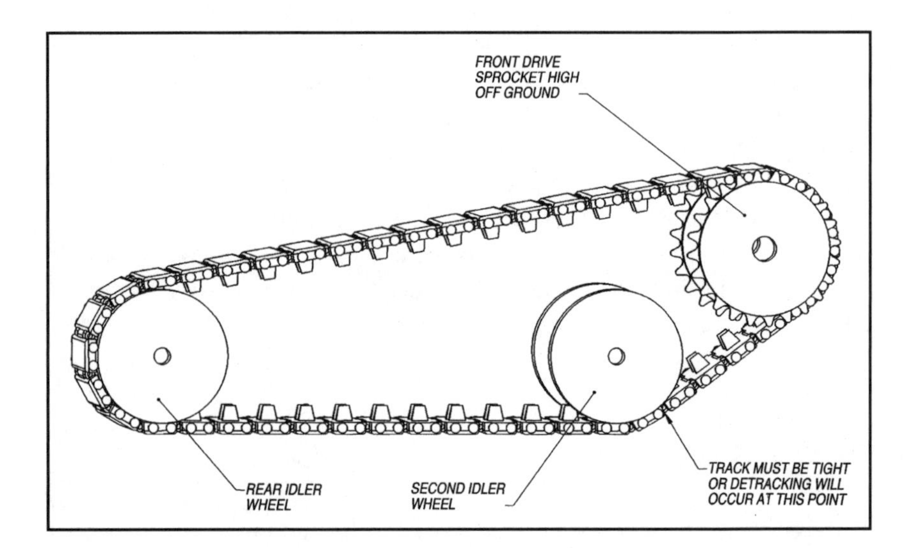

Figure 5-5a–d Various track shapes to improve mobility and robustness

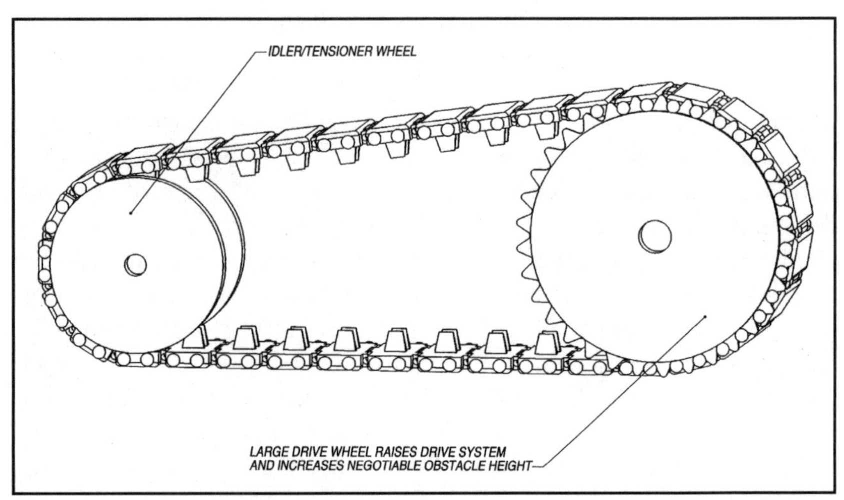

Figure 5-5b

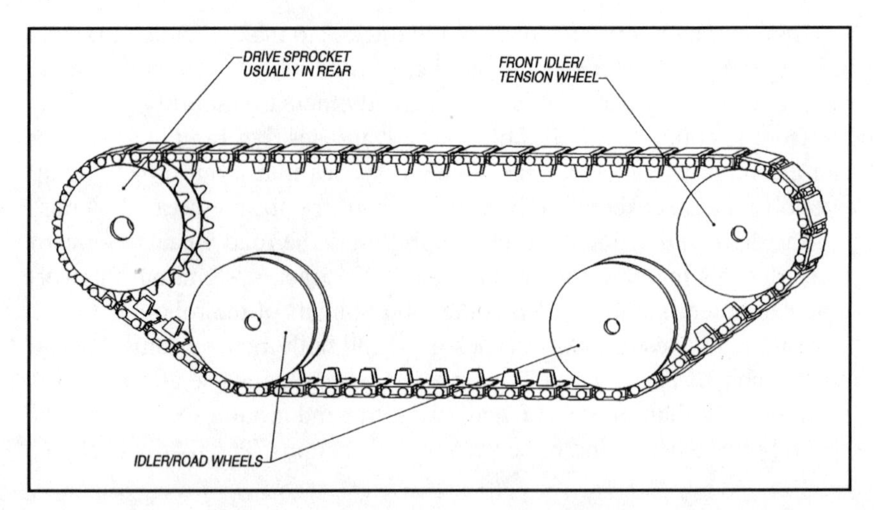

Figure 5-5c

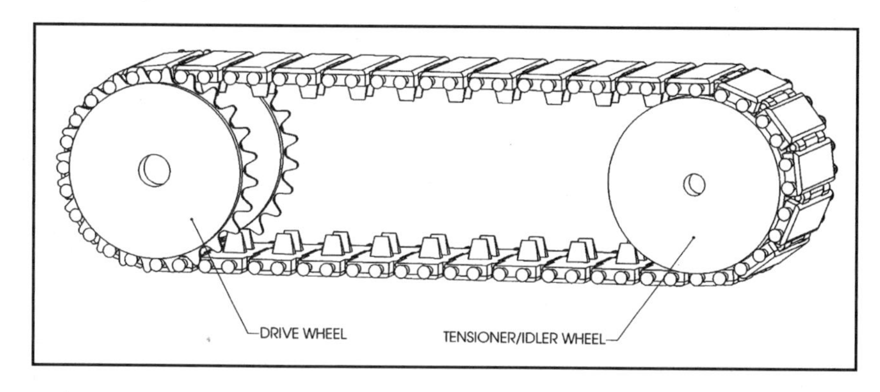

Figure 5-5d

usually mounted directly to the chassis through some common suspension system, but the idler wheel is mounted on an arm that can move through an arc that changes the shape of the front ramp. A second tensioning idler must be incorporated into the track system to maintain tension for all positions of the main arm.

This variability produces very good mobility when system height is included in the equation because the stowed height is relatively small compared to the negotiable obstacle height. The effectively longer track, in addition to a cg shifting mechanism, gives the vehicle the ability to cross wider crevasses. With simple implementations of this concept, the variable geometry track system is a good choice for a drive system for mobile robots. Figure 5-6 (a–b) shows one layout for a variable geometry track system. Many others are possible.

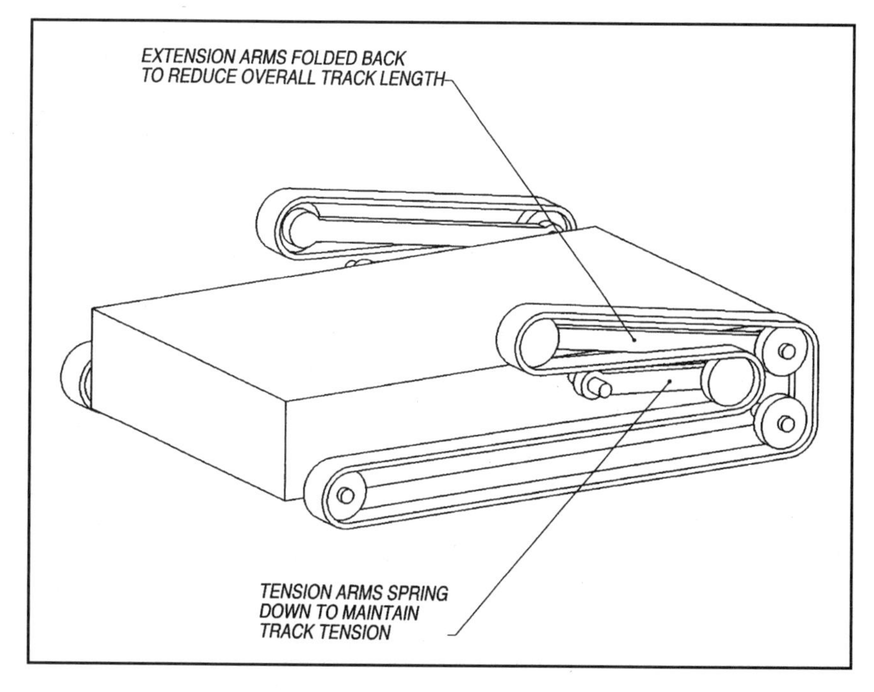

Figure 5-6a–b Variable track system

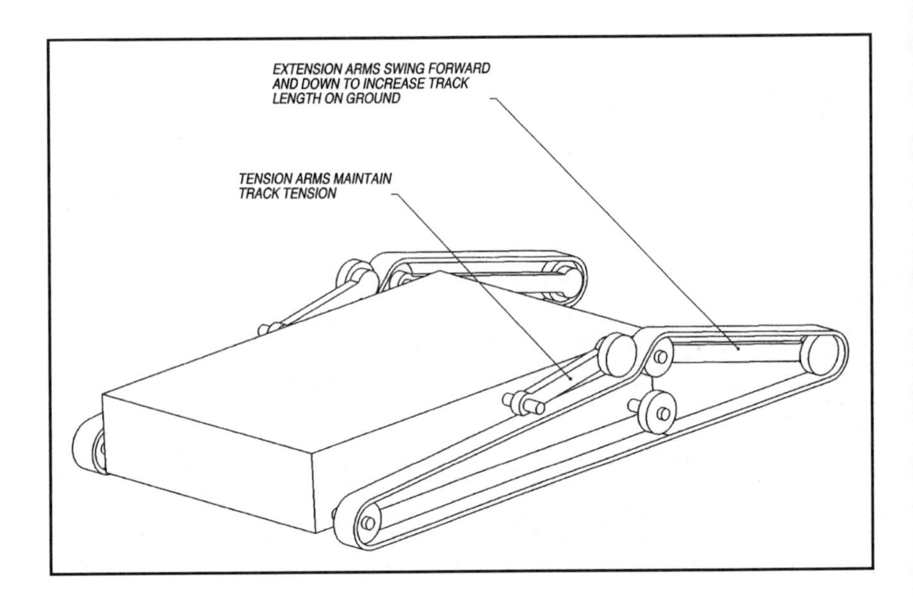

Figure 5-6b

Since it carries both the tension in the track and the drive torque, the drive sprocket (and associated drive mechanism) is the most vulnerable moving part of a track system. They can be located at either the front or rear of the track, though they are usually in the rear to keep them away from the inevitable bumps the front of an autonomous vehicle takes. Raising the sprocket up off the ground removes the sprocket from possible damage when hitting something on the road surface. These modifications result in a common track shape, shown in Figure 5-5c.

A simple method that extends the mobility of a tracked vehicle is to incorporate a ramp into the chassis or body of the vehicle. The static ramp extends in front and above the tracks and slides up obstacles that are taller than the track. This gives the vehicle the ability to negotiate obstacles that are taller than the mobility system using a non-moving part, a neat trick.

TRACK SUSPENSION SYSTEMS

The space between the drive sprocket and idler wheel needs to be uniformly supported on the ground to achieve the maximum benefit of tracks. This can be done in one of several ways. The main differences between these methods is drive efficiency, complexity, and ride characteristics. For especially long tracks, the top must also be supported, but

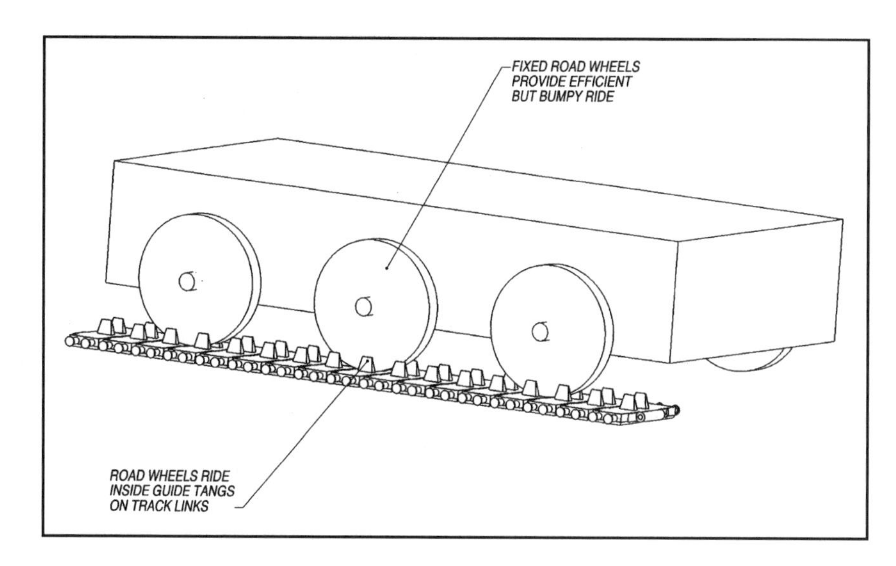

Figure 5-7 Fixed road wheels

this is usually a simple passive roller or two evenly spaced between the drive sprocket and idler.

The main types of ground support methods are

- Guide blades
- Fixed road wheels
- Rocker road wheel pairs
- Road wheels mounted on sprung axles

Guide blades are simple rails that are usually designed to ride in the V shaped guide receivers on the track's links. They can extend continuously from one end to the other, and therefore are the most effective at supporting the track along its whole length. Unfortunately, they are also quite inefficient since there is the long sliding surface that cannot be practically lubricated. They also produce a jarring ride for the rest of the vehicle.

One step up from guide blades is fixed road wheels (Figure 5-7). These are wheels on short stub-shafts solidly mounted to the robot's chassis. The wheels can be small relative to the track, since the thing they roll on is always just the smooth inner surface of the track. They too produce a jarring ride, more so than guide blades, but they are far more efficient. Fixed rollers are a good choice for a robust track system on a robot since ride comfort is not as important, at lower speeds, as on a vehicle carrying a person.

The bumpy ride does hamper track efficiency, however, because the chassis is being moved up and down by the rough terrain. Reducing this

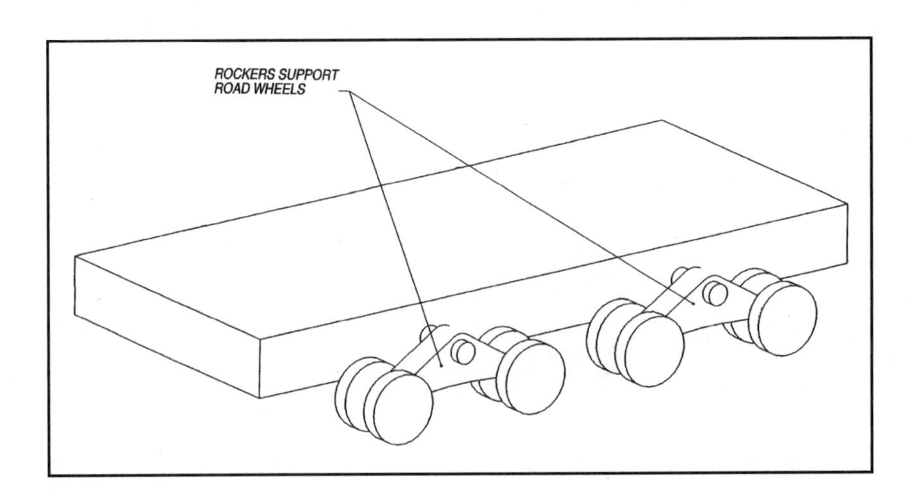

Figure 5-8 Road wheels on rockers

motion is especially beneficial at higher speeds, and the rocker layout used on wheeled vehicles is almost as effective on tracks. The rollers are mounted in pairs on rockers between the drive sprocket and the idler wheel. The rockers (Figure 5-8) allow the track to give a little when traversing bumpy terrain, which reduces vertical motion of the robot chassis. Careful tensioning of the track is essential with movable road wheels.

The most complex, efficient, and smooth ride is produced by mounting the road wheels on sprung axles. There are three main types of suspension systems in common use.

- Trailing arm on torsion spring
- Trailing arm with coil spring
- Leaf spring rocker

The trailing arm on a torsion spring is pictured in Figure 5-9. It is a simple device that relies on twisting a bunch of steel rods, to which the trailing arm is attached at one end. It gets its name because the arms that support the wheel trail behind the point where they attach, through the torsion springs, to the chassis. The road wheels mount to the end of the trailing arms and forces on the road wheel push up on the arm, twisting the steel rods. This system was quite popular in the 1940s and 1950s and was used on the venerable Volkswagen beetle to support the front wheels. It was also used on the Alvis Stalwart, described in more detail in Chapter Four.

You can also support the end of the trailing arm with a coil spring, or even a coil over-shock suspension system that can probably produce the smoothest ride of any track system (Figure 5-10). The shock can also be added to the torsion arm suspension system. The advantage of the coil

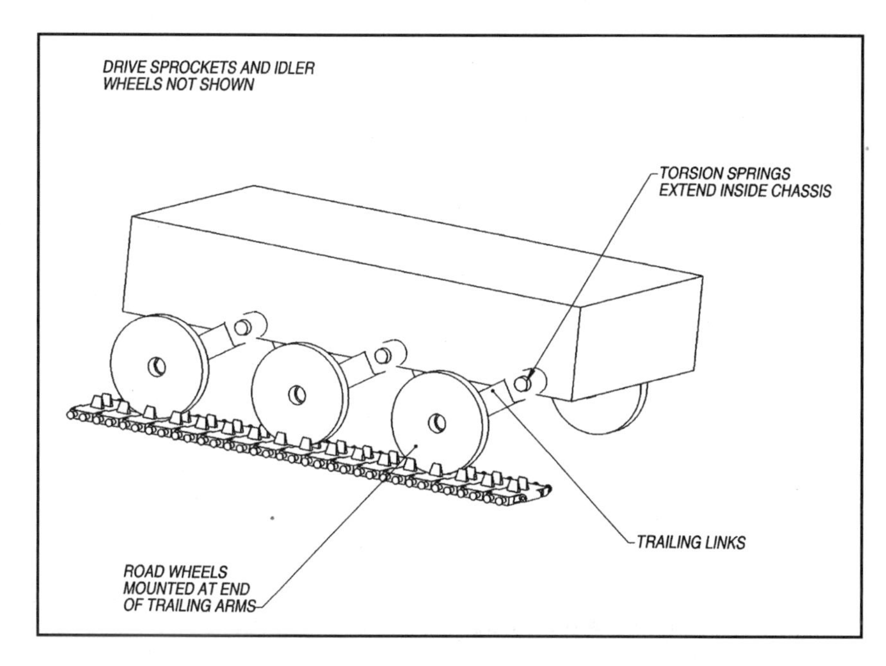

Figure 5-9 Trailing arm

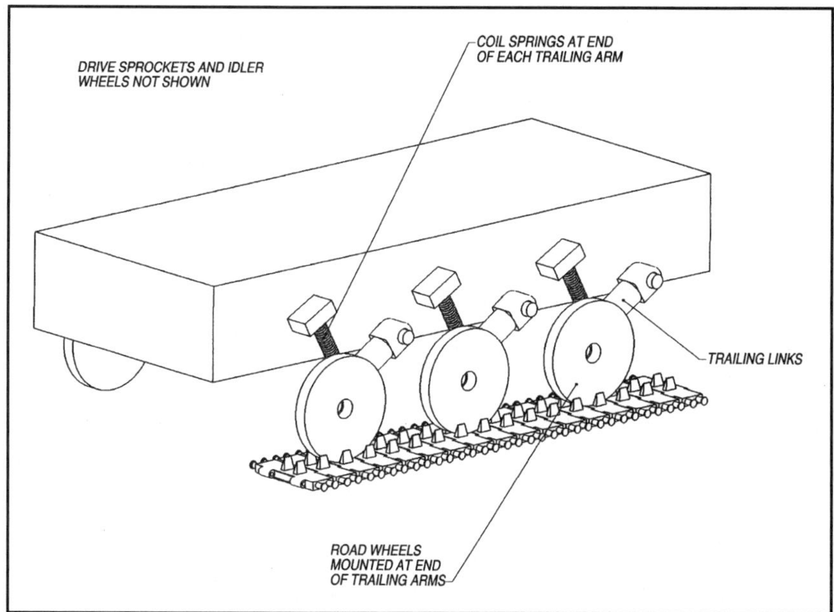

Figure 5-10 Trailing arm and coil springs

spring over the torsion suspension is that the load is supported by the spring very close to the load point, reducing forces and moments in the trailing arm. This can reduce the weight of the suspension system, and puts the system more inside the track's volume rather than inside the chassis.

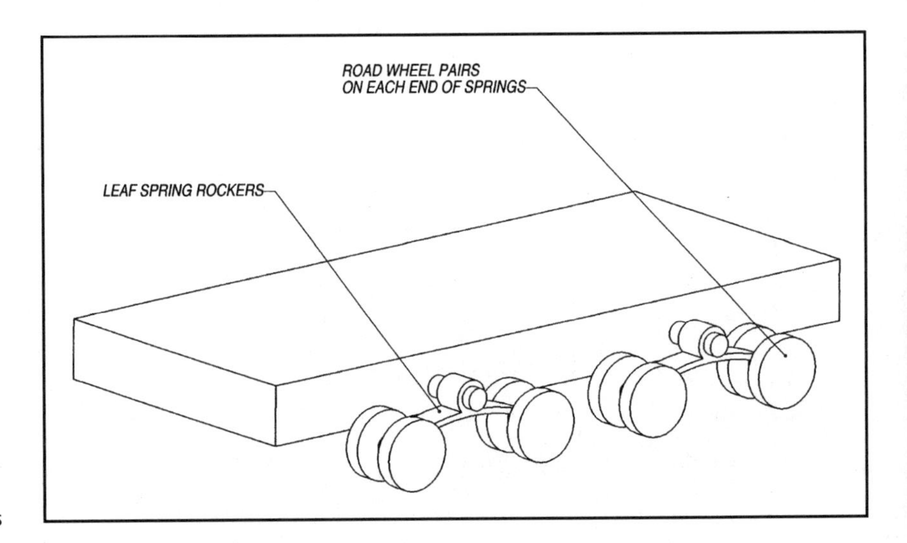

Figure 5-11 Leaf spring rockers

A simple variation of the rocker system is to replace the rockers with leaf springs (Figure 5-11). This eases the shock to the rocker and produces a smoother ride. The springs are usually very stiff since the rocker arm's swinging motion still allows the wheels to make large motions. This system can be retrofitted to rocker arm suspension systems if the current rocker arm does not smooth the ride enough. Having road wheels on both sides of the spring reduces the twisting moment produced by having wheels on only one side. Figure 5-11 shows double wheels.

TRACK SYSTEM LAYOUTS

One-Track Drivetrain

What would seem to be the simplest track layout is one that uses only one track. This layout is actually in existence in at least one form, and mobility can be quite good. The most common commercially available form of a one-track vehicle is the snow mobile. Although these vehicles are designed exclusively for use on snow, replace the skis with wheels, and they can be used on hard surfaces. The track on a snowmobile is quite wide to lower ground pressure as much as practicable, but there is no reason why a narrower track can't be used with the wheeled layout.

Mobility is limited somewhat by two factors: The wheels must be pushed over obstacles and the layout is steered by the Ackerman system

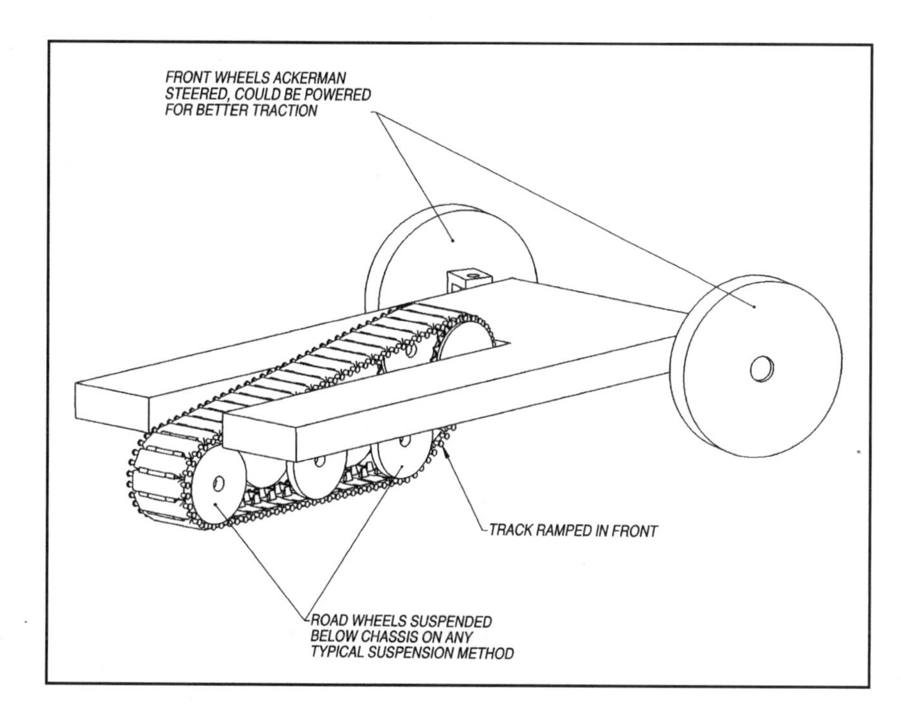

Figure 5-12 One track, two front wheels, Ackerman steer

which prevents turning in place. The first problem can be reduced by powering the wheels. There is no known existence of this layout, but it seems worth investigation. Figure 5-12 shows the typical ramped-front track common on snowmobiles because they normally do not go backwards. A track that is ramped both in front and back would increase mobility. It would be an interesting experiment to build a one-track, two-wheel drive, Ackerman steered robot and test its mobility.

Two-Tracked Drivetrains

The two-track layout is by far the most common. In its basic form, it is simple, easy to understand, and relatively easy to construct. Two tracks are attached to either side of the robot's main chassis, and each are powered by their own motor. Compact designs have the motor mounted substantially inside the track and attached directly to the drive sprocket. Since the drive sprocket must turn at a much lower rpm than the rpm's at which electric motors are most efficient, a speed reduction method almost always needs to be part of the drivetrain. Figure 5-13 shows a two-track layout, with drive motors, gearboxes, fixed track guide blades, and non-ramped tracks. This represents the simplest layout for a tracked vehicle.

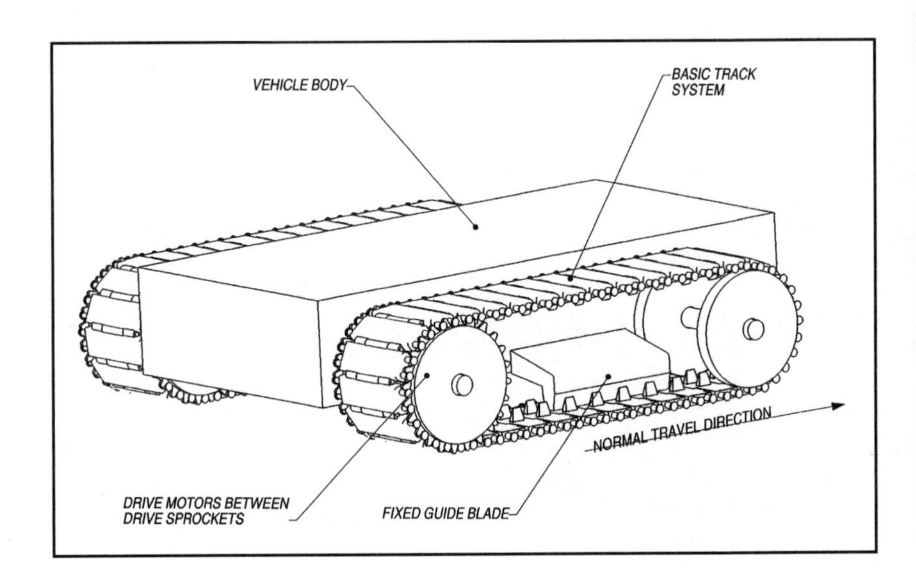

Figure 5-13 Basic two-track layout

Two-Tracked Drivetrains with Separate Steering Systems

A more complex, but less capable, layout is to have the two tracks driven through a differential, and the robot steered by a conventional set of wheels mounted in front of the tracks. This layout came about when large trucks did not have enough traction on unprepared roads and replacing the rear wheels with track systems that took up about the same volume solved that problem. These trucks (Figure 5-14) were called half-tracks. For a mobile robot, this is a less satisfactory layout since it

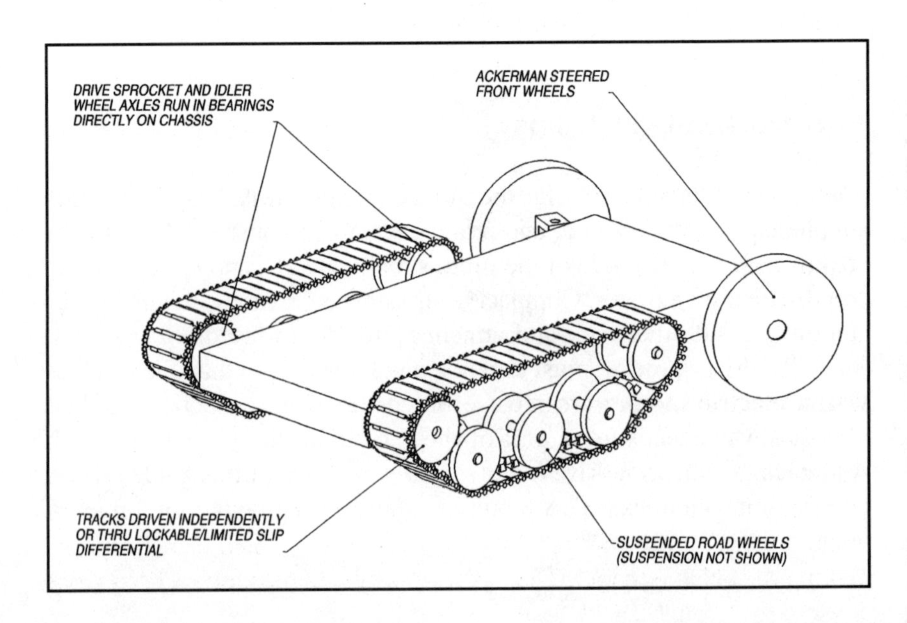

Figure 5-14 The half-track

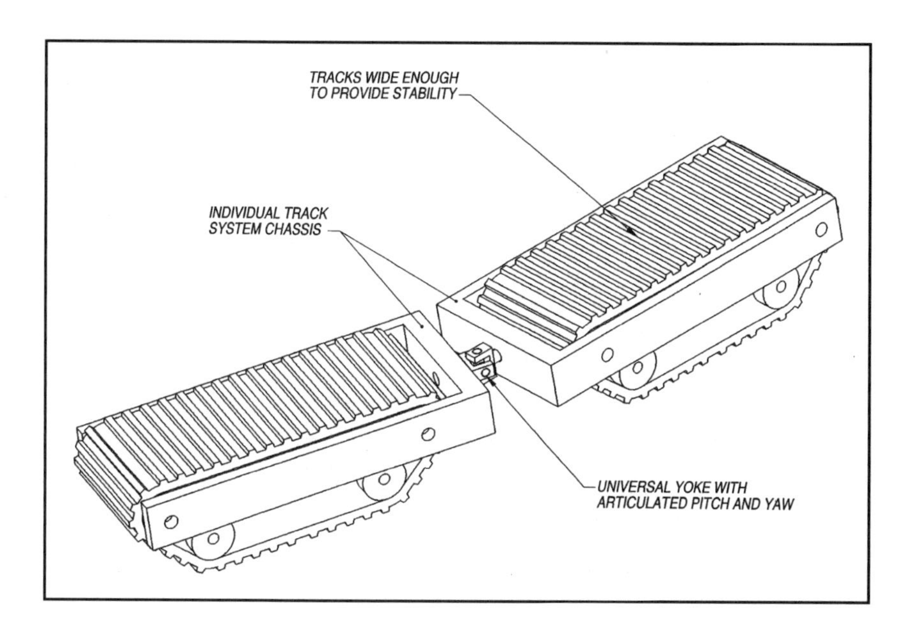

Figure 5-15 Two wide tracks, fore-and-aft

can no longer turn in place like the basic two-track layout can, yet has more moving parts.

An unusual variation of the two-track layout is to place the two tracks inline, one in front of the other (Figure 5-15). Stability is maintained by making the tracks sufficiently wide, and steering is accomplished with an articulated joint between the two tracks. The tracks have to be supported from both sides, like on a snow mobile. Steering power is transmitted through one or two linear or rotary actuators that are part of the articulated joint.

This system also benefits from the trick of making the center of each track a little lower than the ends. Since the tracks are already fore-and-aft, the bowed shape on each track does not produce any wobbling. This system has great mobility, but like the half-track, cannot turn in place. It would probably work very well for vehicles intended for use on snow or sand. The two tracks could either each carry their own chassis, or a single chassis could be attached to the universal joint with the outer ends of the track sections suspended to the chassis with a sprung or active suspension.

Four-Tracked Drivetrains

Adding more tracks would seem to increase the mobility of a tracked vehicle, but there are several problems with this approach. Adding more tracks necessarily means more moving parts, but it also usually means making the vehicle longer. The best layout would be one that adds more

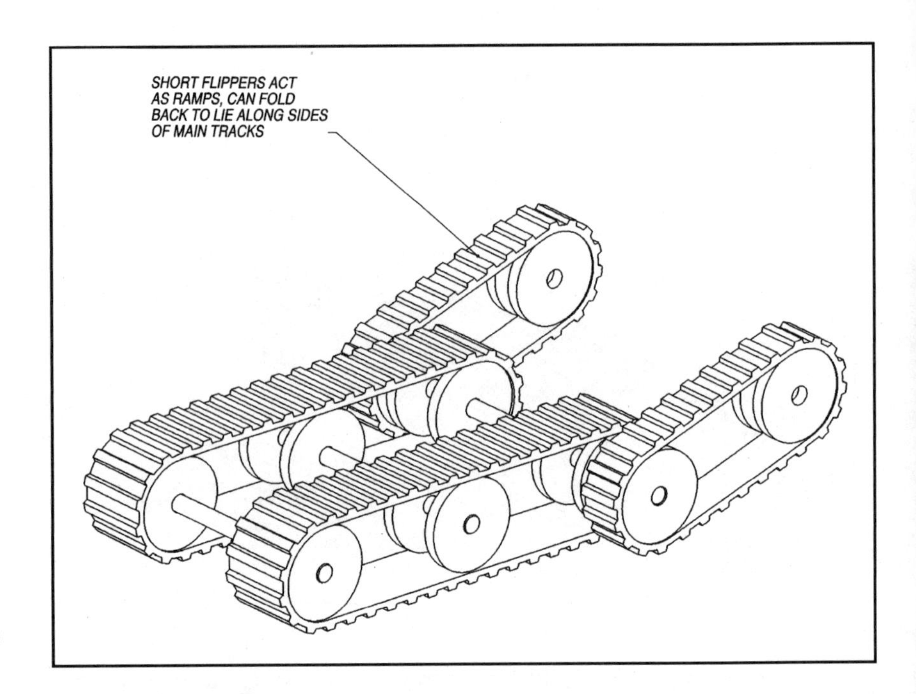

Figure 5-16 iRobot's Urbie, a four-tracked teleoperated robot layout

tracks with the least number of additional moving parts, and keeps the vehicle the same length. This last criteria almost exclusively means a reconfigurable layout, one where the length is longer when that is needed, but can reconfigure into a shorter length when that is needed. This concept has been implemented in a couple of different ways, both of which are patented.

Figure 5-16 shows the general layout of iRobot's Urbie telerobotic platform. This layout uses a third actuator to deploy or stow a pair of flipper-like tracks that rotate around the front idler wheels. They are powered by the same motors that power the main tracks, and always turn with them. This layout represents the simplest form of a four-track vehicle, and has very high mobility.

The center of gravity of Urbie is located ahead of the center of the vehicle so, with the flippers extended, Urbie can cross crevasses that are wider than half the length of the basic vehicle. This clever location of the cg also gives the flippers the ability to flip the robot over if it is overturned. When the flippers are rotated around so they become very large ramps, Urbie can climb over obstacles that are higher than the overall height of the basic track. There are also other functions the flippers can perform unrelated to mobility, like the ability to stand up. This makes the robot much taller and allows a strategically placed camera to see over short walls.

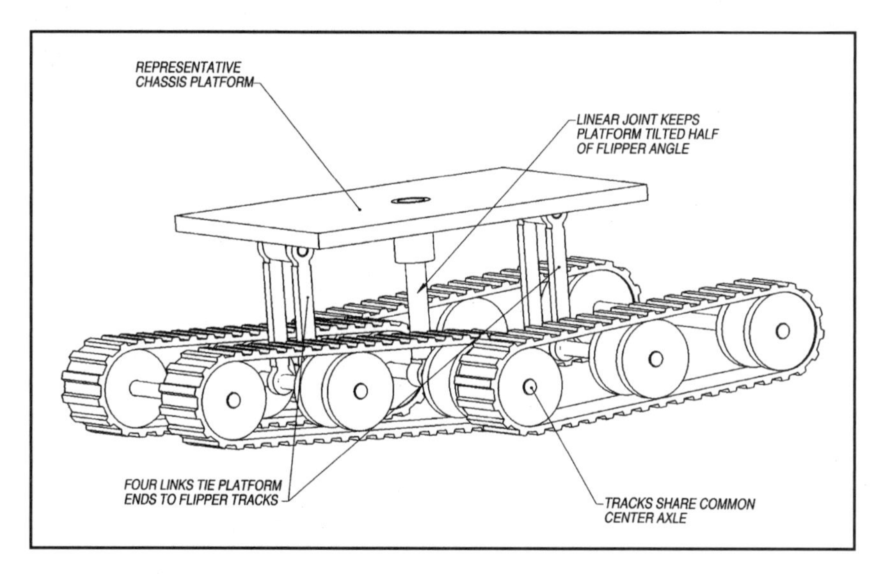

Figure 5-17a Same length flippers, sharing middle axle

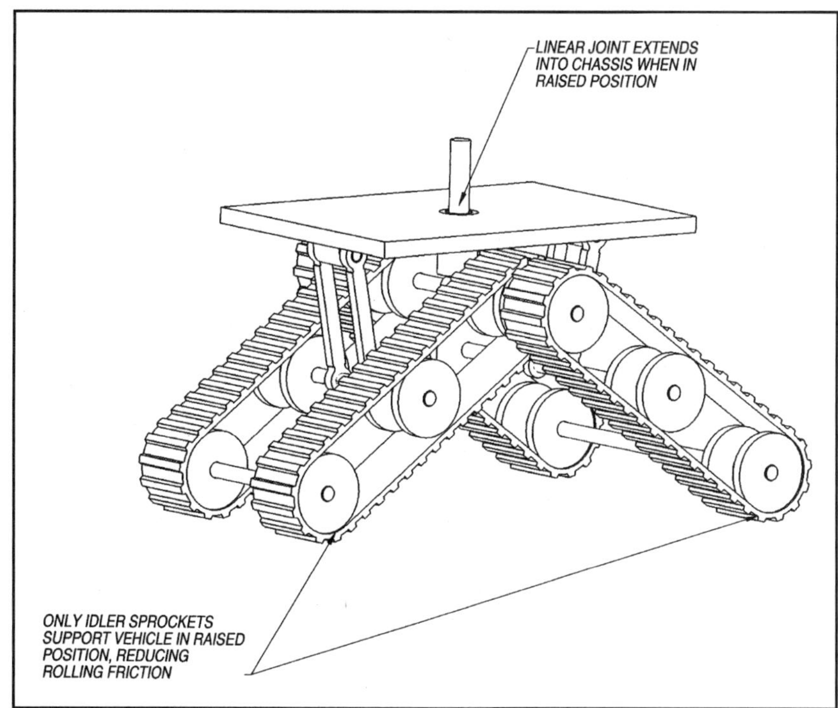

Figure 5-17b

A close relative of Urbie's layout (Figure 5-17a–b) is one where the track pairs are the same size, with the cg located very close to the shared axle of the front and rear tracks. With two actuators to power each of the four tracks independently and a fifth actuator to power the pivot joint a very capable layout results. The main chassis is geared to the shared axle so it is always at the half-angle between the front and rear tracks, which allows it to be raised up yet still be level when folding both tracks down. This trick raises the entire chassis, but it also offloads the weight of the robot from the track guide blades, increasing rolling efficiency when high traction is not needed. This reconfigurable layout combines the high mobility of tracks with good smooth-road rolling efficiency.

There are two basic layouts for four-tracked vehicles. They are both train-like in that there are two two-tracked modules connected by some sort of joint. The two modules must be able to move in several directions relative to each other. They can pitch up and down, yaw left and right, and, ideally, roll (twist).

The simplest connection that allows all three degrees of freedom is the ball joint. If the joint is passive, steering is accomplished in the same way as a two-tracked vehicle, except that now both modules must turn at just the right time to keep skidding between the modules to a minimum. This turns out to be tricky. The ball joint also limits the range of steering angle simply because the socket must wrap around the ball enough to adequately capture it. A universal joint has a greater range of motion, and is easier to use if the joint is to be powered.

The articulated joint, an active universal joint, overcomes the steering problem by allowing the tracks to rotate at whatever speed limits skidding. This steering method makes this layout very agile. The Hagglund Bv206, which uses this layout, is considered nearly unstoppable in almost any terrain from soft snow to steep hills. It is even amphibious, propelled through the water by the tracks. Because it cannot be skid steered, it can't turn in place. Nevertheless, it is a very capable layout. Steering the Hagglund Bv206 is done with a standard steering wheel, which turns the articulated joint and forces the two modules to bend. The tracks are driven through limited slip differentials, allowing the inner and outer tracks in each module to travel at different speeds just like in an Ackerman steered wheeled vehicle.

Six-Tracked Drivetrains

There is at least one track layout (Figure 5-18) incorporated on an existing telerobotic vehicle that uses six tracks. It is an extension of the Urbie design, but was actually invented before Urbie's layout. The two-tracked layout is augmented by flipper tracks on both the front and back, independently tilted, but whose tracks are driven by the main track motors. This layout allows the vehicle to stand up like the one

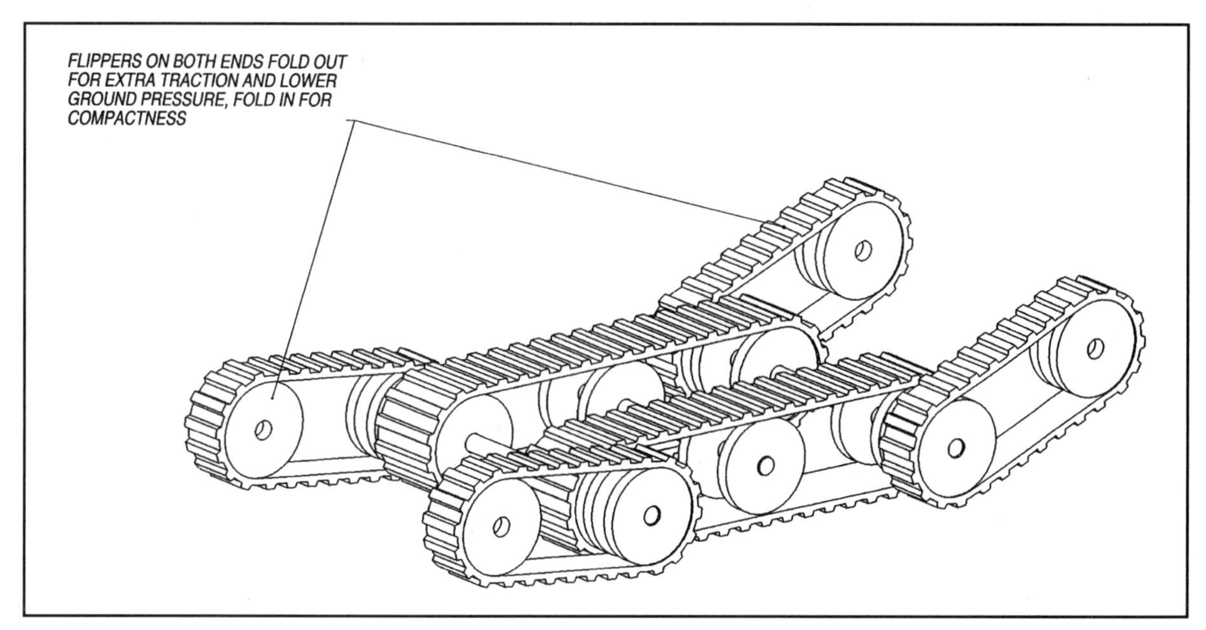

Figure 5-18 Six-tracked, double flippers

shown in Figure 5-17. The double flippers extend the length of the two-tracked base unit by almost a factor of two, facilitating crossing wide crevasses and climbing stairs, yet still being able to turn in place in a small aisle.

Chapter 6 Steering History

The Romans extensively used two wheeled carts, pulled by horses. Pull on the right rein and the horse pulls the cart to the right, and vise versa. The two wheels on the cart were mounted on the same axle, but were attached in a way that each wheel could rotate at whatever speed was needed depending on whether the cart was going straight or around a corner. Carts got bigger and eventually had four wheels, two in front and two in back. It became apparent (though it is unclear if it was the Romans who figured this out) that this caused problems when trying to turn. One or the other set of wheels would skid. The simplest method for fixing this problem was to mount the front set of wheels on each end of an axle that could swivel in the middle (Figure 6-1). A tongue was attached to the axle and stuck out from the front of the vehicle, which in turn was attached to a horse. Pulling on the tongue aligned the front wheels with the turn. The back wheels followed. This method worked well and, indeed, still does for four wheeled horse drawn buggies and carriages.

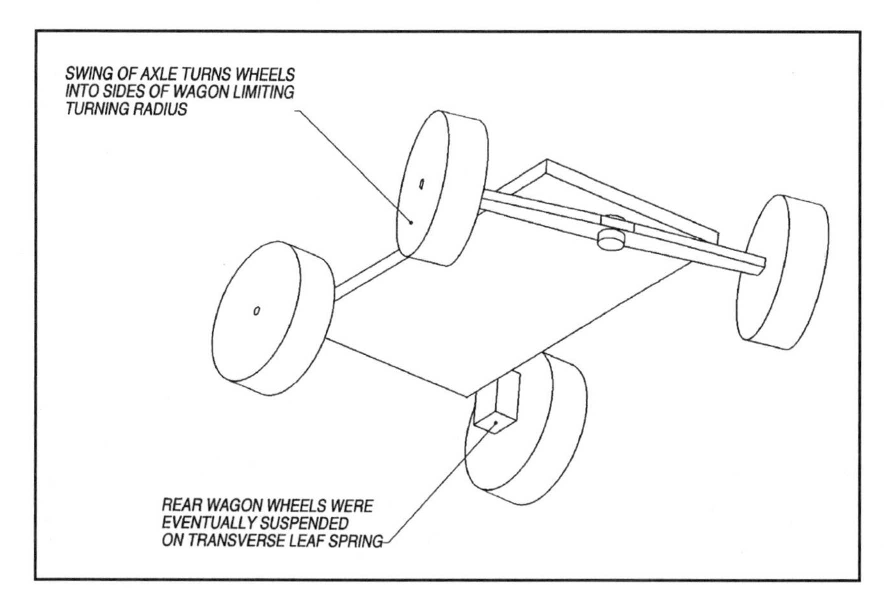

Figure 6-1 Pivot mounted front wheels

In the early 1800s, with the advent of steam engines (and, later, electric motors, gas engines, and diesel engines) this steering method began to show its problems. Vehicles were hard to control at speeds much faster than a few meters per second. The axle and tongue took up a lot of room swinging back and forth under the front of the vehicle. An attempt around this problem was to make the axle long enough so that the front wheels didn't hit the cart's sides when turning, but it was not very convenient having the front wheels wider than the rest of the vehicle.

The first effective fix was to mount the two front wheels on a mechanism that allowed each wheel to swivel closer to its own center. This saved space and was easier to control and it appeared to work well. In 1816, George Lankensperger realized that when turning a corner with the wheels mounted using that geometry the inside wheel swept a different curve than the outside one, and that there needed to be some other mechanical linkage that would allow this variation in alignment. He teamed with Rudolph Ackerman, whose name is now synonymous with this type of steering geometry. Although Ackerman steering is used on almost every human controlled vehicle designed for use on roads, it is actually not well suited for high mobility vehicles controlled by computers, but it feels right to a human and works very well at higher speeds. It turns out there are many other methods for turning corners, some intuitive, some very complex and unintuitive.

STEERING BASICS

When a vehicle is going straight the wheels or tracks all point in the same direction and rotate at the same speed, but only if they are all the same diameter. Turning requires some change in this system. A two-wheeled bicycle (Figure 6-2) shows the most intuitive mechanism for performing this change. Turn the front wheel to a new heading and it rolls in that direction. The back wheel simply follows. Straighten out the front wheel, and the bicycle goes straight again.

Close observation of a tricycle's two rear wheels demonstrates another important fact when turning a corner: the wheel on the inside of the corner rotates slower than the outside wheel, since the inside wheel is going around a smaller circle in the same amount of time. This important detail, shown in Figure 6-3, occurs on all wheeled and tracked vehicles. If the vehicle's wheels are inline, there must be some way to allow the wheels to point in different directions. If there are wheels on either side, they must be able to rotate at different speeds. Any deviation from this

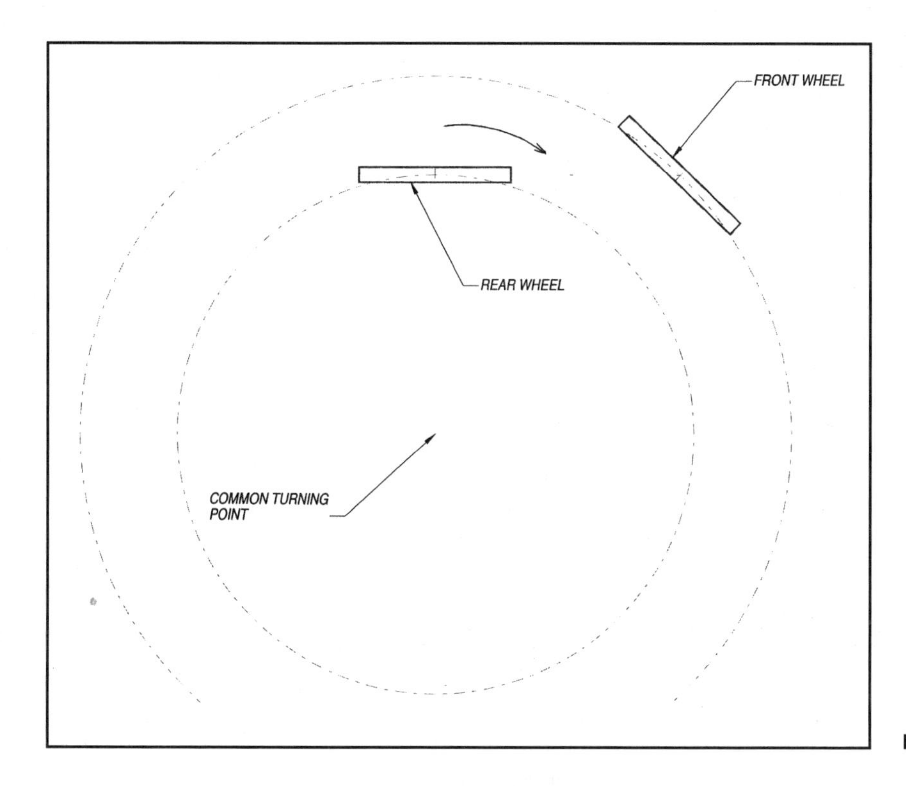

Figure 6-2 Bicycle steering

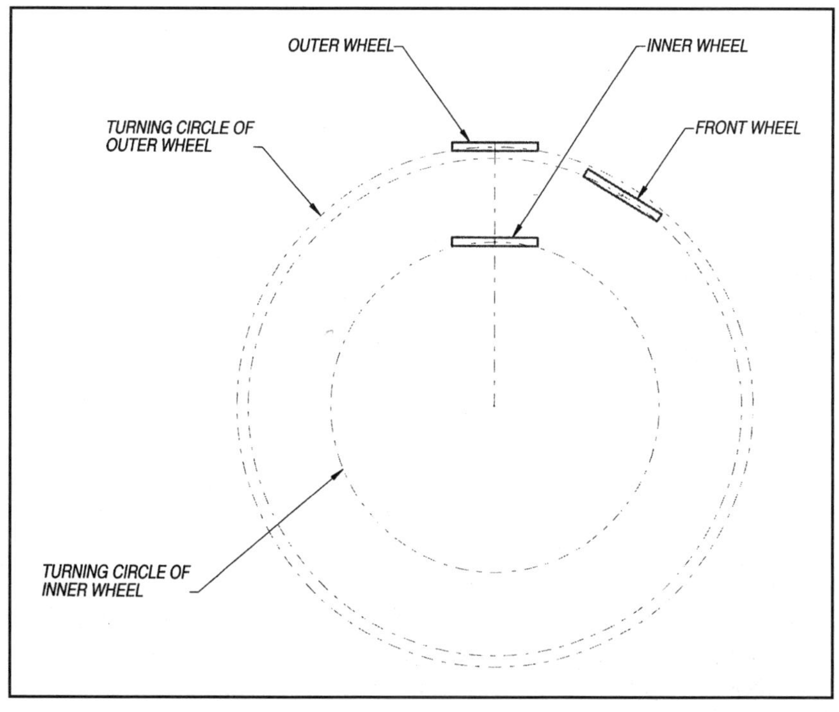

Figure 6-3 Tricycle steering

and some part of the drive train in contact with the ground will have to slide or skid.

Driving straight in one direction requires at least one single direction actuator. A wind-up toy is a good demonstration of this ultra-simple drive system. Driving straight in both directions requires at least one bi-directional actuator or two single-direction actuators. One of those single direction actuators can power either a steering mechanism or a second drive motor. Add one more simple single-direction motor to the wind-up toy, and it can turn to go in any new direction. This shows that the least number of actuators required to travel in any direction is two, and both can be single-direction motors.

In practice, this turns out to be quite limiting, at least partly because it is tricky to turn in place with only two single direction actuators, but mostly because there aren't enough drive and steer options to pick from to get out of a tight spot. Let's investigate the many varieties of steering commonly used in wheeled and tracked robots.

The simplest statically stable vehicle has either three wheels or two tracks, and the simplest power system to drive and steer uses only two single-direction motors. It turns out that there are only two ways to steer these very simple vehicles:

1. Two single-direction motors powering a combined drive/steer wheel or combined drive/steer track with some other passive wheels or tracks
2. Two single-direction motors, each driving a track or wheel (the third wheel on the wheeled layout is a passive swivel caster)

The simplest version of the first steering geometry is a single-wheel drive/steer module mounted on a robot with two fixed wheels. The common tricycle uses this exact layout, but so do some automatic guided vehicles (AGVs) used in automated warehouses. Mobility is limited because there is only one wheel providing the motive force, while dragging two passive wheels. This layout works well for the AGV application because the warehouse's floor is flat and clean and the aisles are designed for this type of vehicle. In an AGV, the drive/steer module usually has a bi-directional steering motor to remove the need to turn the drive wheel past 180° but single direction steer motors are possible. There are many versions of AGVs—the most complicated types have four drive/steer modules. These vehicles can steer with, what effectively amounts to, any common steering geometry; translate in any direction without rotating (commonly called "crabbing"), pseudo-Ackerman steer, turn about any point, or rotate in place with no skidding. Wheel modules

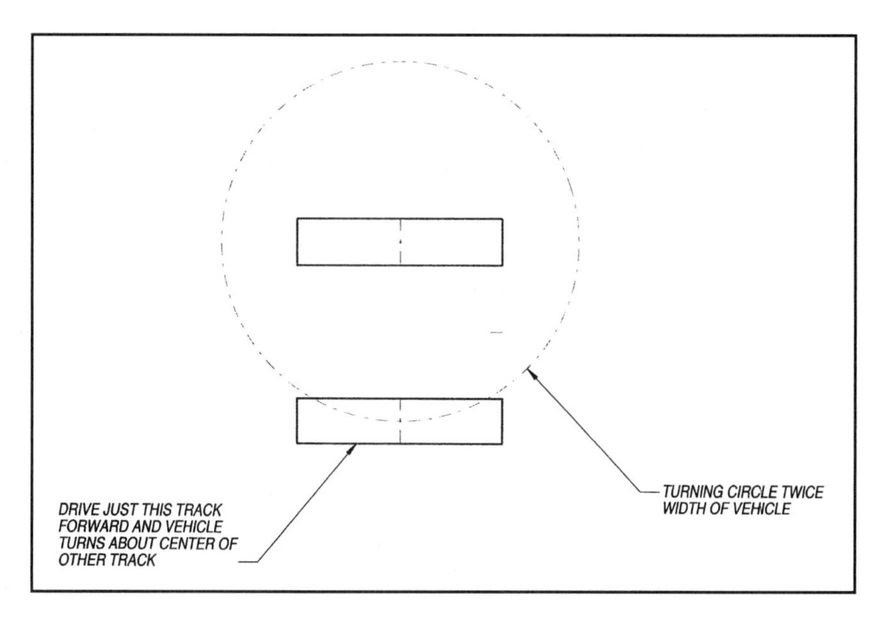

Figure 6-4 Turning about one track

for AGVs are available independently, and come in several sizes ranging from about 30 cm tall to nearly a meter tall.

The second two-single-direction motor steering layout has been successfully tried in research robots and toys, but it doesn't provide enough options for a vehicle moving around in anything but benign environments. It can be used on tracked vehicles, but without being able to drive the tracks backwards, the robot can not turn in place and must turn about one track. Figure 6-4 shows this limitation in turning. This may be acceptable for some applications, and the simplicity of single direction electronic motor-driver may make up for the loss of mobility. The biggest advantage of both of these drive/steering systems is extreme simplicity, something not to be taken lightly.

The Next Step Up

The next most effective steering method is to have one of the actuators bi-directional, and, better than that, to have both bi-directional. The Rug Warrior educational robot uses two bi-directional motors—one at each wheel. This steering geometry (Figure 6-5a, 6-5b) is called differential steering. Varying the relative speed, between the two wheels turns the robot. On some ultra-simple robots, like the Rug Warrior, the third wheel does not even swivel, it simply rolls passively on a fixed axle and skids when the robot makes a turn. Virtually all modern two-tracked

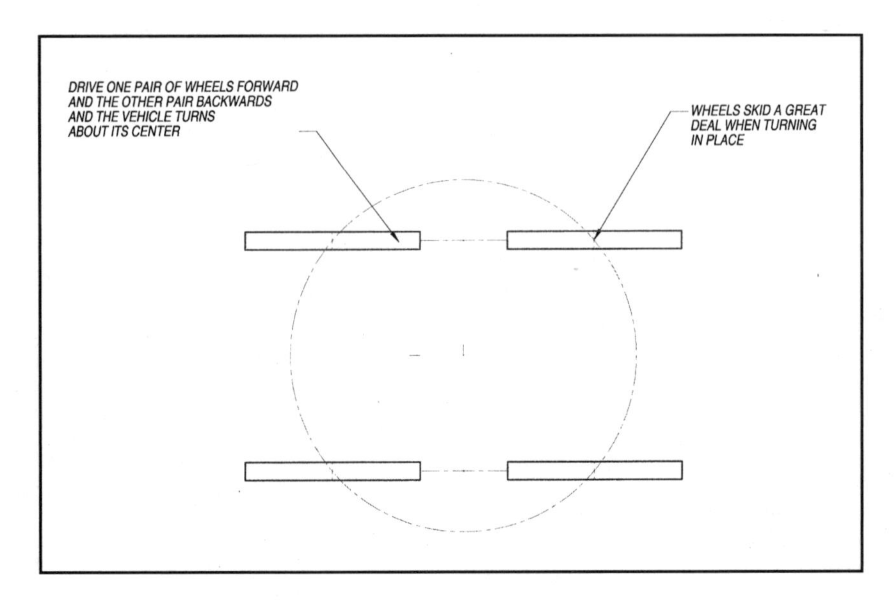

Figure 6-5a Differential steering

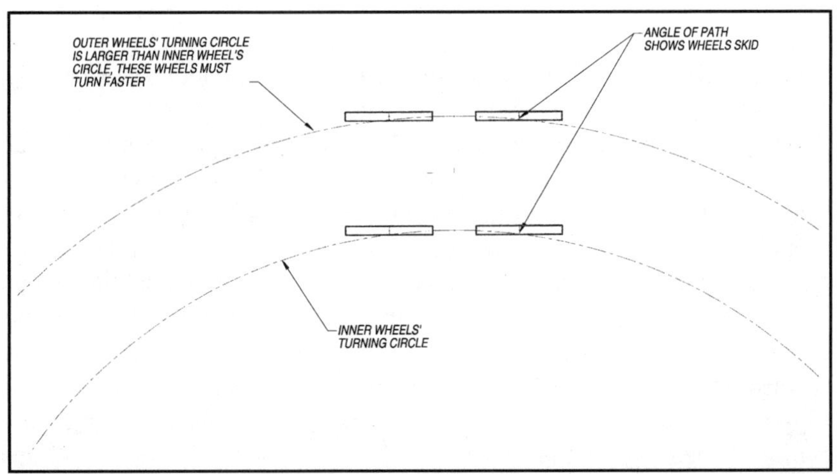

Figure 6-5b

vehicles use this method to steer, while older tracked vehicles would brake a track on one side, slowing down only that track, which turned the vehicle.

As discussed in the chapter on wheeled vehicles, this is also the steering method used on some four-wheel loaders like the well-known Bobcat. One motor drives the two wheels on one side of the vehicle, the other drives the two wheels on the other side. This steering method is so effective and robust that it is used on a large percentage of four-, six-, and even eight-wheeled robots, and nearly all modern tracked vehicles whether autonomous or not. This steering method produces a lot of skidding of the wheels or tracks. This is where the name "skid steer" comes from.

The fact that the wheels or tracks skid means this system is wasting energy wearing off the tires or track pads, and this makes skid steering an inefficient design. Placing the wheels close together or making the tracks shorter reduces this skidding at the cost of fore/aft stability. Six-wheeled skid-steering vehicles can place the center set of wheels slightly below the front and back set, reducing skidding at the cost of adding wobbling. Several all-terrain vehicle manufacturers have made six-wheeled vehicles with this very slight offset, and the concept can be applied to indoor hard-surface robots also. Eight-wheeled robots can benefit from lowering the center two sets of wheels, reducing wobbling somewhat.

The single wheel drive/steer module discussed earlier and shown on a tricycle in Figure 6-6 can be applied to many layouts, and is, in general, an effective mechanism. One drawback is some inherent complexity with powering the wheel through the turning mechanism. This is usually accomplished by putting the drive motor, with a gearbox, inside the wheel. Using this layout, the power to the drive motor is only a couple wires and signal lines from whatever sensors are in the drive wheel. These wires must go through the steering mechanism, which is easier than passing power mechanically through this joint. In some motor-in-wheel layouts, particularly the syncro-drive discussed next, the steering

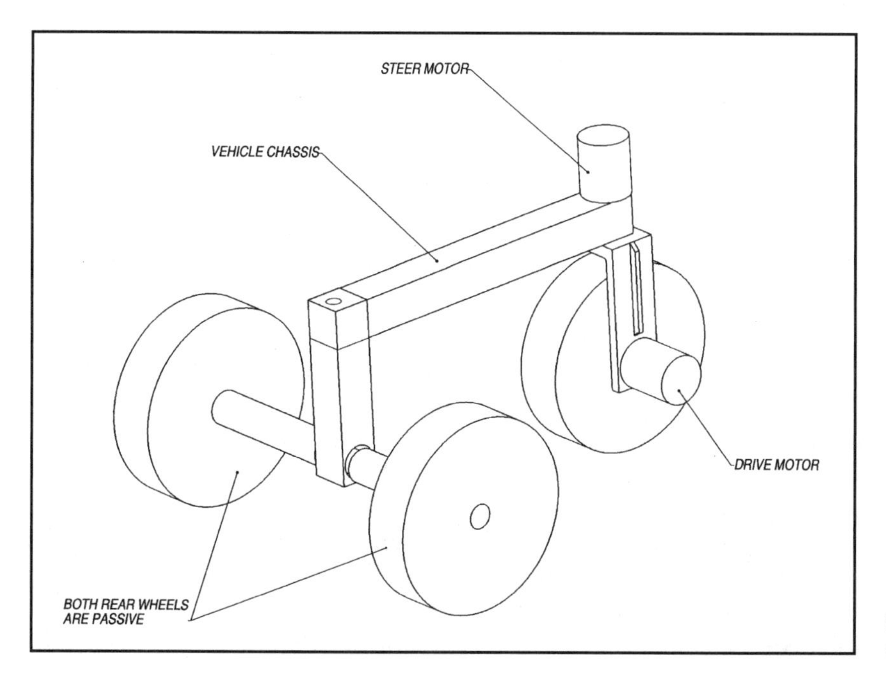

Figure 6-6 Drive/steer module on tricycle

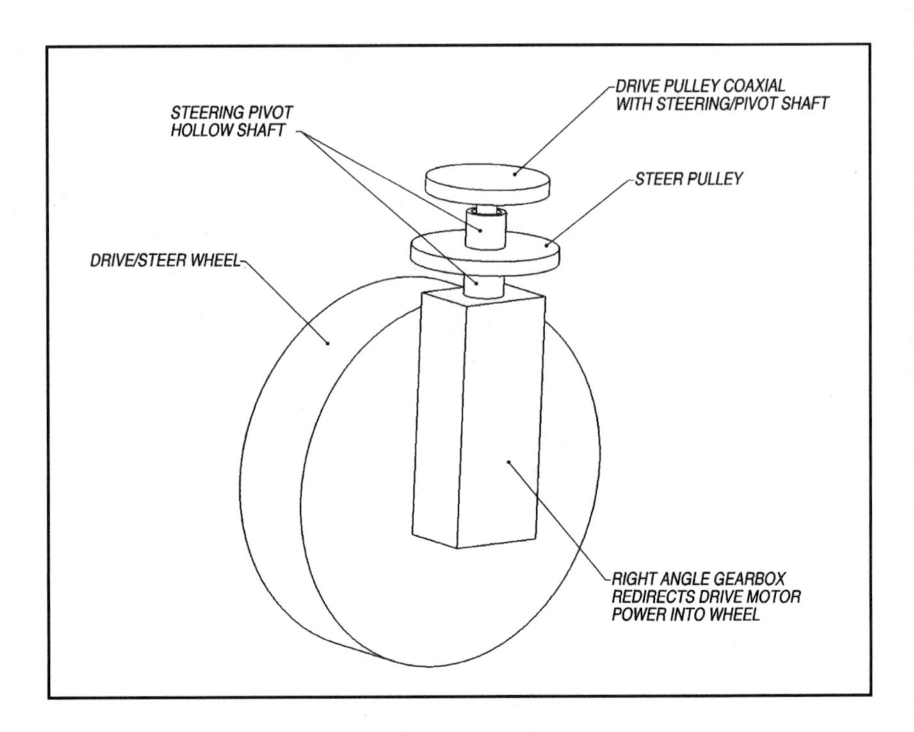

Figure 6-7 Synchronous drive

mechanism must be able to rotate the drive wheel in either direction as much as is needed. This requires an electrical slip ring in the steering joint. Slip rings, also called rotary joints, are manufactured in both standard sizes or custom layouts.

One type of mechanical solution to the problem of powering the wheel in a drive/steer module has been done with great success on several sophisticated research robots and is commonly called a syncro-drive. A syncro-drive (Figure 6-7) normally uses three or four wheels. All are driven and steered in unison, synchronously. This allows fully holonomic steering (the ability to head in any direction without first requiring moving forward). As can be seen in the sketch, the drive motor is directly above the wheel. An axle goes down through the center of the steering shaft and is coupled to the wheel through a right angle gearbox.

This layout is probably the best to use if relying heavily on dead reckoning because it produces little rotational error. Although the dominant dead-reckoning error is usually produced by things in the environment, this system theoretically has the least internal error. The four-wheeled layout is not well suited for anything but flat terrain unless at least one wheel module is made vertically compliant. This is possible, but would produce the complicated mechanism shown in Figure 6-8.

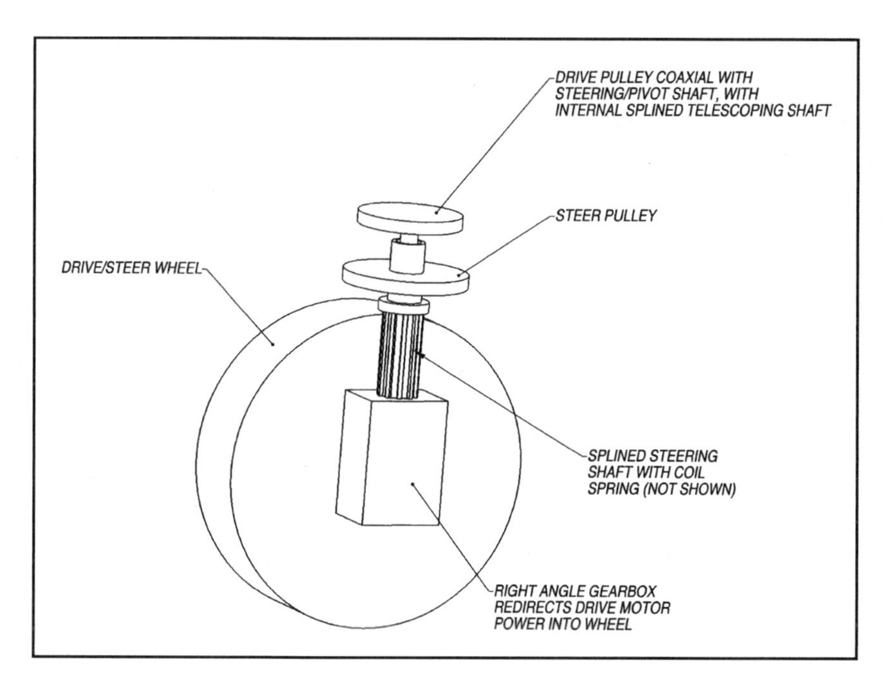

Figure 6-8 Drive/steer module with vertical compliance

All-terrain cycles (ATCs), when they were legal, ran power through a differential to the two rear wheels, and steered with the front wheel in a standard tricycle layout. ATCs clearly pointed out the big weakness of this layout, the tendency to fall diagonally to one side of the front wheel in a tight turn. Mobility was moderately good with a human driver, but was not inherently so.

Quads are the answer to the stability problems of ATCs. Four wheels make them much more stable, and many are produced with four wheel drive, enhancing their mobility greatly although they cannot turn in place. They are, of course, designed to be controlled by humans, who can foresee obstacles and figure out how to maneuver around them. If a mobility system in their size range is needed, they may be a good place to start. They are mass-produced, their price is low, and they are a mature product. Quads are manufactured by a number of companies and are available in many size ranges offering many different mobility capabilities.

As the number of wheels goes up, so does the variety of steering methods. Most are based on variations of the types already mentioned, but one is quite different. In Figure 4-30 (Chapter Four), the vehicle is divided into 2 sections connected by a vertical axis joint. This layout is common on large industrial front-end loaders and provides very good steering ability even though it cannot turn in place. The layout also forces the sections to be rather unusually shaped to allow for tighter turning. Power is transferred to the wheels from a single motor and differentials in the industrial version, but mobility would be increased if each wheel had its own motor.

Chapter 7 Walkers

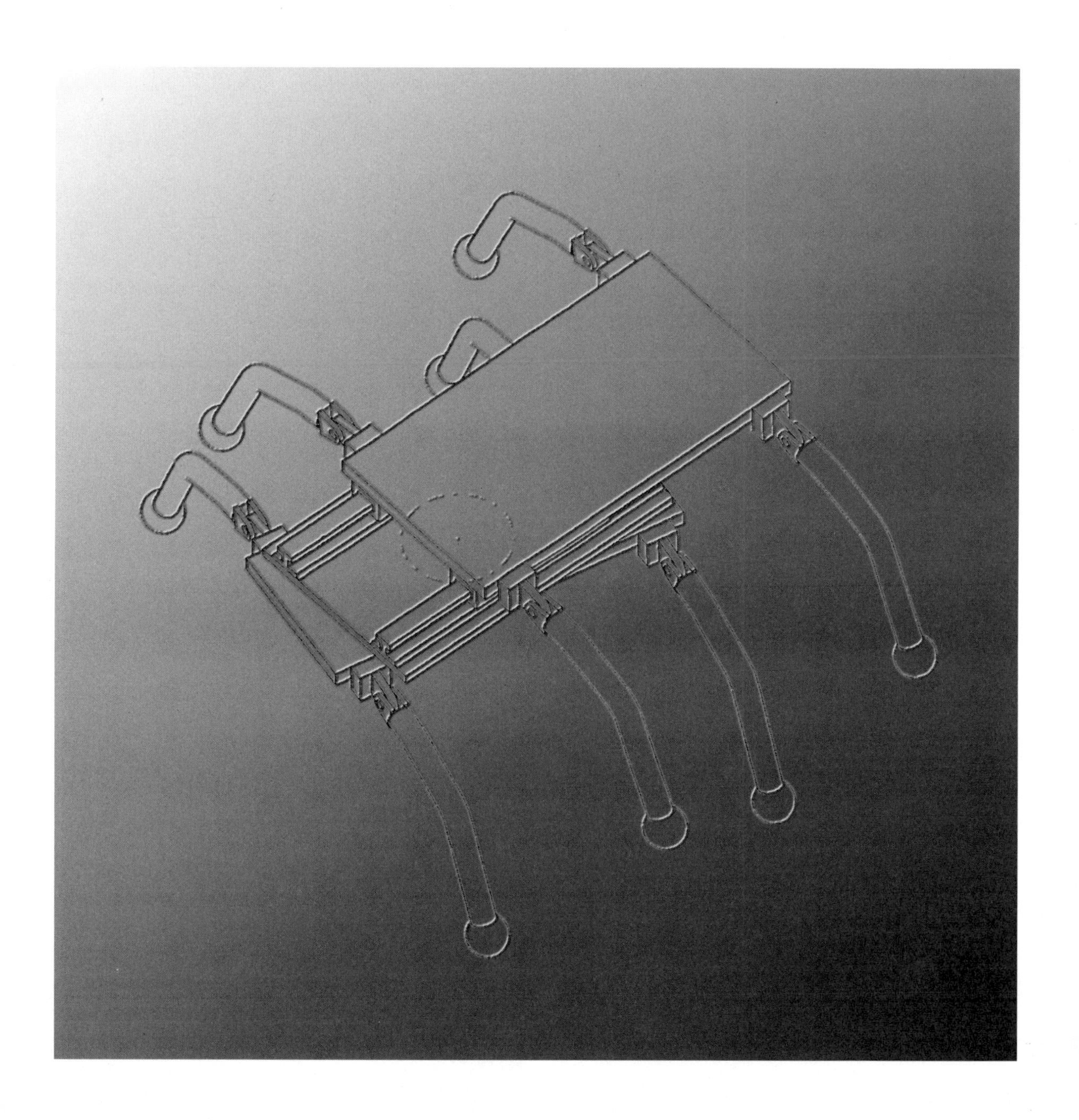

There are no multi-cell animals that use any form of continuously rolling mechanism for propulsion. Every single land animal uses jointed limbs or squirms for locomotion. Walking must be the best way to move then, right? Why aren't there more walking robots? It turns out that making a walking robot is far more difficult than making a wheeled or tracked one. Even the most basic walker requires more actuators, more degrees of freedom, and more moving parts.

Stability is a major concern in walking robots, because they tend to be tall and top heavy. Some types of leg geometries and walking gaits prevent the robot from falling over no matter where in the gait the robot stops. They are statically stable. Other geometries are called "dynamically stable." They fall over if they stop at the wrong point in a step. People are dynamically stable.

An example of a dynamically-stable walker in nature is, in fact, any two-legged animal. They must get their feet in the right place when they want to stop walking to prevent tipping over. Two-legged dinosaurs, humans, and birds are remarkably capable two legged walkers, but any child that has played Red-light/Green-light or Freeze Tag has figured out that it is quite difficult to stop mid-stride without falling over. For this reason, two legged walking robots, whether anthropomorphic (human-like) or birdlike (the knee bends the other way), are rather complicated devices requiring sensors that can detect if the robot is tipping over, and then calculate where to put a foot to stop it from falling.

Some animals with more than two legs are also dynamically stable during certain gait types. Horses are a good example. The only time they are statically stable is when they are standing absolutely still. All gaits they use for locomotion are dynamically stable. When they want to stop, they must plan where to put each foot to prevent falling over. When a horse's shoe needs to be lifted off the ground, it is a great effort for the horse to reposition itself to remain stable on three hooves, even though it is already standing still. Cats, on the other hand, can walk with a gait that allows them to stop at any point without tipping over. They do not need to plan in advance of stopping. This is called statically-stable independent leg walking. Elephants are known to use this technique when crossing streams or difficult terrain. They stand on three legs while the forth leg is moved around until it finds, by feel, a suitable place to set down.

These examples demonstrate that four-legged walkers can have geometries that are either dynamically or statically stable or both. Animals have highly developed sensors, a highly evolved brain, and fantastically high power-density muscles, that allow this variety of motion control. Practical walking robots, because of the limitations of sensors, processors, and fast acting powerful actuators, usually end up being statically stable with two to eight legs.

The design of dynamically-stable, walking mobile robots requires an extensive knowledge of fairly complicated sensors, balance, high-level math, fast-acting actuators, kinematics, and dynamics. This is all beyond the scope of this book. The rest of this chapter will focus on the second major category, statically-stable walkers.

LEG ACTUATORS

First, let's look and leg geometries and actuation methods. There are three major techniques for moving legs on a mobile robot.

- Linear actuators (hydraulic, pneumatic, or electric)
- Direct-drive rotary
- Cable driven

Hydraulics is not covered in this book, but linear motion can be done effectively using two other methods, pneumatic and electric. Pneumatic cylinders come in practically any imaginable size and have been used in many walking robot research projects. They have higher power density than electric linear actuators, but the problem with pneumatics is that the compressed air tank takes up a large volume.

Linear actuators have the advantage that they can be used directly as the leg itself. The body of the actuator is mounted to the robot's chassis or another actuator, and the end of the extending segment has a foot attached to it. This concept has been used to make robots that use Cartesian and cylindrical coordinate walkers. These layouts are not covered in this book, but the reader is urged to investigate them since they can simplify the development of the control code of the robot.

Direct-drive rotary actuators usually have to be custom designed to get torque outputs high enough to rotate the walker's joints. They have low power density and usually make the walker's joints look unnaturally large. They are very easy to control accurately and facilitate a modular design since the actuator can be thought of conceptually and physically as the complete joint. This is not true of either linear actuated or cable driven joints.

Cable-driven joints have the advantage that the actuators can be located in the body of the robot. This makes the limbs lighter and smaller. In applications where the leg is very long or thin, this is critical. They are somewhat easy to implement, but can be tricky to properly tension to get good results. Cable management is a big job and can consume many hours of debug time.

LEG GEOMETRIES

Walking robots use legs with from one to four degrees of freedom (DOF). There are so many varieties of layouts only the basic designs are discussed. It is hoped the designer will use these as a starting point from which to design the geometry and actuation method that best suits the application.

The simplest leg has a single joint at the hip that allows it to swing up and down (Figure 7-1). This leg is used on frame walkers and can be actuated easily by either a linear or rotary actuator. Since the joint is already near the body, using a cable drive is unnecessary. Notice that all the legs shown in the following figures have ball shaped feet. This is necessary because the orientation of the foot is not controlled and the ball gives the same contact surface no matter what orientation it is in. A second method to surmount adding orientation controlled feet is to mount the foot on the end of the leg with a passive ball joint.

The following four figures show two-DOF legs with the different actuation methods. These figures demonstrate the different attributes of the actuation method. Figure 7-2 shows that linear actuators make the legs much wider in one dimension but are the strongest of the three. Figure 7-3 shows a mechanism that keeps the second leg segment vertical as it is raised and lowered. The actuator can be replaced with a passive link, making this a one-DOF leg whose second segment doesn't swing out as much as the leg shown in Figure 7-2.

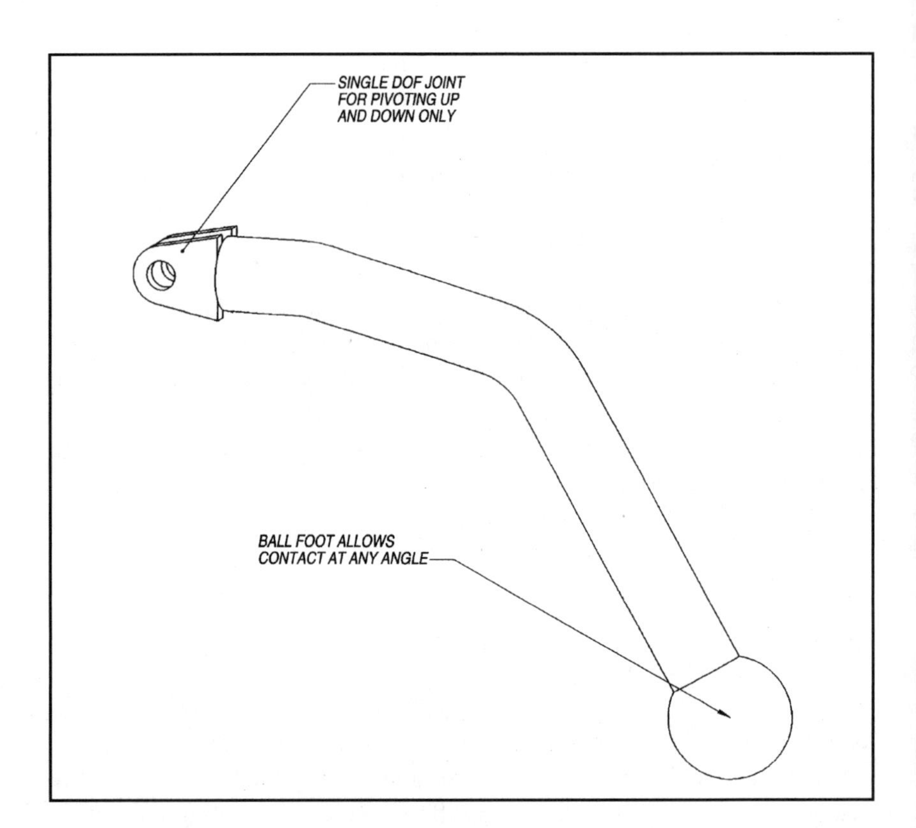

Figure 7-1 One-DOF leg for frame walkers

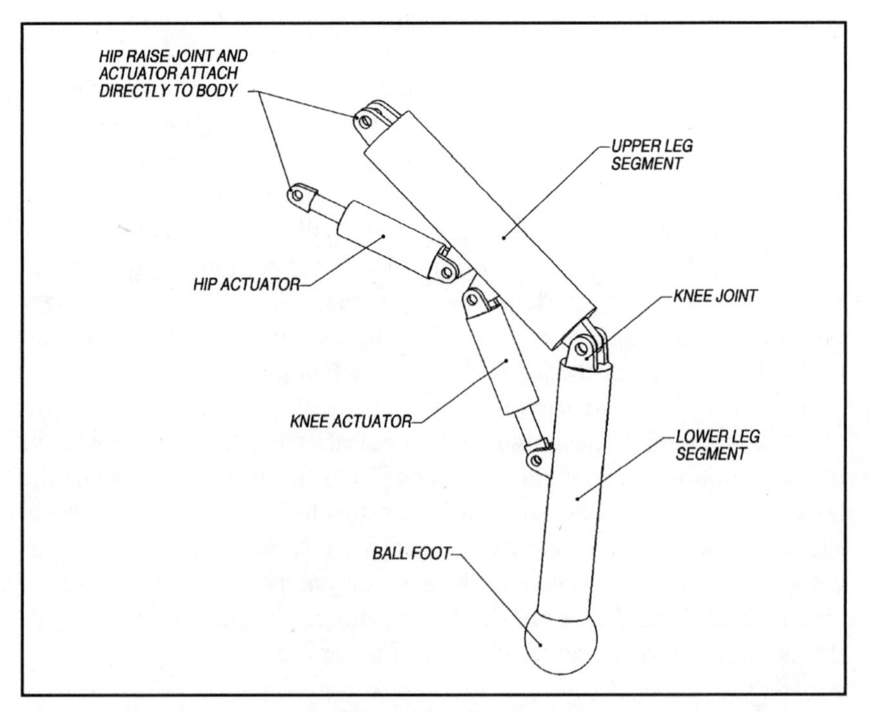

Figure 7-2 Two-DOF leg using linear actuators

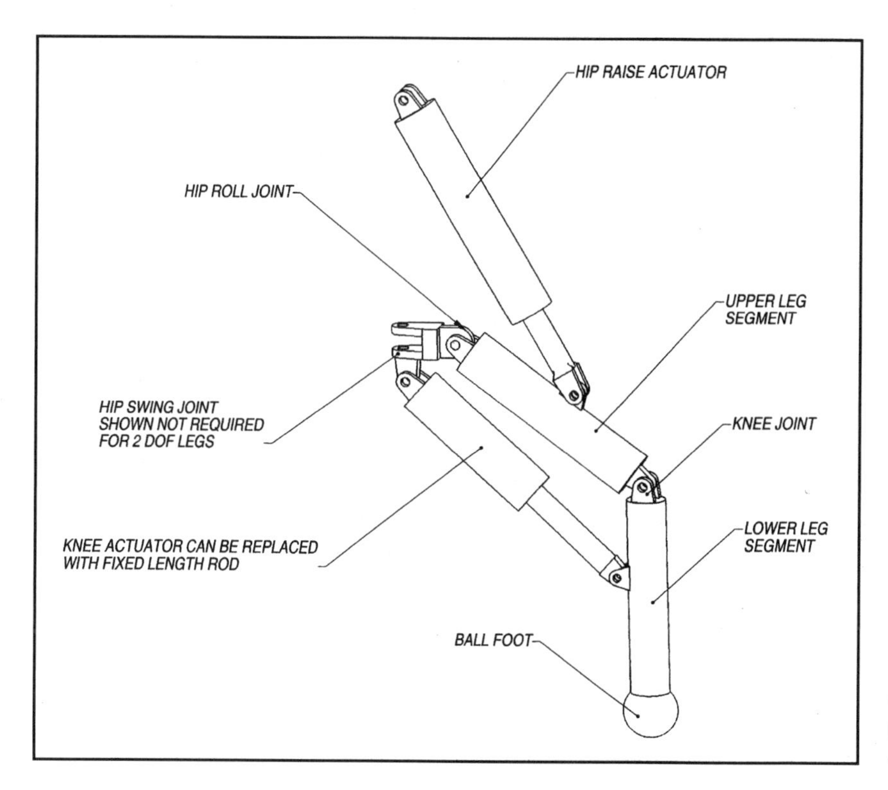

Figure 7-3 Two-DOF leg using linear actuators with chassis-mounted knee actuator

Rotary actuators (Figure 7-4) are the most elegant, but make the joints large. The cable driven layout (Figure 7-5) takes up the least volume and has no exposed actuators. Both of these methods are common, mostly because they use motors in a simple configuration, rather than linear actuators. Their biggest drawback is that they need to be big to get enough power to be useful. iRobot's Genghis robot used two hobby servos bolted together, acting as rotary actuators, to get a very effective two-axis hip joint. This robot, and several others like it, use simple straight legs. These simple walker layouts are useful preliminary tools for those interested in studying six-legged walking robots.

To turn the two-DOF linear actuator layout into a three-DOF, a universal joint can be added at the hip joint. This is controlled with an actuator attached horizontally to the chassis. Figure 7-6 shows a simple design for this universal hip joint. The order of the joints (swing first, then raise; or raise first, then swing) makes a big difference in how the foot location is controlled and should be carefully thought out and prototyped before building the real parts.

The three-DOF rotary actuator leg (Figure 7-7) adds a knee joint to the Genghis layout for improved dexterity and mobility. There are many varieties of this layout that change the various lengths of the segments

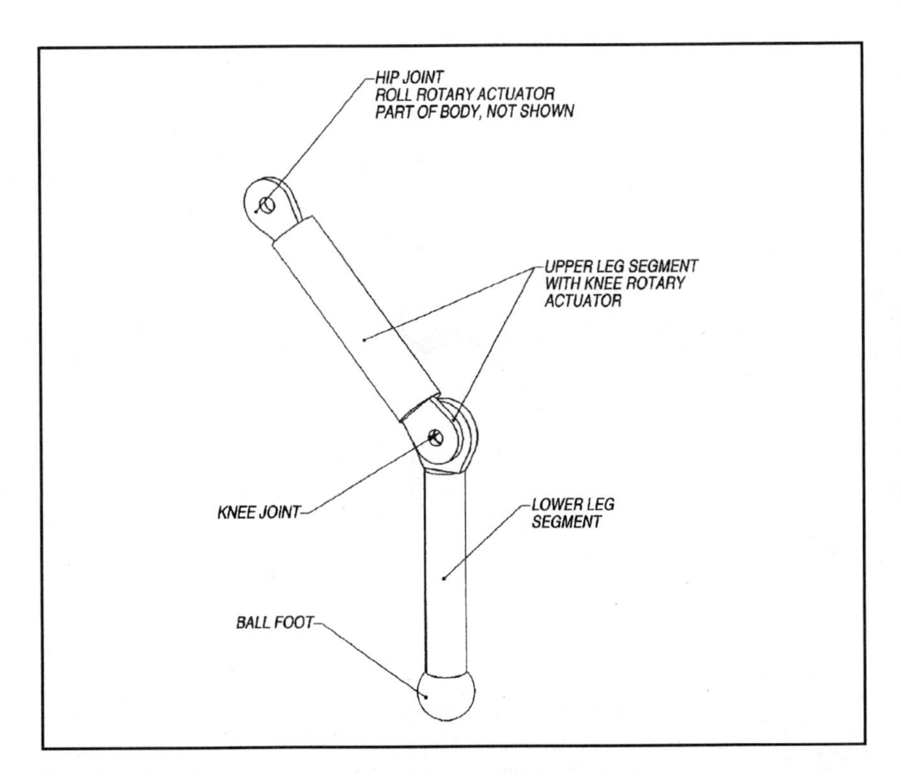

Figure 7-4 Two-DOF leg using rotary actuators

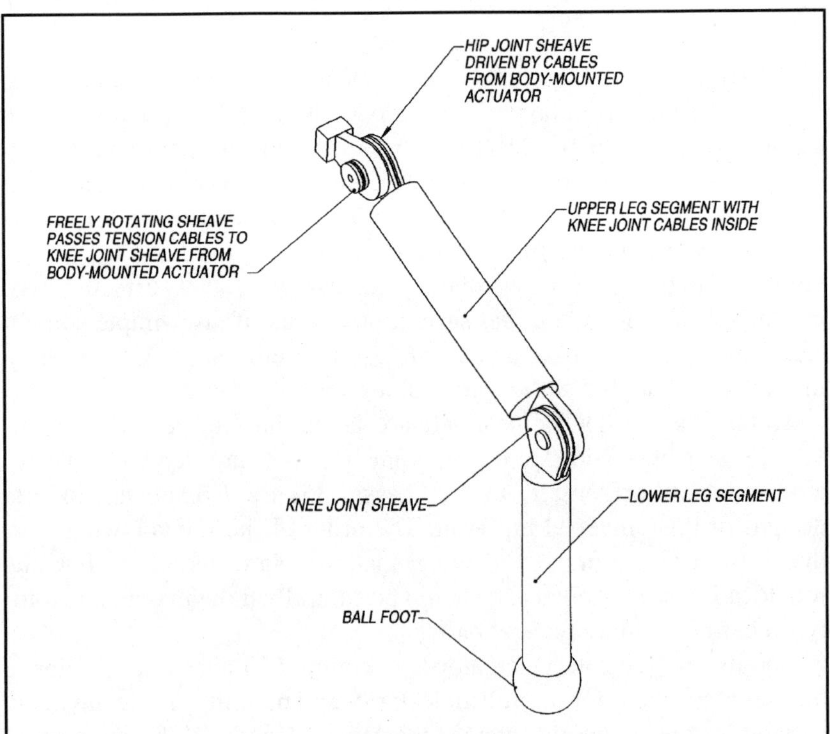

Figure 7-5 Two-DOF leg using cable driven actuators

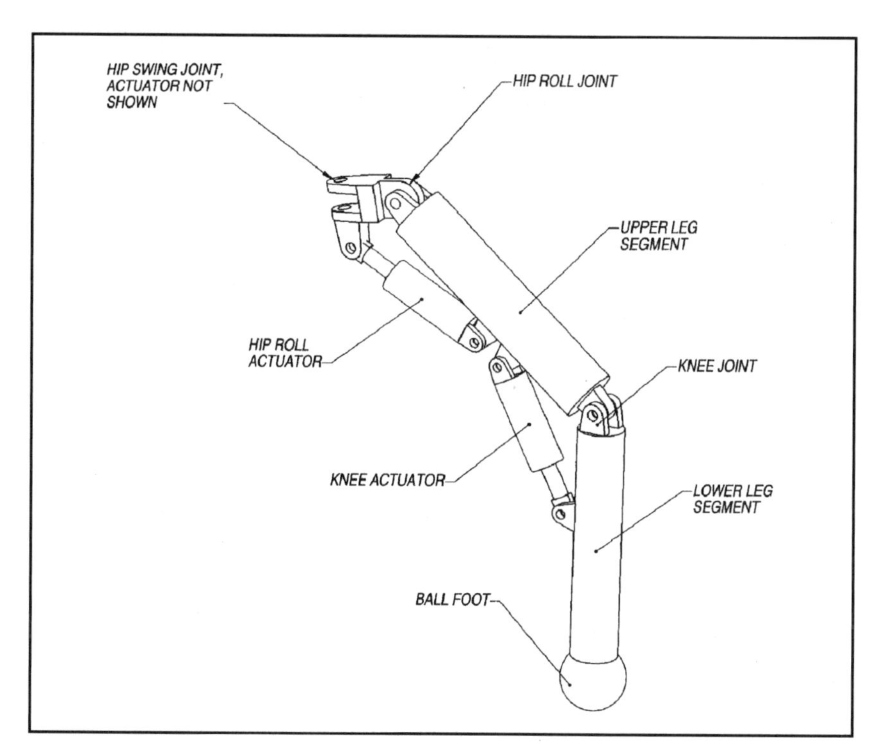

Figure 7-6 Three-DOF leg using linear actuators

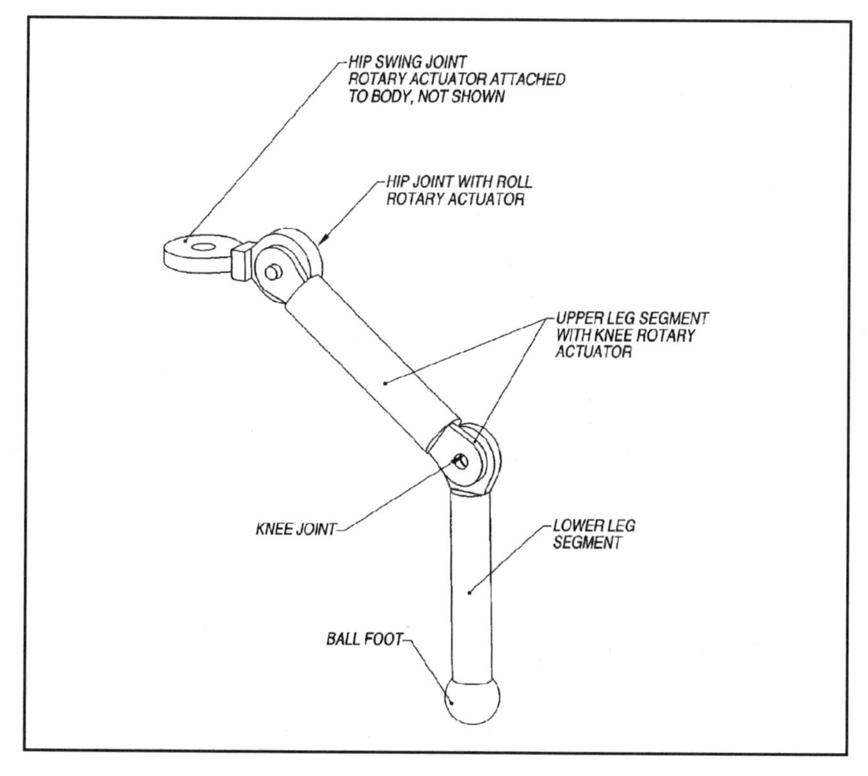

Figure 7-7 Three-DOF leg using rotary actuators

and the relative location of each actuator. It is quite difficult to drive a two-DOF hip joint with cables, but it can be done. The general layout would look much like what is shown in Figure 7-7.

WALKING TECHNIQUES

Statically-stable walkers are easier to implement than dynamically-stable walkers. A method used to group statically-stable walkers is the technique used to move the legs. There are three useful sub groups: wave walking, independent leg walking, and frame walking. Wave walking is what animals with many legs use, like millipedes. Independent leg walking is used by just about every four, six, and eight-leg walker, although some simplify things by moving their legs in groups for certain speeds or motions. Frame walking exists in nature in the form of an inchworm, and is the simplest of the three, but to have high mobility still requires many actuators. As we shall see, frame walking can be a very effective mobility method for a mobile robot.

Wave Walking

Centipedes and millipedes use a walking technique that must be mentioned, although it is simple in concept, for walking robots, it is less efficient than other methods. The robot lifts its rear-most set of legs and swings them forward and sets them down, then the next set of legs is moved similarly. When the front-most set of legs is moved, the whole robot chassis is moved forward relative to the legs. The process can be smoothed out some by averaging the position of the body as each set of legs moves forward. This technique can be used with six- or more-legged robots, but is not very common in robots because of the large numbers of joints and actuators.

Independent Leg Walking

Virtually all other legged animals in nature that don't use wave walking can control each leg independently. Some animals are better than others, but the ability is there. Figures 7-8 and 7-9 show four- and six-legged walkers with three rotary-actuated joints in each leg. An eight-leg layout would have no less than 24 actuators. The four- and six-legged versions

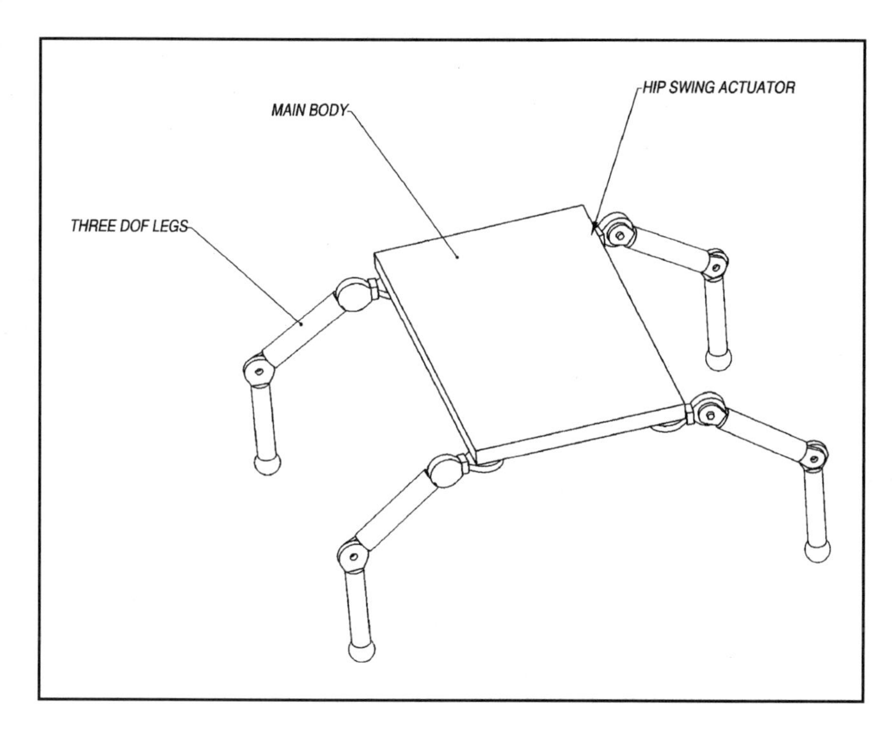

Figure 7-8 Independent leg walker, four legs, twelve DOF

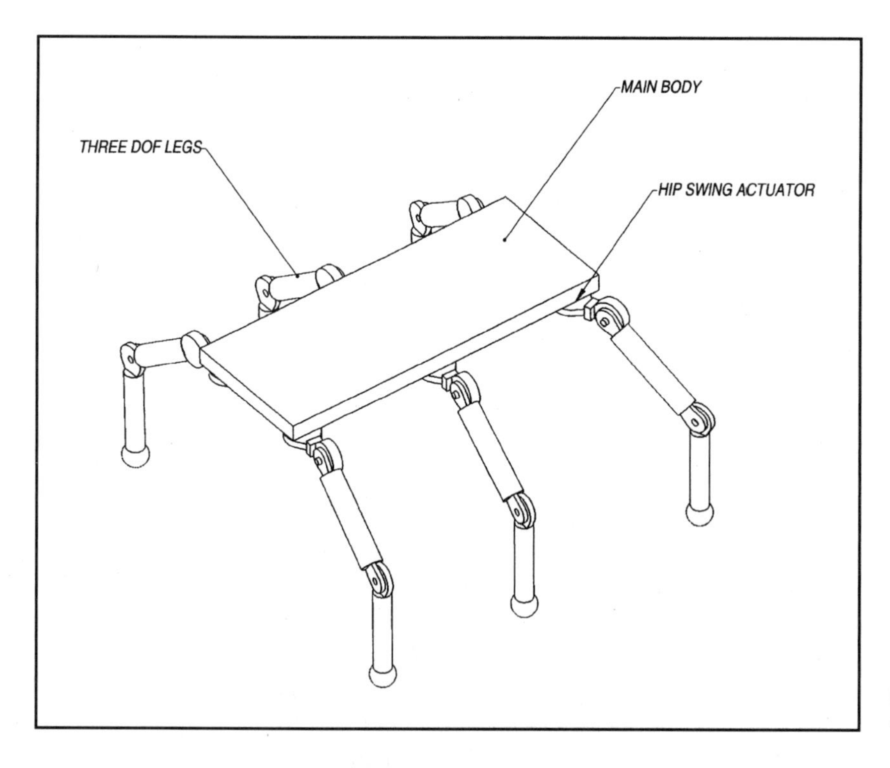

Figure 7-9 Independent leg walker, six legs, eighteen DOF

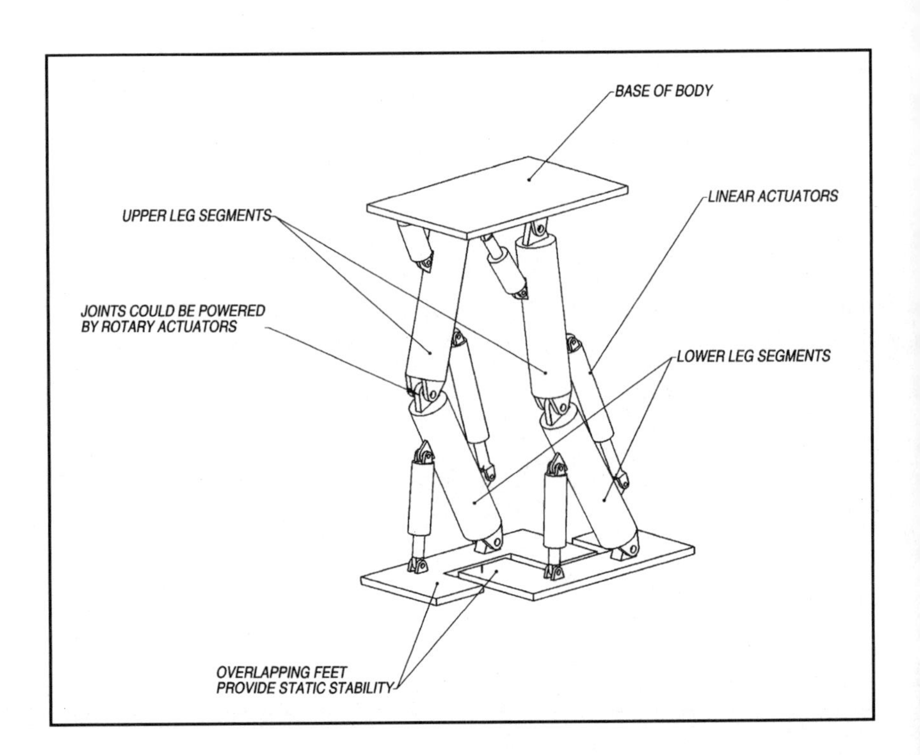

Figure 7-10 Extra wide feet provide two-legged stability

theoretically have very high mobility. Many research robots have been built that use four or six legs and are impressively agile, if very slow.

Although it would seem impossible to build a two-legged statically-stable robot, there is a trick that toys and some research robots use that gives the robot the appearance of being dynamically stable when they are actually statically stable. The trick is to have feet that are large enough to hold the robot upright on one foot without requiring the foot to be in exactly the right place. In effect, foot size reduces the required accuracy of foot placement so that the foot can be placed anywhere it can reach and the robot will not fall over.

The wide feet must also prevent tipping over sideways and are so wide that they overlap each other and must be carefully shaped and controlled so they don't step on each other. Two-legged walking, with oversized and overlapping feet, is simply picking up the back foot, bringing it forward, and putting it down. The hip joints require a second DOF in addition to swinging fore and aft, to allow rotation for turning. Each leg must have at least three DOF, and usually requires four. The layout shown in Figure 7-10 can only walk in a straight line because it lacks the hip rotation joint. Notice that even with only two legs and no ability to turn, this layout requires six actuators to control its six degrees of freedom.

This layout provides a good educational tool to learn about walking. Although in the final implementation it may have eight DOF and its knees bend backwards, it is familiar to the designer. One leg at a time can be built, tested, and debugged and then both attached to a simple plate for a chassis.

Frame Walking

The third general technique for walking with a legged robot is frame walking. Frame walking relies on the robot having two major sections, each with their own set of legs, both sections statically stable. Walking is accomplished by raising the legs of one frame, traversing that frame forward relative to the frame whose legs are still on the ground, and then setting the legs down. The other frame's legs are then raised and traversed forward.

The coupling between the two frames usually has a second rotating DOF to facilitate turning, rather than by adding a rotation in each leg. Figure 7-11 shows a mechanism for traversing and turning the two parts of the body. In nature, an inchworm uses a form of frame walking. The two frames are the front and back sections of the worm. The coupling is the leg-less section in between. In the case of the inchworm, the coupling has many degrees of freedom, but two is all that is required if the legs each have their own ability to move up and down. Unfortunately for robot designers, the inchworm also has the ability to grasp with its claw-

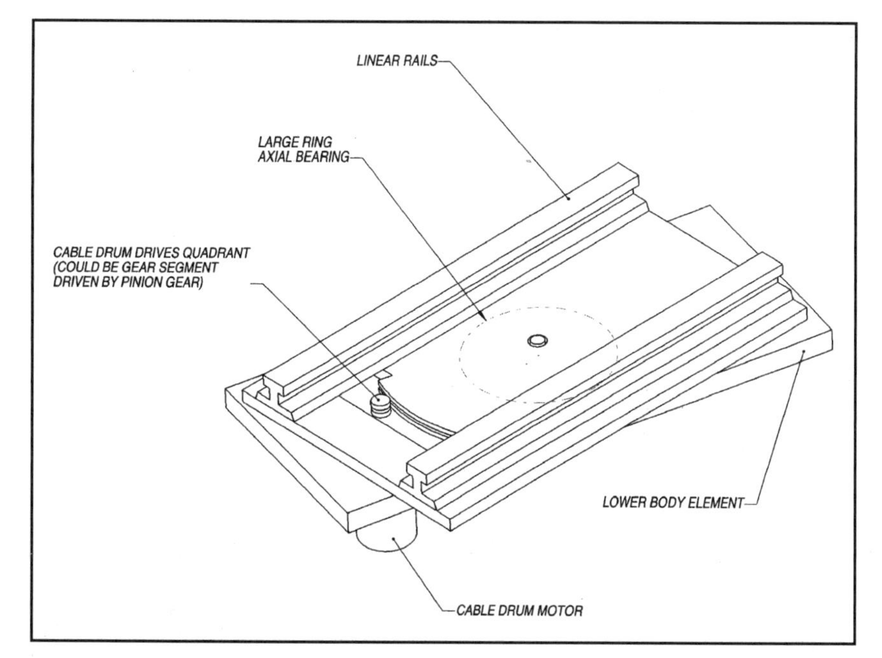

Figure 7-11 Mechanism for frame traversing and rotating

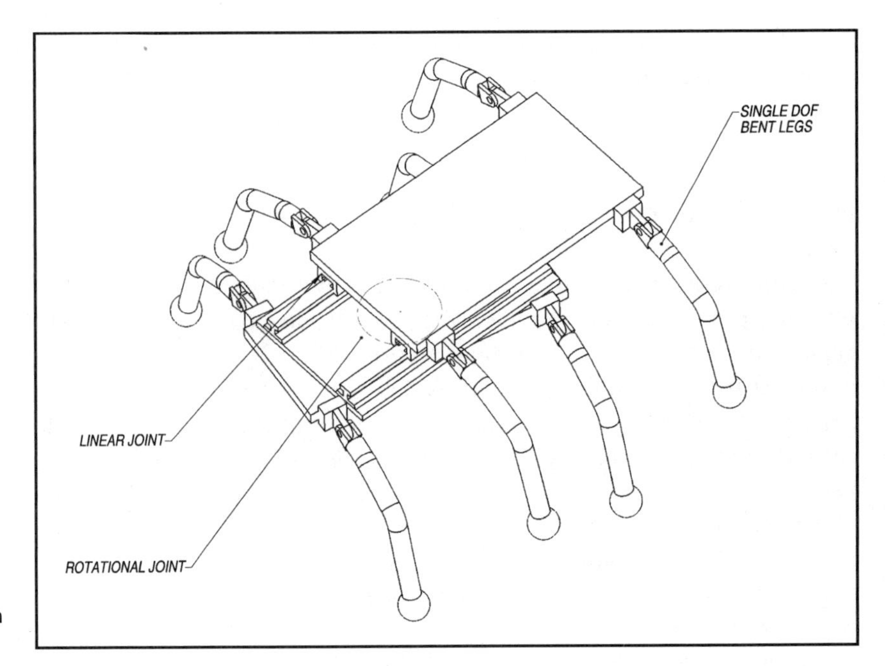

Figure 7-12 Traversing/rotating frame eight-leg frame walker with single-DOF legs

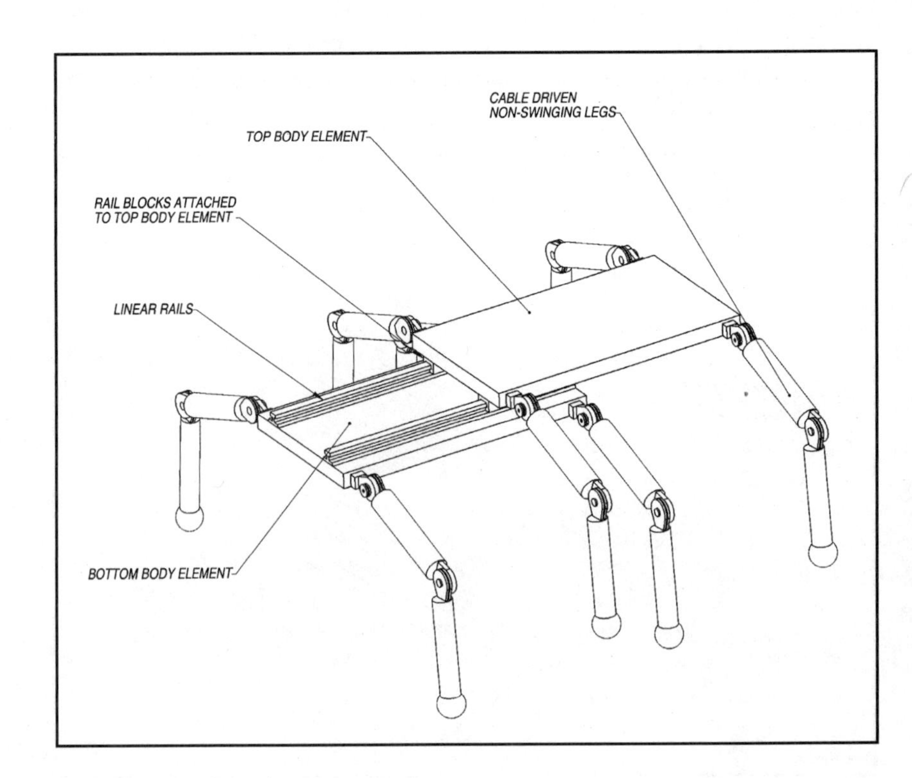

Figure 7-13 Eight-leg frame walker with two-DOF legs

like feet, making it quasi-statically stable. Figure 7-12 shows an implementation of the traversing/rotating frame with simple one-DOF legs. This layout has 10 DOF. Figure 7-13 removes the rotating joint, which forces placing a second joint in each leg to be able to turn. The actuator count goes up to 17 with this layout. The advantage of having the second joint in each leg is the ability to place each leg in the most optimum point to maintain traction and stability. Still, 17 actuators is a lot to control and maintain.

A six-leg tripod-gait frame walker could, however, have just three degrees of freedom, all in one joint between the two frames. This joint would have a linear motion for traversing, a rotary motion for steering, and a vertical motion to lift one frame and then the other. Mobility would suffer with such a simple platform because the robot would lack the ability to stand level on uneven terrain. Perhaps the best is a six-leg tripod-gait frame walker with one linear DOF in each leg and two in the coupling, bringing the total DOF to eight. Figure 7-14 shows just such a layout, perhaps the best walking layout to start with if designing a walking robot.

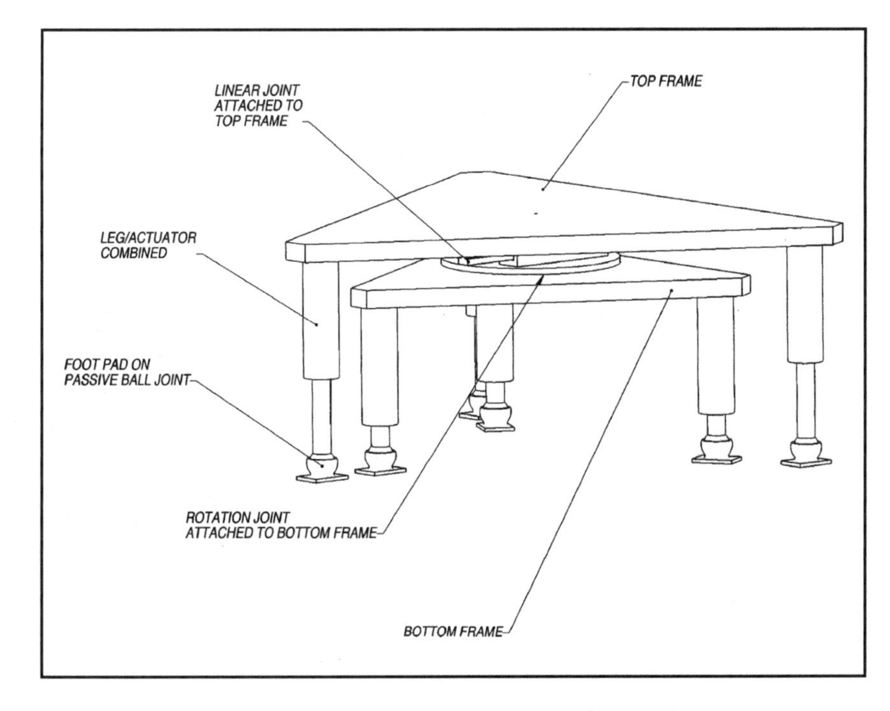

Figure 7-14 Six-legged tripod-frame walker with single-DOF legs

ROLLER-WALKERS

A special category of walkers is actually a hybrid system that uses both legs and wheels. Some of these types have the wheels mounted on fixed legs; others have the wheels mounted on legs that have one or two degrees of freedom. There doesn't seem to be any widely accepted term for these hybrids, but perhaps roller walkers will suffice.

A commercially available roller walker has one leg with a wheel on its end, and two jointed legs with no wheels, each with three DOF. The machine is a logging machine that can stand level even on very steep slopes. Although this machine looks ungainly with its long legs with a wheel on one of them, it is quite capable. Because of its slow traverse speed, it is transported to a job sight on the back of a special truck.

Wheels on legs can be combined to form many varieties of roller walkers. Certain terrain types may be more easily traversed with this unusual mobility system. The concept is gaining wider appeal as it becomes apparent a hybrid system can combine the better qualities of wheeled and legged robots. If contemplating designing a roller walker, it may be more effective to think of the mobility system as a wheeled vehicle with the wheels mounted on jointed appendages rather than a walking vehicle with wheels. The biggest limitation of walkers is still top speed. This limitation is easily overcome by wheels. A big limitation of a wheeled vehicle is getting over obstacles that are higher than the wheels. The ability to raise a wheel, or reconfigure the vehicle's geometry to allow a wheel to easily drive up a high object, reduces this limitation.

There are several researchers working on roller walkers. There are no figures included here, but the reader is urged to investigate these web sites:

http://mozu.mes.titech.ac.jp/
http://www.aist.go.jp/MEL/mainlab/rob/rob08e.html

FLEXIBLE LEGS

A trick taken from animals and being tested in mobility labs is the use of flexible-leg elements. A compliant member can sometimes be used to great advantage by reducing the requirement for exact leg placement. They are simple, extremely robust mobility systems that use independent leg-walking techniques. A simple version of this concept is closer to a wheeled robot than a walker. The tires are replaced with several long flexible arms, like whiskers, extending out from the wheel. This increases their ability to deal with large perturbations in the environ-

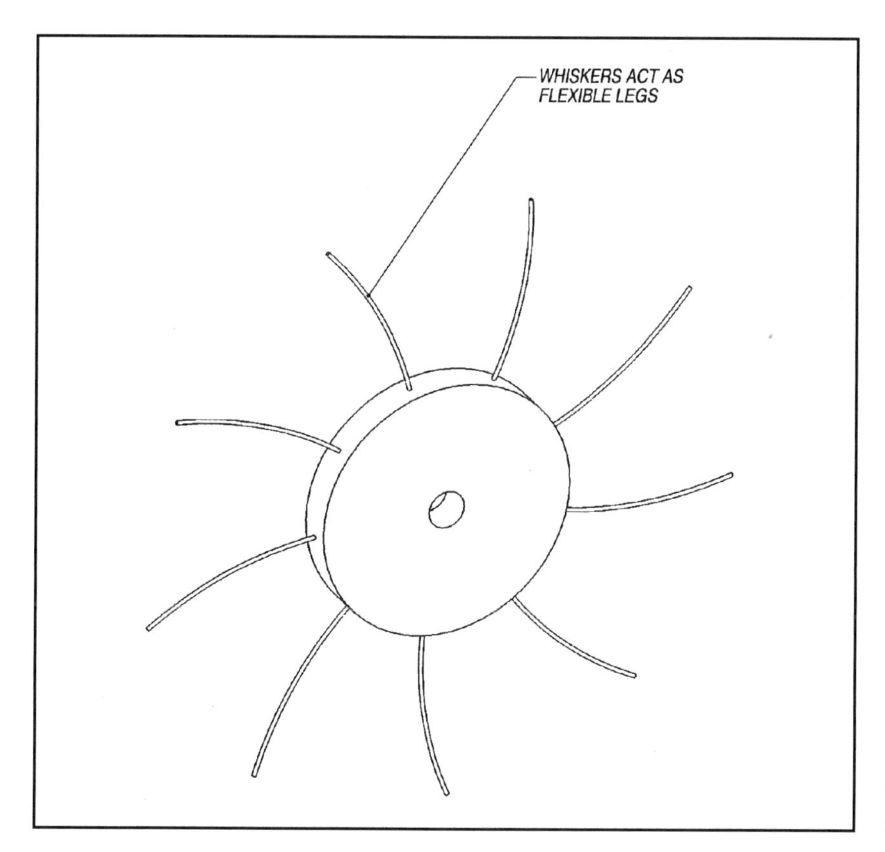

Figure 7-15 Whisker-wheeled roller walker

ment, but decreases efficiency. They have very high mobility, able to climb steps nearly as high as the legs are long. Robotics researchers are working on small four- and six-wheel leg robots that use this concept with very good results. Figure 7-15 shows the basic concept. A variation of this design extends the whisker legs more axially than radially. This idea is taken from studying cockroaches whose legs act like paddles when scrambling over bumpy terrain.

If walking is being considered as the mobility system for an autonomous robot, there are several things to remember.

- Using a statically-stable design requires far less expertise in several fields of engineering and will therefore dramatically increase the chances of success.
- Frame walking is easier to implement than wave- or independent-leg walking.
- Studies have shown six legs are optimal for most applications.
- Rotary joints are usually more robust.

Walkers have inherently more degrees of freedom, which increases complexity and debug time. As will be investigated in the chapter on mobility, walkers deal with rugged terrain very well, but may not actually be the best choice for a mobility system. Roller walkers offer the advantages of both walking and rolling and in a well thought out design may prove to be very effective.

Walkers have been built in many varieties. Some are variations on what has been presented here. Some are totally different. In general, with the possible exception of the various roller walkers, they share two common problems, they are complicated and slow. Nature has figured out how to make high-density actuators and control many of them at a time at very high speed. Humans have figured out how to make the wheel and its close cousin, the track. The fastest land animal, the cheetah, has been clocked at close to 100km/hr. The fastest land vehicle has hit more than seven times that speed. Contrarily, a mountain goat can literally run along the face of a steep cliff and a cockroach can scramble over terrain that has obstacles higher than itself, and can do so at high speed. There are no human-made locomotion devices that can even come close to a goat's or cockroach's combined speed and agility.

Nature has produced what is necessary for survival, but nothing more. Her most intelligent product has not yet been able to produce anything that can match the mobility of several of her most agile products. Perhaps someday we will. For the person just getting started in robotics, or for someone planning to use a robot to do a practical task, it is suggested to start with a wheeled or tracked vehicle because of their greater simplicity. For a mechanical engineer interested in designing a complex mechanism to learn about statics, dynamics, strength of materials, actuators, kinematics, and control systems, a walking robot is an excellent tool.

Chapter 8 Pipe Crawlers and Other Special Cases

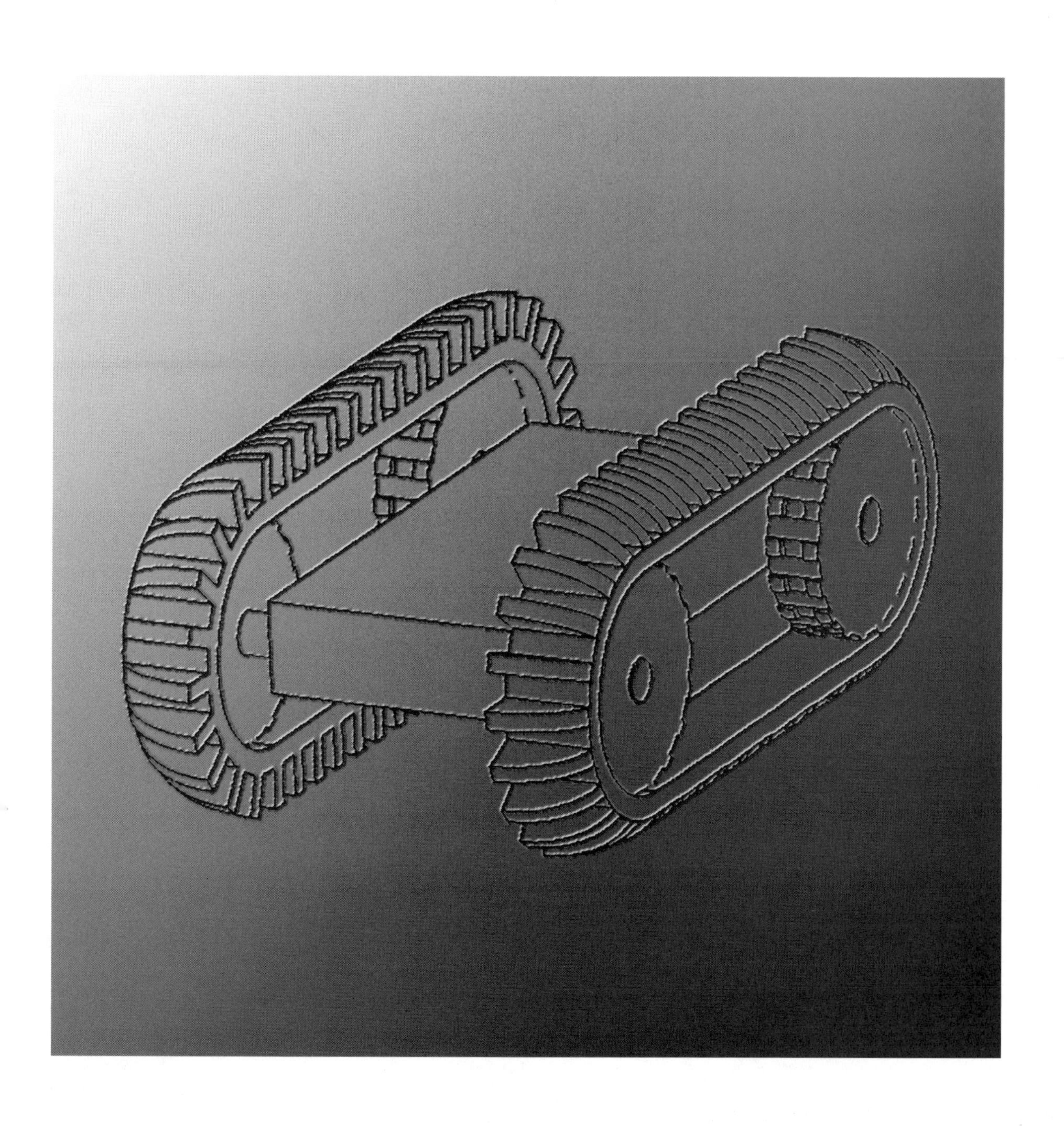

There are many less obvious applications for mobile robots. One particularly interesting problem is inspecting and repairing pipelines from the inside. Placing a robot inside a pipe reduces and, sometimes, removes the need to dig up a section of street or other obstruction blocking access to the pipe. The robot can be placed inside the pipe at a convenient location by simply separating the pipe at an existing joint or valve. These pipe robots, commonly called pipe crawlers, are very special designs due to the unique environment they must work in. Pipe crawlers already exist that inspect, clean, and/or repair pipes in nuclear reactors, water mains under city streets, and even down five-mile long oil wells.

Though the shape of the environment may be round and predictable, there are many problems facing the locomotion system of a pipe crawler. The vehicle might be required to go around very sharp bends, through welded, sweated, or glued joints. Some pipes are very strong and the crawlers can push hard against the walls for traction, some are very soft like heating ducts requiring the crawler to be both light and gentle. Some pipes transport slippery oil or very hot water. Some pipes, like water mains and oil pipelines, can be as large as several meters in diameter; other pipes are as small as a few centimeters. Some pipes change size along their length or have sections with odd shapes.

All these pipe types have a need for autonomous robots. In fact, pipe crawling robots are frequently completely autonomous because of the distance they must travel, which can be so far that it is nearly impossible to drag a tether or communicate by radio to the robot when it is inside the pipe. Other pipe crawlers do drag a tether which can place a large load on the crawler, forcing it to be designed to pull very hard, especially while going straight up a vertical pipe. All of these problems place unusual and difficult demands on the crawler's mechanical components and locomotion system.

End effectors on these types of robots are usually inspection tools that measure wall thickness or cameras to visually inspect surface conditions. Sometimes mechanical tools are employed to scrape off surface rust or other corrosion, plug holes in the pipe wall, or, in the case of oil wells, blow holes in the walls. These effectors are not complex mechanically and this chapter will focus on the mobility systems required for unusual environments and unusual methods for propulsion including external pipe walking and snakes.

The pipe crawler mechanisms shown in the following figures give an overview of the wide variety of methods of locomoting inside a pipe. Choosing between one and the other must be based on the specific attributes of the pipe and the material it transports, and if the robot has to work in-situ or in a dry pipe. In addition to those shown in this book, there are many other techniques and layouts for robots designed to move about in pipes or tanks.

HORIZONTAL CRAWLERS

Moving along horizontal pipes is very similar to driving on level ground. The crawler must still be able to steer to some degree because it must negotiate corners in the pipes, but also because it must stay on the bottom of the pipe or it may swerve up the walls and tip over. There are many horizontal pipe crawlers on the market that use the four-wheeled skid-steer principle, but tracked drives are also common. The wheels of wheeled pipe crawlers are specially shaped to conform to the round shape of the pipe walls, on tracked crawlers the treads are tilted for the same reason. These vehicles' suspension and locomotion systems are frequently quite simple. Figures 8-1 and 8-2 show two examples.

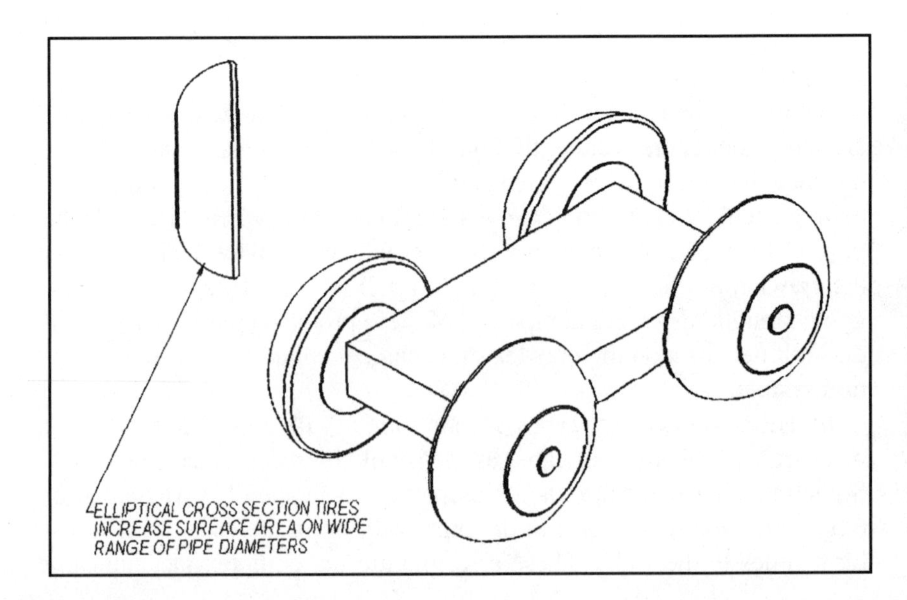

Figure 8-1 Four-wheeled horizontal pipe crawler

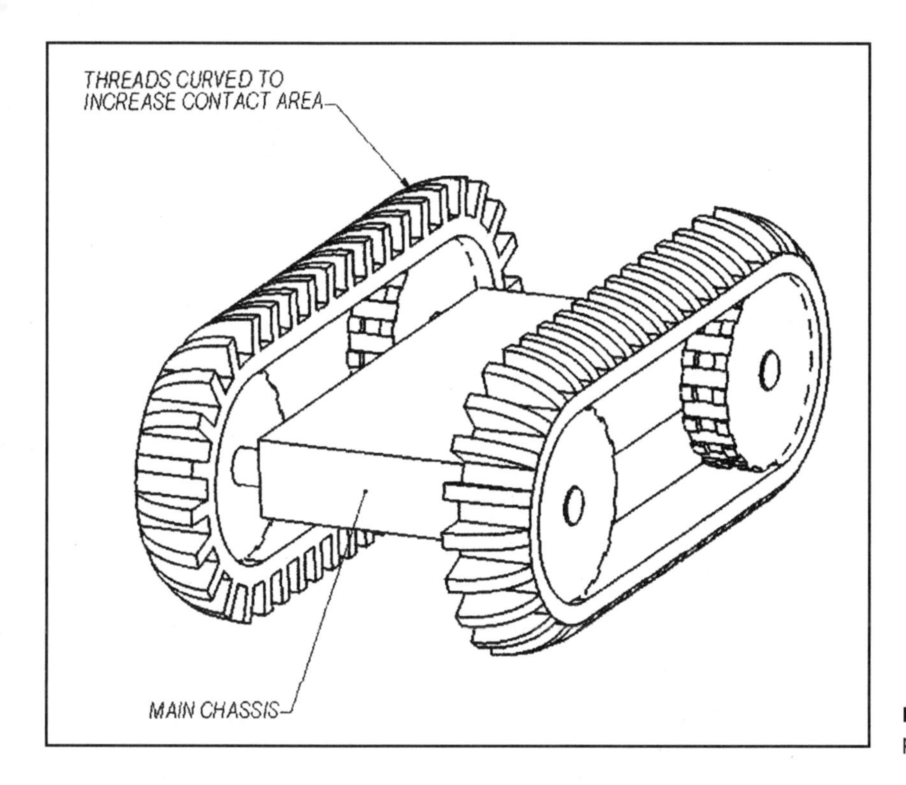

Figure 8-2 Two-track horizontal pipe crawler

VERTICAL CRAWLERS

Robotic vehicles designed to travel up vertical pipe must have some way to push against the pipe's walls to generate enough friction. There are two ways to do this, reaching across the pipe to push out against the pipe's walls, or putting magnets in the tires or track treads. Some slippery nonferrous pipes require a combination of pushing hard against the walls and special tread materials or shapes. Some pipes are too soft to withstand the forces of tires or treads and must use a system that spreads the load out over a large area of pipe.

There is another problem to consider for tethered vertical pipe crawlers. Going straight up a vertical pipe would at first glance seem simple, but as the crawler travels through the pipe, it tends to corkscrew because of slight misalignment of the locomotors or deformities on the pipe's surface. This corkscrewing winds up the tether, eventually twisting and damaging it. One solution to this problem is to attach the tether to the chassis through a rotary joint, but this introduces another degree of freedom that is both complex and expensive. For multi-section crawlers, a better solution is to make one of the locomotor sections steerable by a small amount.

Traction Techniques for Vertical Pipe Crawlers

There are at least four tread treatments designed to deal with the traction problem.

- spikes, studs, or teeth
- magnets
- abrasives or nonskid coating
- high-friction material like neoprene

Each type has its own pros and cons, and each should be studied carefully before deploying a robot inside a pipe because getting a stuck robot out of a pipe can be very difficult. The surface conditions of the pipe walls and any active or residual material in the pipe should also be investigated and understood well to assure the treatment or material is not chemically attacked.

Spiked, studded, or toothed wheels or treads can only be used where damage to the interior of the pipe can be tolerated. Galvanized pipe would be scratched leading to corrosion, and some hard plastic pipe material might stress crack along a scratch. Their advantage is that they can generate very high traction. Spiked wheels do find use in oil wells, which can stand the abuse. They require the crawler to span the inside of the pipe so they can push against opposing walls.

The advantage of magnetic wheels is that the wheels pull themselves against the pipe walls; the disadvantage is that the pipe must be made of a ferrous metal. Magnets remove the need to have the locomotion system provide the force on the walls, which reduces strain on the pipe. They also have the advantage that the crawler can be smaller since it no longer must reach across the whole of a large pipe. Use of magnetic wheels is not limited to pipe crawlers and should be considered for any robot that will spend most of its life driving on a ferrous surface.

Tires made of abrasive impregnated rubber hold well to iron and plastic pipe, but these types loose effectiveness if the abrasive is loaded with gunk or worn off. Certain types of abrasives can grip the surface of clean dry pipes nearly as well as toothed treads, and cause less damage.

High-friction rubber treads work in many applications, but care must be taken to use the right rubber compound. Some rubbers maintain much of their stickiness even when wet, but others become very slippery. Some compounds may also corrode rapidly in fluids that might be found in pipes. They cause no damage to pipe walls and are a simple and effective traction technique.

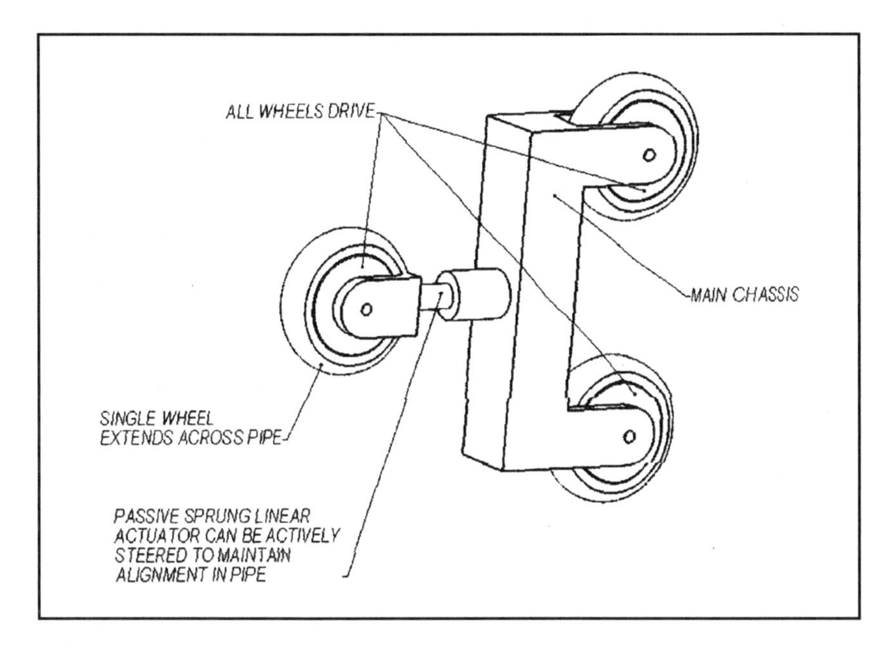

Figure 8-3 Basic three-wheeled

Wheeled Vertical Pipe Crawlers

Wheeled pipe crawlers, like their land-based cousins, are the simplest type of vertical pipe crawlers. Although these types use wheels and not tracks, they are still referred to as pipe crawlers. Practical layouts range from three to six or more wheels, usually all driven for maximum traction on frequently very slippery pipe walls.

Theoretically, crawling up a pipe can be done with as little as one actuator and one passive sprung joint. Figure 8-3 shows the simplest layout required for moving up vertical pipe. This design can easily get trapped or be unable to pass through joints in the pipe and can even be stopped by large deformities on the pipe walls.

The next best layout adds a fourth wheel. This layout is more capable, but there are situations in certain types of pipes and pipe fittings in which it too can become trapped, see Figure 8-4. The center linear degree of freedom can be actuated to keep the vehicle aligned in a pipe.

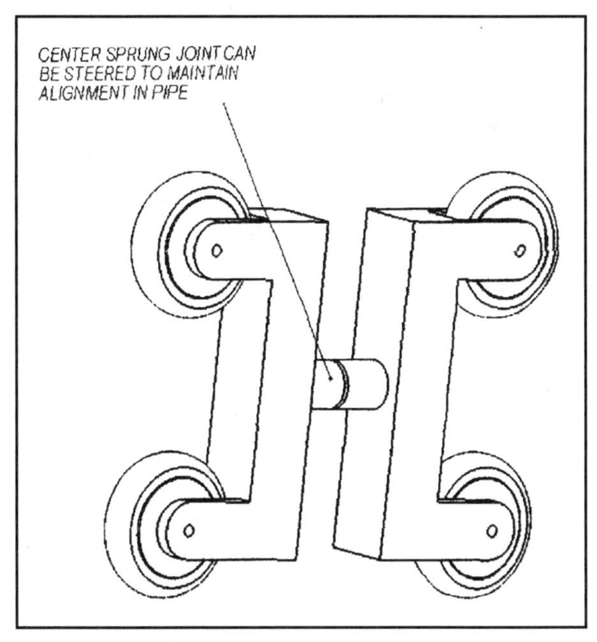

Figure 8-4 Four-wheeled, center steer

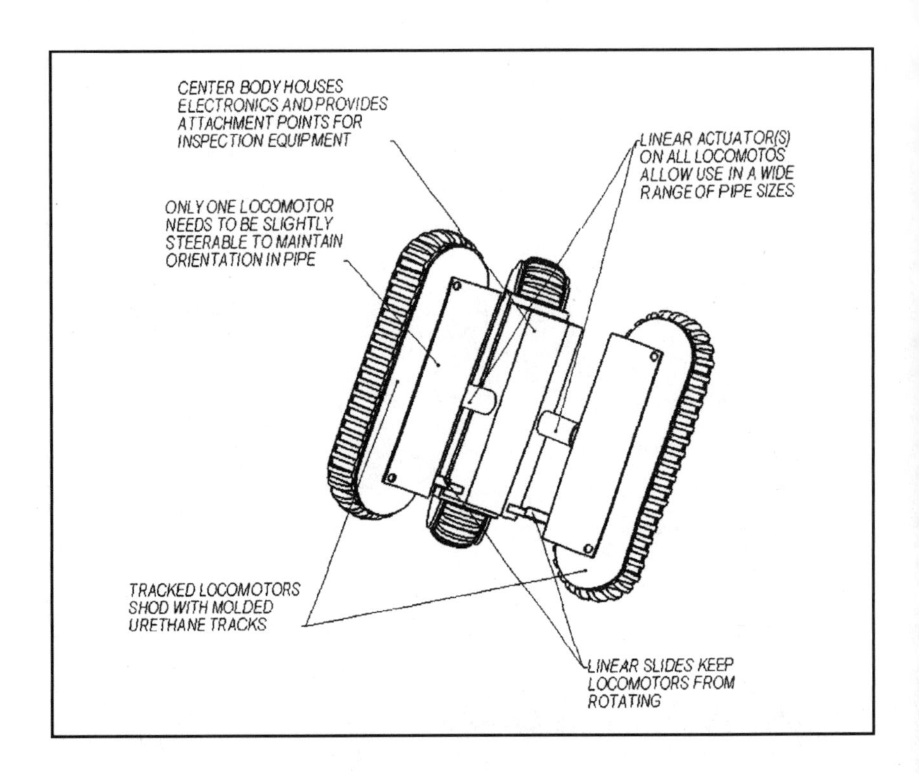

Figure 8-5 Three locomotors, spaced 120° apart

TRACKED CRAWLERS

Wheeled crawlers work well in many cases, but tracks do offer certain advantages. They exert much less pressure on any given spot due to their larger footprint. This lower pressure tends to scratch the pipe less. Spreading out the force of the mechanism that pushes the locomotor sections against the walls also means that the radial force itself can be higher, greatly increasing the slip resistance of the vehicle. Figure 8-5 shows the very common three-locomotor tracked pipe crawler.

OTHER PIPE CRAWLERS

For pipes that cannot stand high internal forces, another method must be used that further spreads the forces of the crawler over a larger area. There are at least two concepts that have been developed. One uses balloons, the other linear extending legs.

The first is a unique concept that uses bladders (balloons) on either end of a linear actuator, that are filled with air or liquid and expand to push out against the pipe walls. The rubber bladders cover a very large section of the pipe and only low pressure inside the bladder is required to

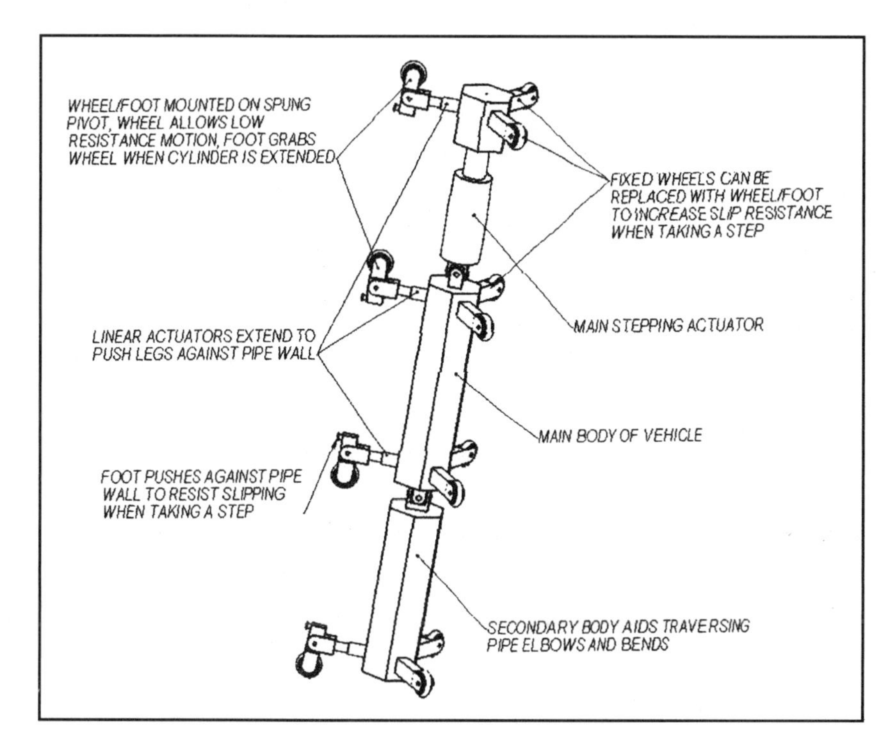

Figure 8-6 Inchworm multi-section roller walker

get high forces on the pipe walls, generating high-friction forces. Steering, if needed, is accomplished by rotating the coupling between the two sections.

This coupling is also the inchworm section, and forward motion of the entire vehicle is done by retracting the front bladder, pushing it forward, expanding it, retracting the rear section, pulling it towards the front section, expanding it, then repeating the whole process. Travel is slow, and this concept does not deal well with obstructions or sharp corners, but the advantage of very low pressures on the pipe walls may necessitate using this design. A concept that uses this design was proposed for moving around in the flexible Kevlar pipes of the Space Shuttle.

Another inchworm style pipe crawler has a seemingly complex shape, but this shape has certain unusual advantages. The large pipes inside nuclear reactor steam pipes have sensors built into the pipes that extend in from the inner walls nearly to the center of the pipe. These sensor wells are made of the same material as the pipe, usually a high-grade stainless steel, but cannot be scraped by the robot. The robot has to have a shape that can get around these protrusions. An inchworm locomotion vehicle consisting of three sections, each with extendable legs, provides great mobility and variable geometry to negotiate these obstacles. Figure 8-6 shows a minimum layout of this concept.

EXTERNAL PIPE VEHICLES

There are some applications that require a vehicle to move along the outside of a pipe, to remove unwanted or dangerous insulation, or to move from one pipe to another in a process facility cluttered with pipes. CMU's asbestos removing external pipe walker, BOA, is just such a vehicle. Though not a robot according to this book's definition, it is still worth including because it shows the wide range of mobility systems that true robots might eventually have to have to move in unexpected environments. BOA is a frame walker. Locomotion is accomplished by moving and clamping one set of grippers on a pipe, extending another set ahead on the pipe, and grasping the pipe with a second set of grippers.

RedZone Robotics' Tarzan, an in-tank vertical pipe walking arm, is an example of a very unusual concept proposed to move around inside a tank filled with pipes. This vehicle is similar to the International Space Station's maintenance arm in that it moves from one pipe to another, on the outside of the pipes. Unlike the ISS arm, Tarzan must work against the force of gravity. Since Tarzan is not autonomous, it uses a tether to get power and control signals from outside the tank. The arm is all-hydraulic, using both rotary actuators and cylinders. All together, there are 18 actuators. Imagine the complexity of controlling 18 actuators and managing a tether all on an arm that is walking completely out of view inside a tank filled with a forest of pipes!

SNAKES

In nature, there is a whole class of animals that move around by squirming. This has been applied to robots with a little success, especially those intended to move in all three dimensions. Almost by definition, squirming requires many actuators, flexible members, and/or clever mechanisms to couple the segments. The advantage is that the robot is very small in cross section, allowing it to fit into very complex environments, propelling itself by pushing on things. The disadvantage is that the number of actuators and high moving parts count.

There are many other unusual locomotion methods, and many more are being developed in the rapidly growing field of mobile robots. The reader is encouraged to search the web to learn more of these varied and sometimes strange solutions to the problem of moving around in uncommon environments like inside and outside pipes, inside underground storage tanks, even, eventually, inside the human body.

Chapter 9 Comparing Locomotion Methods

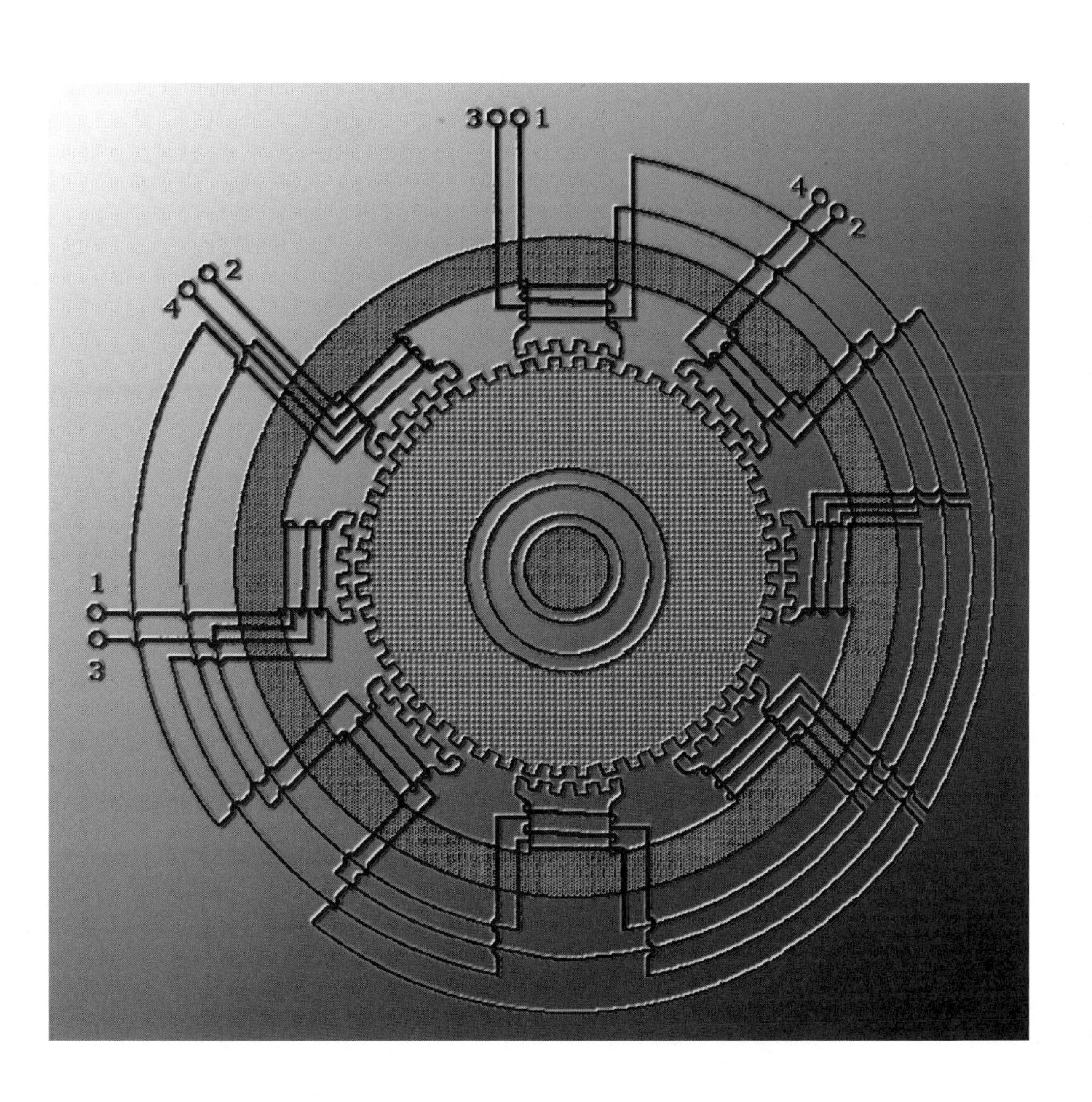

WHAT IS MOBILITY?

Now that we have seen many methods, mechanisms, and mechanical linkages for moving around in the environment, let's discuss how to compare them. A standardized set of parameters will be required, but this comparison implies that we must first answer the question: What is mobility? Is it defined by how big an obstacle the mobility system can get over, or is it how steep a slope it can climb? Perhaps it is how well, or even if, it climbs stairs? What about how deep a swamp it can get through or how wide a crevasse it can traverse? Is speed part of the equation?

The answer would seem to be all of these things, but how can we compare the mobility of an autonomous diesel powered 40-ton bulldozer to a double "A" battery powered throwable two-wheeled tail-dragger robot the size of a soda can? That seems inherently impossible. There needs to be some way to even the playing field so it is the effectiveness of the mobility system that is being compared regardless of its size. In this chapter, we'll investigate several ways of comparing mobility systems starting with a detailed discussion of ways of describing the mobility system itself. Then, the many mobility challenges the outdoor environment presents will be investigated. A set of mobility indexes that provide an at-a-glance comparison will be generated, and finally a practical specific-case comparison method will be discussed.

THE MOBILITY SYSTEM

To level the playing field, the mobility systems being compared have to be scaled to be effectively the same size. This means that there needs to be a clear definition of size. Since most robots are battery powered, energy efficiency must also be included in the comparison because there are advantages of shear power in overcoming some obstacles that battery powered vehicles simply would not have. This limited available power in most cases also limits speed. In some situations, simply going at an obstacle fast can aid in getting over it. For simplicity and because of the relatively low top speeds of battery powered robots, forward momentum is not included as a comparison of mobility methods in this book.

One last interesting criteria that bears mentioning is the vehicle's shape. This may not seem to have much bearing on mobility, and indeed in most situations it does not. However, for environments that are crowded with obstacles that cannot be driven over, where getting around things is the only way to proceed, a round or rounded shape is easier to maneuver. The round shape allows the vehicle to turn in place even if it is against a tree trunk or a wall. This ability does not exist for vehicles that are nonround. The nonround shaped vehicle can get quite inextricably stuck in a blind alley in which it tries to turn around. For most outdoor environments, simply rounding the corners somewhat is enough to aid mobility. In some environments (very dense forests or inside buildings) a fully round shape will be advantageous.

Size

Overall length and height of the mobility system directly affect a vehicle's ability to negotiate an obstacle, but width has little affect, so size is, at least, mostly length and height. The product of the overall length and height, the elevation area, seems to give a good estimate of this part of its size, but there needs to be more information about the system to accurately compare it to others. The third dimension, width, seems to be an important characteristic of size because a narrower vehicle can potentially fit through smaller openings or turn around in a narrower alley. It is, however, the turning width of the mobility system that is a better parameter to compare.

For some obstacles, just being taller is enough to negotiate them. For other obstacles, being longer works. A simple way to compare these two parameters together would be helpful. A length/height ratio or elevation area would be useful since it reduces the two parameters down to one. The length/height ratio gives an at-a-glance idea of how suited a system is to negotiating an environment that is mostly bumps and steps or one that is mostly tunnels and low passageways.

Width has little effect on getting over or under obstacles, but it does affect turning radius. It is mostly independent of the other size parameters, since the width can be expanded to increase the usable volume of the robot without affecting the robot's ability to get over or under obstacles. Since turning in place is the more critical mobility trait related to width, the right dimension to use is the diagonal length of the system. This is set by the expected minimum required turning width as determined by environmental constraints. It may, however, be necessary to make the robot wider for other reasons, like simply adding volume to the robot. A rule of thumb to use when figuring out the robot's width is to make it about 62 percent of the length of the robot.

The components of the system each have their own volume, and moving parts sweep out a sometimes larger volume. These pieces of the robot are independent of the function of the robot, but take up volume. Including the volume of the mobility system's pieces is useful. As will be seen later, weight is critical, so the total mass of the mobility system's components needs to be included. Since mass is directly related (roughly, since materials have different densities) to the volume of a given part, and volume is easier to calculate and visualize, volume negates any need to include mass.

Efficiency

Another good rule of thumb when designing anything mechanical is that less weight in the structure and moving parts is always better. This rule applies to mobile vehicles. If there were no weight restriction and little or no size restriction, then larger and therefore heavier wheels, tracks, or legs would allow a vehicle to get over more obstacles. However, weight is important for several reasons.

- The vehicle can be transported more easily.
- It takes less of its own power to move over difficult terrain and, especially, up inclines.
- Maintenance that requires lifting the vehicle is easier to perform and less dangerous.
- The vehicle is less dangerous to people in its operating area.

For all these reasons, smaller and lighter suspension and drive train components are usually the better choice for high mobility vehicles.

There are three motions in which the robot moves: fore/aft, turn, and up/down, and each requires a certain amount of power. The three axes of a standard coordinate system are labeled X, Y, and Z, but for a mobile robot, these are modified since most robot's turn before moving sideways. The robot's motions are commonly defined as traverse, turn, and climb. A robot can be doing any one, two, or all three at the same time, but the power requirements of each is so different that they can easily be listed independently by magnitude. Climbing uses the most power and turning in place usually requires more power than moving forwards or backwards. This does not apply to all mobility systems but is a good general rule.

THE ENVIRONMENT

Moving around in the relatively benign indoor environment is a simple matter, with the notable exception of staircases. The systems in this book mostly focus on systems designed for the unpredictable and highly varied outdoor environment, an environment that includes large variations in temperature, ground cover, topography, and obstacles. This environment is so varied, that only a small percentage of the problems can be listed, or the number of comparison parameters would become much too large.

Hot and cold may not seem related to mobility, but they are in that the mobility system must be efficient so it doesn't create too much heat and damage itself or nearby components when operating in a desert. The mobility system must not freeze up or jam from ice when operating in loose snow or freezing rain. As for ground cover, the mobility system might have to deal with loose dry sand, which can get everywhere and rapidly wear out bearings, or operate in muddy water. It might also have to deal with problematic topography like steep hills, seemingly impassable nearly vertical cliffs, chasms, swamps, streams, or small rivers. The mobility system will almost definitely have to travel over some or all of those topographical challenges. In addition, there are the more obvious obstacles like rocks, logs, curbs, pot holes, random bumps, stone or concrete walls, railroad rails, up and down staircases, tall wet grass, and dense forests of standing and fallen trees.

This means that the mobility system's effectiveness should be evaluated using the aforementioned parameters. How does it handle sand or pebbles? Is its design inherently difficult to seal against water? How steep an incline can it negotiate? How high an obstacle, step, or bump can it get over or onto? How wide a chasm can it cross? Somehow, all these need to be simplified to reduce the wide variety down to a manageable few.

The four categories of temperature, ground cover, topography, and obstacles can be either defined clearly or broken up into smaller more easily defined subcategories without ending up with an unmanageably large list. Let's look at each one in greater detail.

Thermal

Temperature can be divided simply into the two extremes of hot and cold. Hot relates to efficiency. A more efficient machine will have fewer problems in hot climates, but better efficiency, more importantly, means battery powered robots will run longer. Cold relates to pinch points, which can collect snow and ice, causing jamming or stalling. A useful pair of temperature-related terms to think about in a comparison of mobility systems would then be efficiency and pinch points.

Ground Cover

Ground cover is more difficult to define, especially in the case of sand, because it can't be scaled. Sand is just sand no matter what size the vehicle is (except for tiny robots of course), and mud is still mud. Driving on sand or mud would then be a function of ground pressure, the maximum force the vehicle can exert on a wheel, track, or foot, divided by the area that a supporting element places on the ground. Lower ground pressure reduces the amount the driving element sinks, thereby reducing the amount of power required to move that element. Higher ground pressure is helpful in only two cases: towing a heavy load behind the robot, and climbing steep slopes.

Robots are infrequently required to be tow trucks, but this may change as the variety of tasks they are put to widens. Climbing hills, though, is a common task. The effect of ground pressure on hill climbing can be overcome with careful tread design (independent of the mobility system), which combines the benefits of low ground pressure with high traction. Lower ground pressure should be considered to indicate a more capable mobility system.

The theory that sand and mud are not scalable can't be applied to grass however, because tall field grass really is significantly larger than short lawn grass. Grass seems benign, but it is strong enough when bunched up to throw tracks, stall wheeled vehicles, and trip walkers. These problems can be roughly related to ground pressure since a lighter pressure system would tend to ride higher on wet grass, reducing its tangling problems. The problems caused by grass, then, can be assumed to be effectively covered by the ground pressure category.

Topography

Topography can be scaled to any size making it very simple to include. It can be defined by angle of slope. The problem with angle of slope, though, is that it can be more a function of the friction of the material and the tread shape of whatever is in ground contact, than a function of the geometry of the mobility system. There are some geometries that are easier to control on steep slopes, and there are some walkers, climbers really, that can climb slopes that a wheeled or tracked vehicle simply could not get up. Negotiable slope angle is therefore important, but it should be assumed that the material in ground contact is the same no matter what type of mobility system is used.

Obstacles

Obstacles can also be scaled, but they create a special case. The effectiveness of the mobility system could be judged almost entirely by how high, relative to its elevation area, an obstacle it can negotiate. Obstacle negotiation is a little more complicated than that but it can be simplified by dividing it into three subcategories.

- Mobility system overall height to negotiable obstacle height
- System length to negotiable obstacle height
- System elevation area to negotiable obstacle height

The comparison obstacle parameters can be defined to be the height of a square step the system can climb onto and the height of a square topped wall the system can climb over without high centering, or otherwise becoming stuck.

An inverted obstacle, a chasm, is also significant. Negotiable chasm width is mostly a function of the mobility system's length, but some clever designs can vary their length somewhat, or shift their center of gravity, to facilitate crossing wider chasms. For systems that can vary their length, negotiable chasm width should be compared using the system's shortest overall length. For those that are fixed, use the overall length.

Another facet of obstacle negotiation is turning width. This is important because a mobility system with a small turning radius is more likely to be able to get out of or around confining situations. Turning width is not directly a function of vehicle width, but is defined as the narrowest alley in which the vehicle can turn around. This is in contrast to ratings given by some manufacturers that give turning radius as the radius of a circle defined as the distance from the turning point to the center of the vehicle width. This can be misleading because a very large vehicle that turns about its center can be said to have a zero-radius turning width.

A turning ability parameter must also show how tightly a vehicle can turn around a post, giving some idea how well it could maneuver in a forest of closely spaced trees. There are, then, two width parameters, alley width and turning-around-a-post width.

COMPLEXITY

A more nebulous comparison criteria that must be included in an evaluation of any practical mechanical device is its inherent complexity. A common method for judging complexity is to count the number of moving parts or joints. Ball or roller bearings are usually counted as one part of a joint although there may be 10s of balls or rollers moving inside the bearing. A problem with this method is that some parts, though moving, have very small forces on them or operate in a relatively hazard-free environment and, so, last a very long time, sometimes even longer than nonmoving parts in the same system.

A second method is to count the number of actuators since their number relates to the number of moving parts and they are the usually the source of greatest wear. The drawback of this method is that it ignores passive moving parts like linkages that may well cause problems or wear out before an actively driven part does. The first method is probably a better choice because robots are likely to be moving around in completely unpredictable environments and any moving part is equally susceptible to damage by things in the environment.

Speed and Cost

There are two other comparison parameters that could be included in a comparison of mobility methods. They are velocity of the moving vehicle and cost of the locomotion system. Moving fast over rough and unpredictable terrain places large and complex loads on a suspension system. These loads are difficult to calculate precisely because the terrain can be so unpredictable. Powerful computer simulation programs can predict a suspension system's performance with a moderate degree of accuracy, but the suspension system still must always be tested in the real world. Usually the simulation program's predictions are proven inaccurate to a significant degree. It is too difficult to accurately predict and design for a specific level of performance at speeds not very far above eight m/s to have any useful meaning. It is assumed that slowing is an acceptable way to increase mobility, and that slowing can be done with any suspension design. Mobility is not defined as getting over an obstacle at a certain speed; it is simply getting over the obstacle at whatever speed works.

Cost can be related to size, weight, and complexity. Fewer, smaller, and lighter parts are usually cheaper. The design time to get to the simplest, lightest design that meets the criteria may be longer, but the end cost will usually be less. Since cost is closely related to size, weight, and complexity, it does not need to be included in a comparison of suspension and drive train methods.

THE MOBILITY INDEX COMPARISON METHOD

Another, perhaps simpler, method is to create an index of the mobility design's capabilities listed as a set of ratios relating the mobility system's length, height, width, and possibly complexity, to a small set of terrain parameters. The most useful set would seem to be obstacle height, crevasse width, and narrowest alley in which the vehicle can turn around. Calculating the vehicle's ground pressure would cover mobility in sand or mud. The pertinent ratios would be:

- Step/Elevation Area: Negotiable step height divided by the elevation area of mobility system
- Step/System Height: Highest negotiable wall or platform, whichever is shorter, divided by mobility system height
- Crevasse/System Length: Negotiable crevasse width divided by vehicle length (in the case of variable geometry vehicles, the shortest length of the mobility system)
- System Width/Turning diameter: Vehicle width divided by outermost swept diameter of turning circle
- System Width/Turning-Around-a-Post Width: Vehicle width divided by width of path it sweeps when turning around a very thin post
- Ground Pressure

These are all set up so that a higher ratio number means theoretically higher mobility. No doubt, some mobility system designs will have very high indexes in some categories, and low indexes in others. Having a single Mobility Index for each mobility system design would be convenient, but it would be difficult to produce one that describes the system's abilities with enough detail to be useful. These six, however, should give a fairly complete at-a-glance idea of how well a certain design will perform in many situations.

THE PRACTICAL METHOD

A third way to compare mobility systems that may work well for a designer working on a specific robot design, is to calculate the total volume of everything on the robot not related to the mobility system (including the power supply), and define this volume with a realistic ratio of length, width, and height. A good place to start for the size ratios is to make the width 62 percent of the length, and the height one quarter of the length. This box represents the volume of everything the mobility system must carry.

The next step is to define the mobility requirements, allowing substantial leeway if the operating environment is not well known. The basic six parameters discussed above are a good place to start.

- Step or wall height
- Minimum tunnel height
- Crevasse width
- Maximum terrain slope
- Minimum spacing of immovable objects
- Maximum soil density

All of these need to be studied carefully to aid in determining the most effective mobility system layout to use. The more time spent doing this study, the better the mobility system choice will match the terrain's requirements.

When this study is completed, selecting and designing the mobility system is then a combination of scaling the system to the robot's box size, and meeting the mobility constraints. It should be remembered that this process will include several iterations, trial and error, and perseverance to guarantee that the best system is being incorporated. The more information that can be obtained about the operating environment, the more likely the robot will be successful. In the end, one of the more capable and versatile mobility systems, like the six-wheeled rocker bogie or the four-tracked front-flipper layouts will probably work well enough even without complete knowledge of the environment.

A generic rule of thumb for mobility system design can be extracted from the investigations done in this chapter. Relative to the size and weight of the vehicle the mobility system is carrying, make the mobility system big, light, slow, low (or movable) cg, and be sure it has sufficient treads. If all these are maximized, they will make your robot a high mobility robot.

Chapter 10 Manipulator Geometries

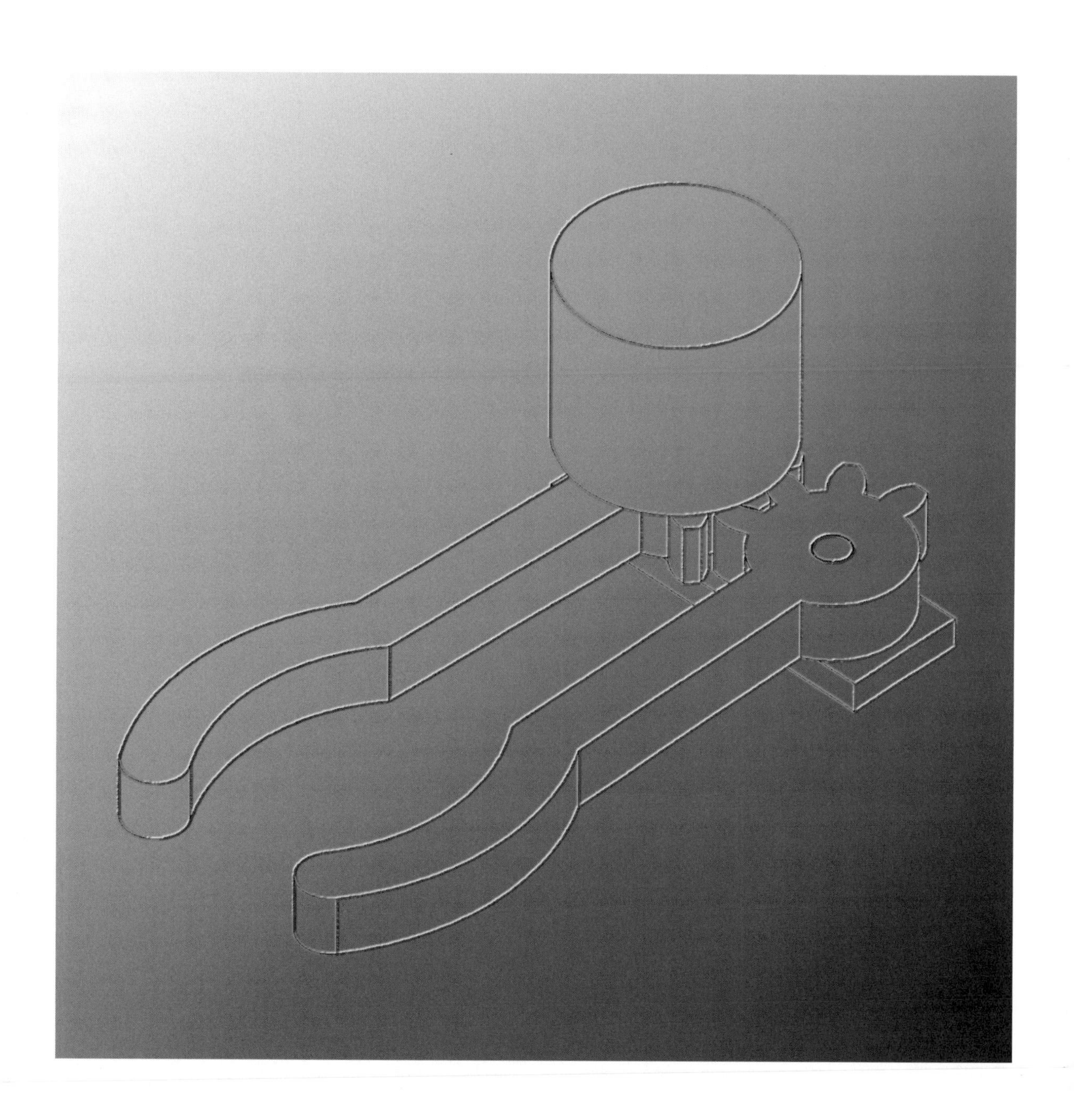

Manipulator is a fancy name for a mechanical arm. A manipulator is an assembly of segments and joints that can be conveniently divided into three sections: the arm, consisting of one or more segments and joints; the wrist, usually consisting of one to three segments and joints; and a gripper or other means of attaching or grasping. Some texts on the subject divide manipulators into only two sections, arm and gripper, but for clarity the wrist is separated out as its own section because it performs a unique function.

Industrial robots are stationary manipulators whose base is permanently attached to the floor, a table, or a stand. In most cases, however, industrial manipulators are too big, use a geometry that is not effective on a mobile robot, or lack enough sensors -(indeed many have no environmental sensors at all) to be considered for use on a mobile robot. There is a section covering them as a group because they demonstrate a wide variety of sometimes complex manipulator geometries. The chapter's main focus, however, will be on the three general layouts of the arm section of a generic manipulator, and wrist and gripper designs. A few unusual manipulator designs are also included.

It should be pointed out that there are few truly autonomous manipulators in use except in research labs. The task of positioning, orienting, and doing something useful based solely on input from frequently inadequate sensors is extremely difficult. In most cases, the manipulator is teleoperated. Nevertheless, it is theoretically possible to make a truly autonomous manipulator and their numbers are expected to increase dramatically over the next several years.

POSITIONING, ORIENTING, HOW MANY DEGREES OF FREEDOM?

In a general sense, the arm and wrist of a basic manipulator perform two separate functions, positioning and orienting. There are layouts where the wrist or arm are not distinguishable, but for simplicity, they are treated as separate entities in this discussion. In the human arm, the shoulder and elbow do the gross positioning and the wrist does the orienting. Each joint allows one degree of freedom of motion. The theoretical minimum number of degrees of freedom to reach to any location in the work envelope and orient the gripper in any orientation is six; three for location, and three for orientation. In other words, there must be at least three bending or extending motions to get position, and three twisting or rotating motions to get orientation.

Actually, the six or more joints of the manipulator can be in any order, and the arm and wrist segments can be any length, but there are only a few combinations of joint order and segment length that work effectively. They almost always end up being divided into arm and wrist. The three twisting motions that give orientation are commonly labeled pitch, roll, and yaw, for tilting up/down, twisting, and bending left/right respectively. Unfortunately, there is no easy labeling system for the arm itself since there are many ways to achieve gross positioning using extended segments and pivoted or twisted joints. A generally excepted generic description method follows.

A good example of a manipulator is the human arm, consisting of a shoulder, upper arm, elbow, and wrist. The shoulder allows the upper arm to move up and down which is considered one DOF. It allows forward and backward motion, which is the second DOF, but it also allows rotation, which is the third DOF. The elbow joint gives the forth DOF. The wrist pitches up and down, yaws left and right, and rolls, giving three DOFs in one joint. The wrist joint is actually not a very well designed joint. Theoretically the best wrist joint geometry is a ball joint, but even in the biological world, there is only one example of a true full motion ball joint (one that allows motion in two planes, and twists 360°) because they are so difficult to power and control. The human hip joint is a limited motion ball joint.

On a mobile robot, the chassis can often substitute for one or two of the degrees of freedom, usually fore/aft and sometimes to yaw the arm left/right, reducing the complexity of the manipulator significantly. Some special purpose manipulators do not need the ability to orient the gripper in all three axes, further reducing the DOF. At the other extreme, there are arms in the conceptual stage that have more than fifteen DOF.

To be thorough, this chapter will include the geometries of all the basic three DOF manipulator arms, in addition to the simpler two DOF arms specifically for use on robots. Wrists are shown separately. It is left to you to pick and match an effective combination of wrist and arm geometries to solve your specific manipulation problem. First, let's look at an unusual manipulator and a simple mechanism—perhaps the simplest mechanism for creating linear motion from rotary motion.

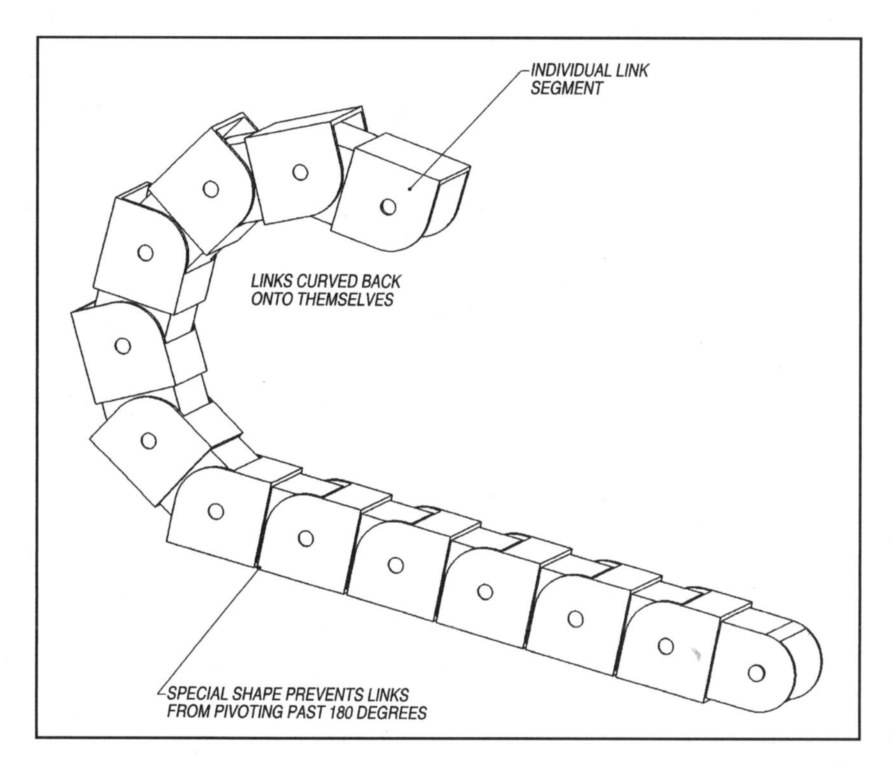

Figure 10-1 E-chain

E-Chain

An unusual chain-like device can be used as a manipulator. It is based on a flexible cable bundle carrier called E-Chain and has unique properties. The chain can be bent in only one plane, and to only one side. This allows it to cantilever out flat creating a long arm, but stored rolled up like a tape measure. It can be used as a one-DOF extension arm to reach into small confined spaces like pipes and tubes. Figure 10-1 shows a simplified line drawing of E-chain's allowable motion.

Slider Crank

The slider-crank (Figure 10-2) is usually used to get rotary motion from linear motion, as in an internal combustion engine, but it is also an efficient way to get linear motion from the rotary motion of a motor/gearbox. A connecting rod length to / crank radius ratio of four to one produces nearly linear motion of the slider over most of its stroke and is, therefore, the most useful ratio. Several other methods exist for creating

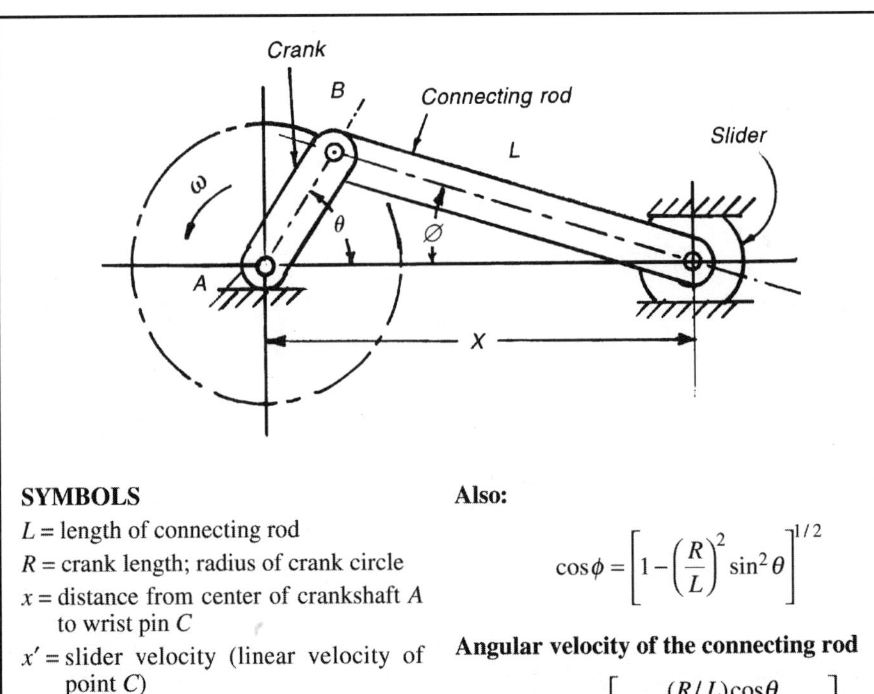

SYMBOLS

L = length of connecting rod

R = crank length; radius of crank circle

x = distance from center of crankshaft A to wrist pin C

x' = slider velocity (linear velocity of point C)

x'' = slider acceleration

θ = crank angle measured from dead center (when slider is fully extended)

ϕ = angular position of connecting rod; $\phi = 0$ when $\theta = 0$

ϕ' = connecting-rod angular velocity = $d\phi/dt$

ϕ'' = connecting-rod angular acceleration = $d^2\phi/dl^2$

ω = constant crank angle velocity

Displacement of slider

$\boldsymbol{x = L \cos \phi + R \cos \phi}$

Also:

$$\cos\phi = \left[1-\left(\frac{R}{L}\right)^2 \sin^2\theta\right]^{1/2}$$

Angular velocity of the connecting rod

$$\phi' = \omega\left[\frac{(R/L)\cos\theta}{[1-(R/L)^2\sin^2\theta]^{1/2}}\right]$$

Linear velocity of the piston

$$\frac{x'}{L} = -\omega\left[1+\frac{\phi'}{\omega}\right]\left(\frac{R}{L}\right)\sin\theta$$

Angular acceleration of the connecting rod

$$\phi'' = \frac{\omega^2(R/L)\sin\theta[(R/L)^2-1]}{[1-(R/L)^2\sin^2\theta]^{3/2}}$$

Slider acceleration

$$\frac{x''}{L} = -\omega^2\left(\frac{R}{L}\right)\left[\cos\theta+\frac{\phi''}{\omega^2}\sin\theta+\frac{\phi'}{\omega}\cos\theta\right]$$

Figure 10-2 Slider Crank

linear motion from rotary, but the slider crank is particularly effective for use in walking robots.

The motion of the slider is not linear in velocity over its full range of motion. Near the ends of its stroke the slider slows down, but the force produced by the crank goes up. This effect can be put to good use as a clamp. It can also be used to move the legs of walkers. The slider crank should be considered if linear motion is needed in a design.

In order to put the slider crank to good use, a method of calculating the position of the slider relative to the crank is helpful. The equation for calculating how far the slider travels as the crank arm rotates about the motor/gearbox shaft is: $x = L \cos \varnothing + r \cos \varnothing$.

ARM GEOMETRIES

The three general layouts for three-DOF arms are called Cartesian, cylindrical, and polar (or spherical). They are named for the shape of the volume that the manipulator can reach and orient the gripper into any position—the work envelope. They all have their uses, but as will become apparent, some are better for use on robots than others. Some use all sliding motions, some use only pivoting joints, some use both. Pivoting joints are usually more robust than sliding joints but, with careful design, sliding or extending can be used effectively for some types of tasks.

Pivoting joints have the drawback of preventing the manipulator from reaching every cubic centimeter in the work envelope because the elbow cannot fold back completely on itself. This creates dead spaces—places where the arm cannot reach that are inside the gross work volume. On a robot, it is frequently required for the manipulator to fold very compactly. Several manipulator manufacturers use a clever offset joint design depicted in Figure 10-3 that allows the arm to fold back on itself

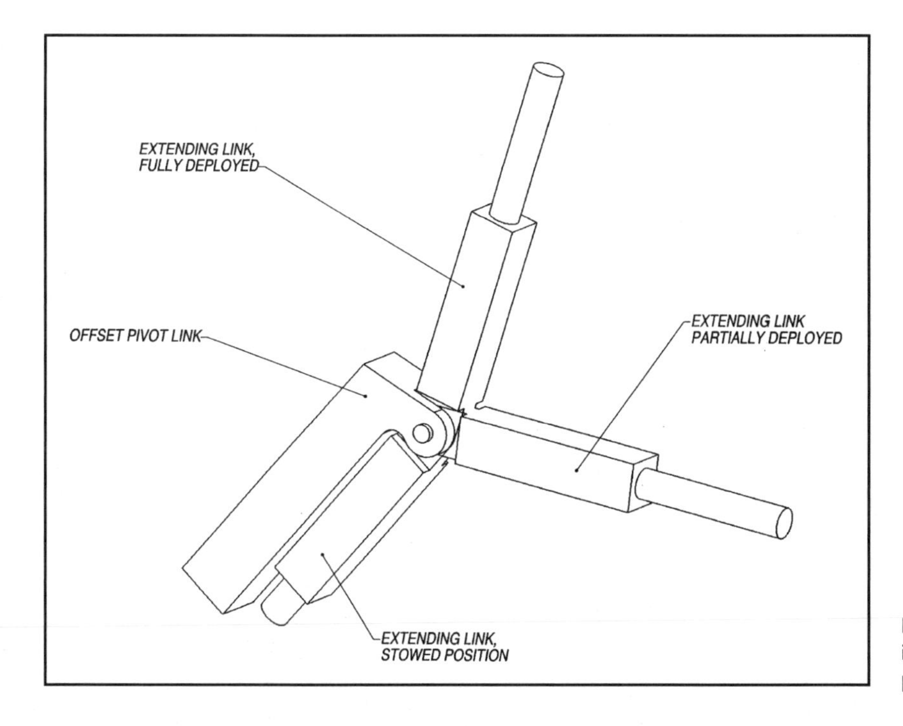

Figure 10-3 Offset joint increases working range of pivoting joints

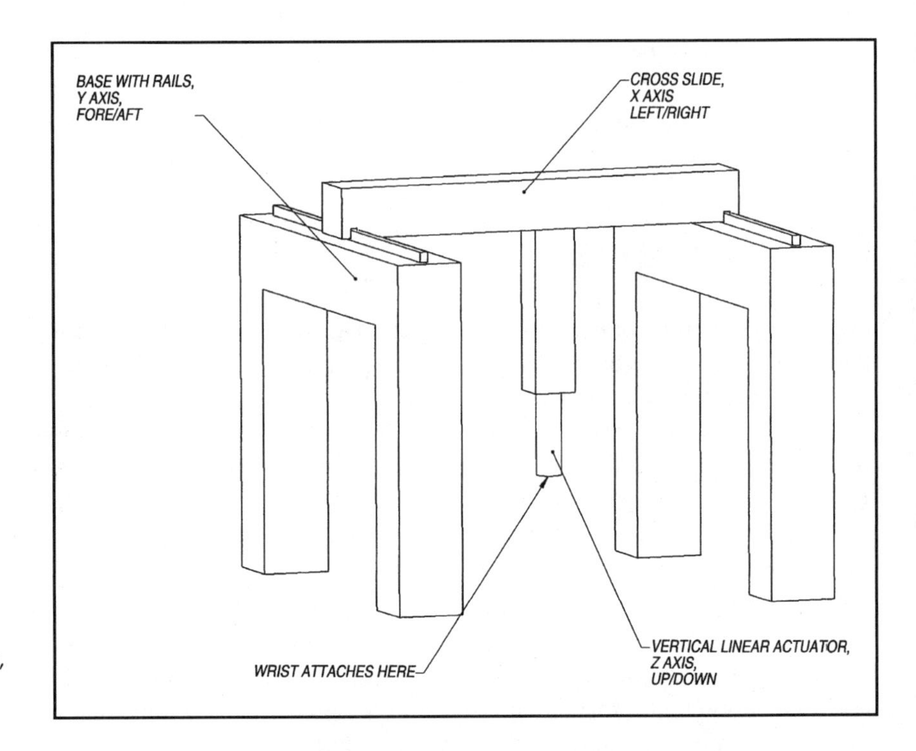

Figure 10-4 Gantry, simply supported using tracks or slides, working from outside the work envelope.

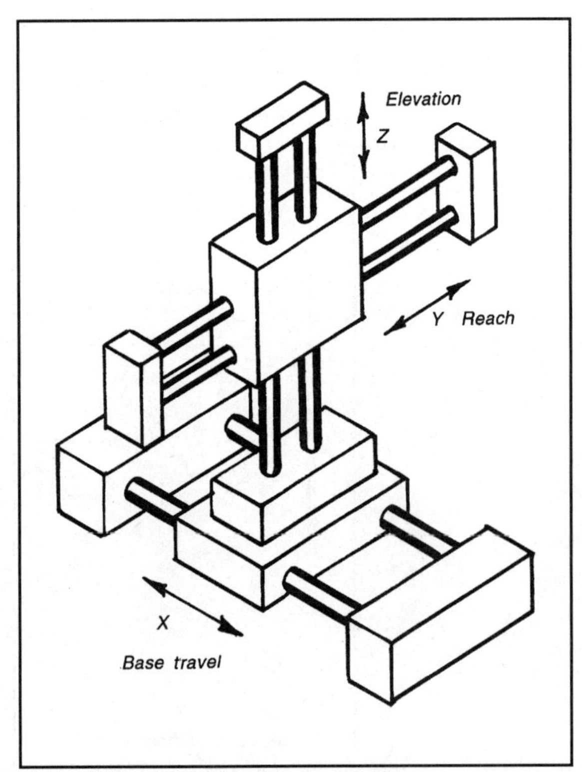

Figure 10-5 Cantilevered manipulator geometry

180°. This not only reduces the stowed volume, but also reduces any dead spaces. Many industrial robots and teleoperated vehicles use this or a similar design for their manipulators.

CARTESIAN OR RECTANGULAR

On a mobile robot, the manipulator almost always works beyond the edge of the chassis and must be able to reach from ground level to above the height of the robot's body. This means the manipulator arm works from inside or from one side of the work envelope. Some industrial gantry manipulators work from outside their work envelope, and it would be difficult indeed to use their layouts on a mobile robot. As shown in Figure 10-4, gantry manipulators are Cartesian or rectangular manipulators. This geometry looks like a three dimensional XYZ coordinate system. In fact, that is how it is controlled and how the working end moves around in the work envelope. There are two basic layouts based on how the arm segments are supported, gantry and cantilevered.

Mounted on the front of a robot, the first two DOF of a cantilevered Cartesian manipulator can move left/right and up/down; the Y-axis is not necessarily needed on a mobile robot because the robot can move back/forward. Figure 10-5 shows a cantilevered layout with three DOF. Though not the best solution to the problem of working off the front of a robot, it will work. It has the benefit of requiring a very simple control algorithm.

CYLINDRICAL

The second type of manipulator work envelope is cylindrical. Cylindrical types usually incorporate a rotating base with the first segment able to telescope or slide up and down, carrying a horizontally telescoping segment. While they are very simple to picture and the work envelope is fairly intuitive, they are hard to implement effectively because they require two linear motion segments, both of which have moment loads in them caused by the load at the end of the upper arm.

In the basic layout, the control code is fairly simple, i.e., the angle of the base, height of the first segment, and extension of the second segment. On a robot, the angle of the base can simply be the angle of the chassis of the robot itself, leaving the height and extension of the second segment. Figure 10-6 shows the basic layout of a cylindrical three-DOF manipulator arm.

A second geometry that still has a cylindrical work envelope is the SCARA design. SCARA means Selective Compliant Assembly Robot Arm. This design has good stiffness in the vertical direction, but some compliance in the horizontal. This makes it easier to get close to the right location and let the small compliance take up any misalignment. A SCARA manipulator replaces the second telescoping joint with two vertical axis-pivoting joints. Figure 10-7 shows a SCARA manipulator.

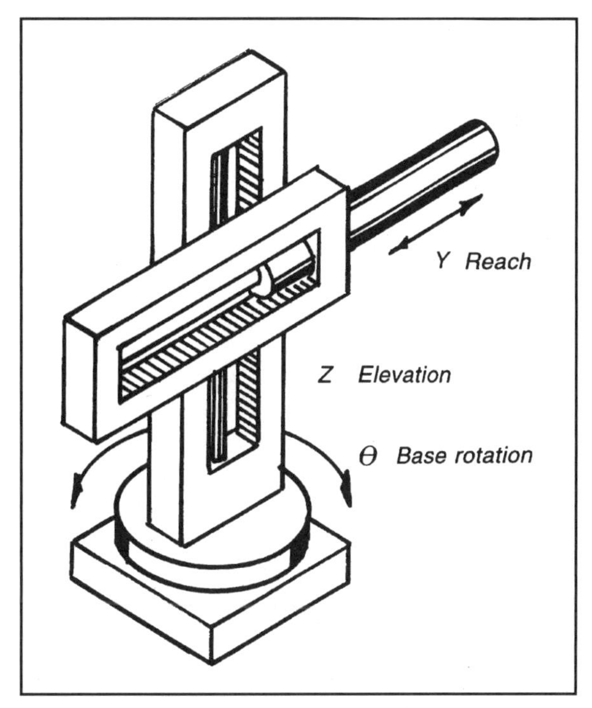

Figure 10-6 Three-DOF cylindrical manipulator

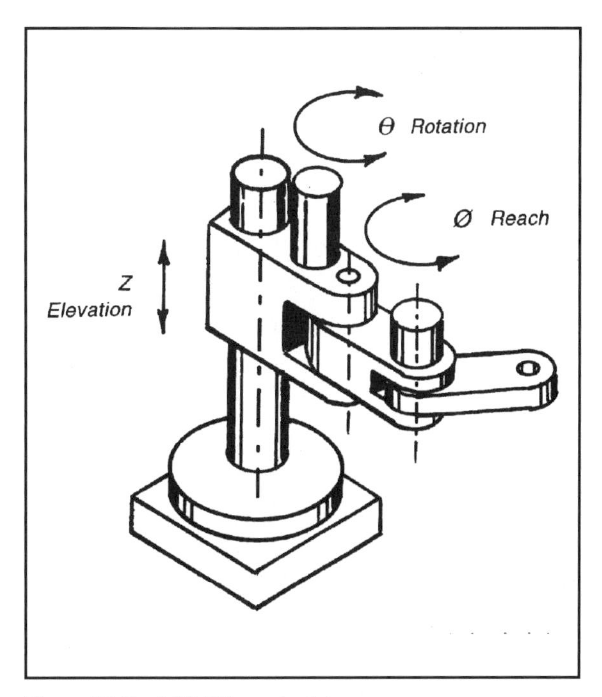

Figure 10-7 A SCARA manipulator

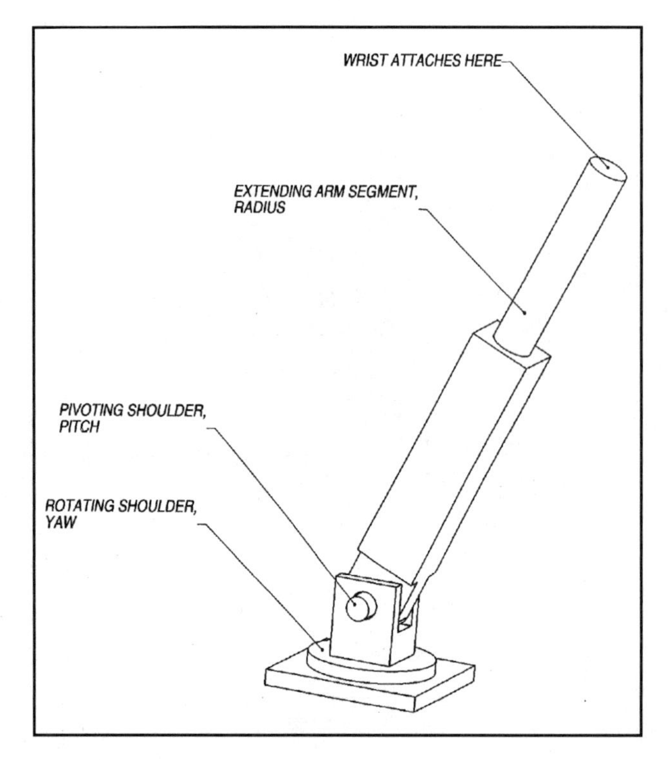

POLAR OR SPHERICAL

The third, and most versatile, geometry is the spherical type. In this layout, the work envelope can be thought of as being all around. In reality, though, it is difficult to reach everywhere. There are several ways to layout an arm with this work envelope. The most basic has a rotating base that carries an arm segment that can pitch up and down, and extend in and out (Figure 10-8). Raising the shoulder up (Figure 10-9) changes the envelope somewhat and is worth considering in some cases. Figures 10-10, 10-11, and 10-12 show variations of the spherical geometry manipulator.

Figure 10-8 Basic polar coordinate manipulator

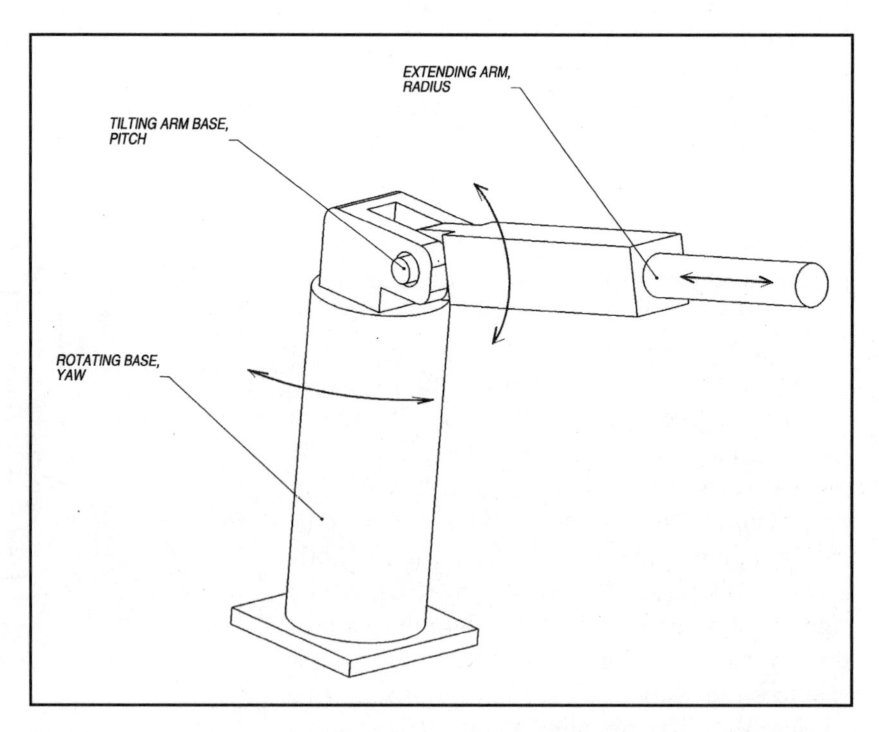

Figure 10-9 High shoulder polar coordinate manipulator with offset joint at elbow

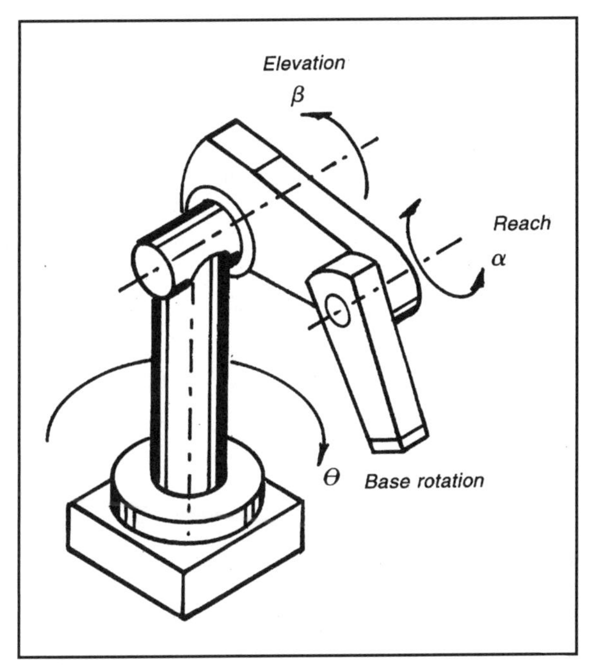

Figure 10-10 High shoulder polar coordinate manipulator with overlapping joints

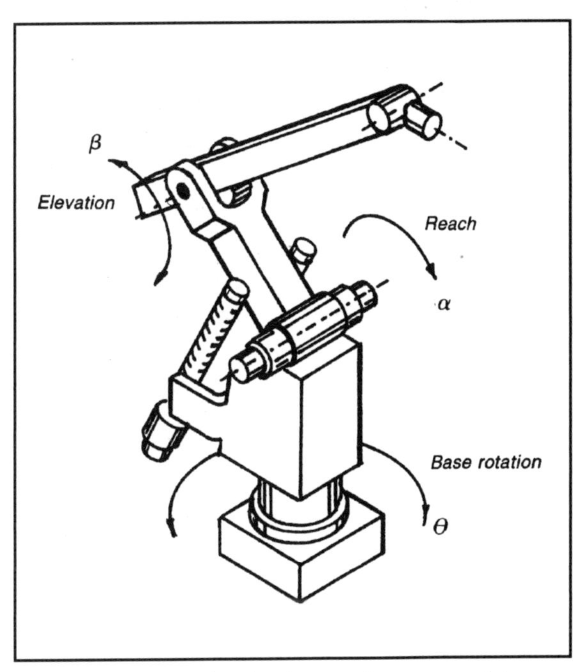

Figure 10-11 Articulated polar coordinate manipulator

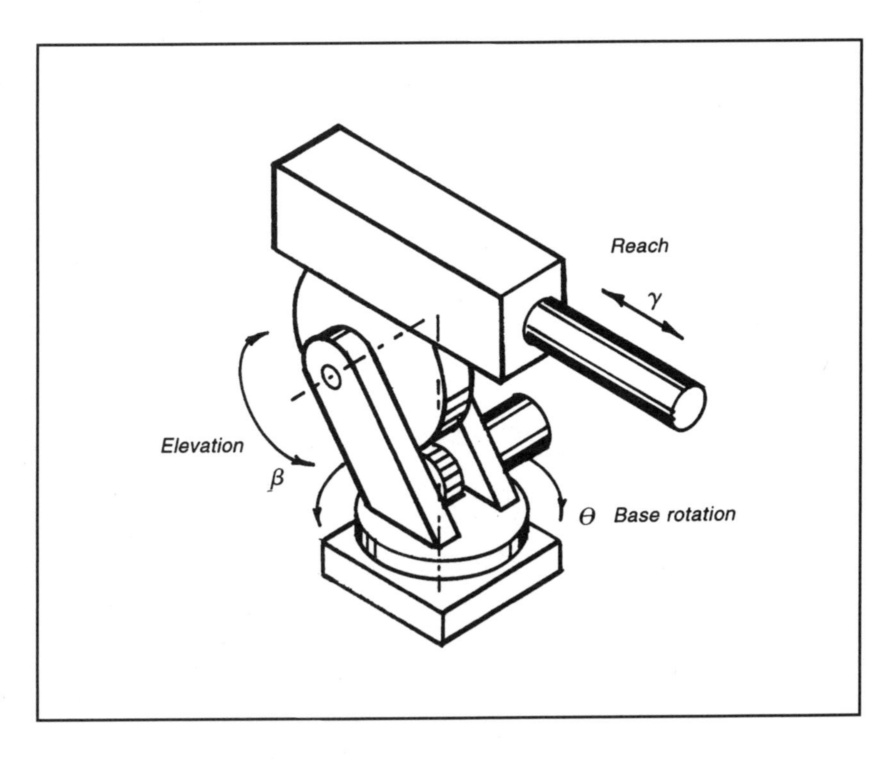

Figure 10-12 Gun turret polar coordinate manipulator

THE WRIST

The arm of the manipulator only gets the end point in the right place. In order to orient the gripper to the correct angle, in all three axes, a second set of joints is usually required—the wrist. The joints in a wrist must twist up/down, clockwise/counter-clockwise, and left/right. They must pitch, roll, and yaw respectively. This can be done all-in-one using a ball-in-socket joint like a human hip, but controlling and powering this type is difficult.

Most wrists consist of three separate joints. Figures 10-13, 10-14, and 10-15 depict one, two, and three-DOF basic wrists each building on the previous design. The order of the degrees of freedom in a wrist has a large effect on the wrist's functionality and should be chosen carefully, especially for wrists with only one or two DOF.

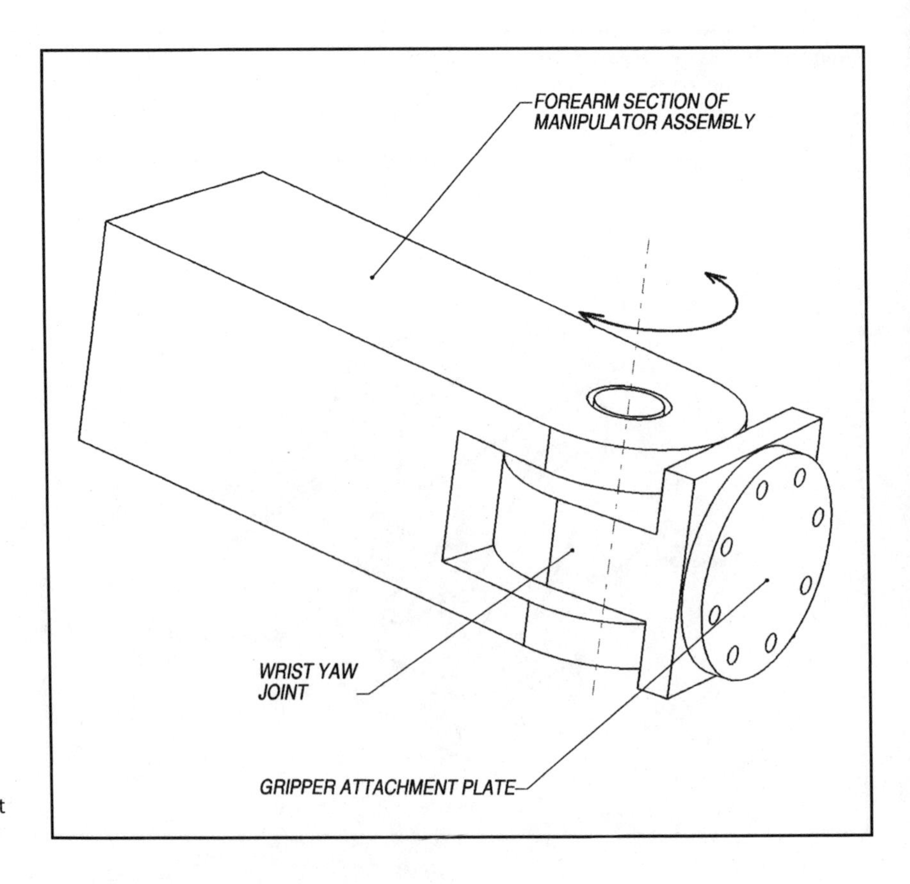

Figure 10-13 Single-DOF wrist (yaw)

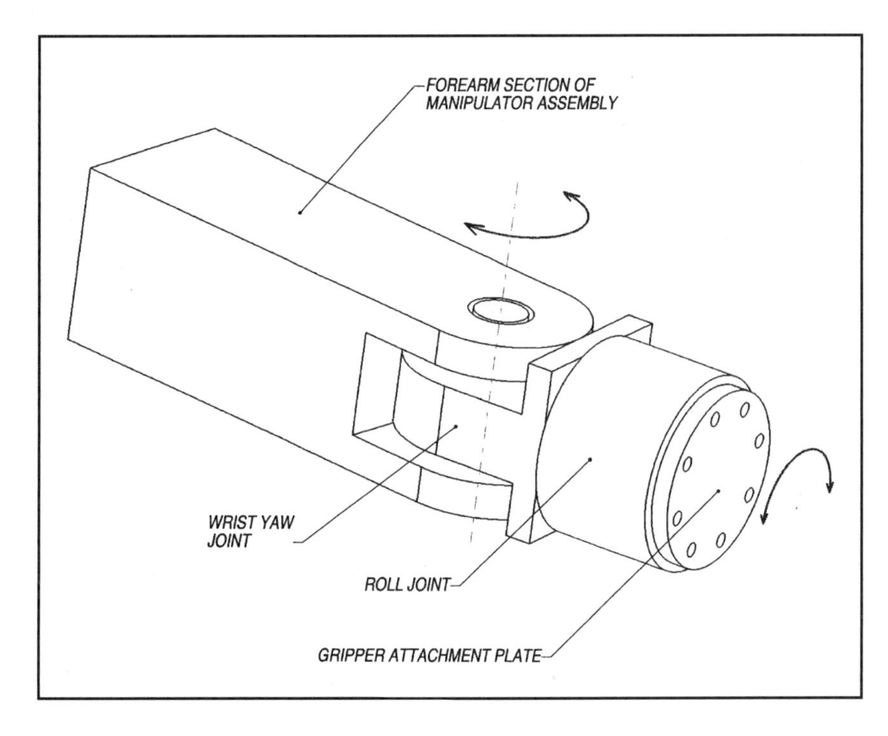

Figure 10-14 Two-DOF wrist (yaw and roll)

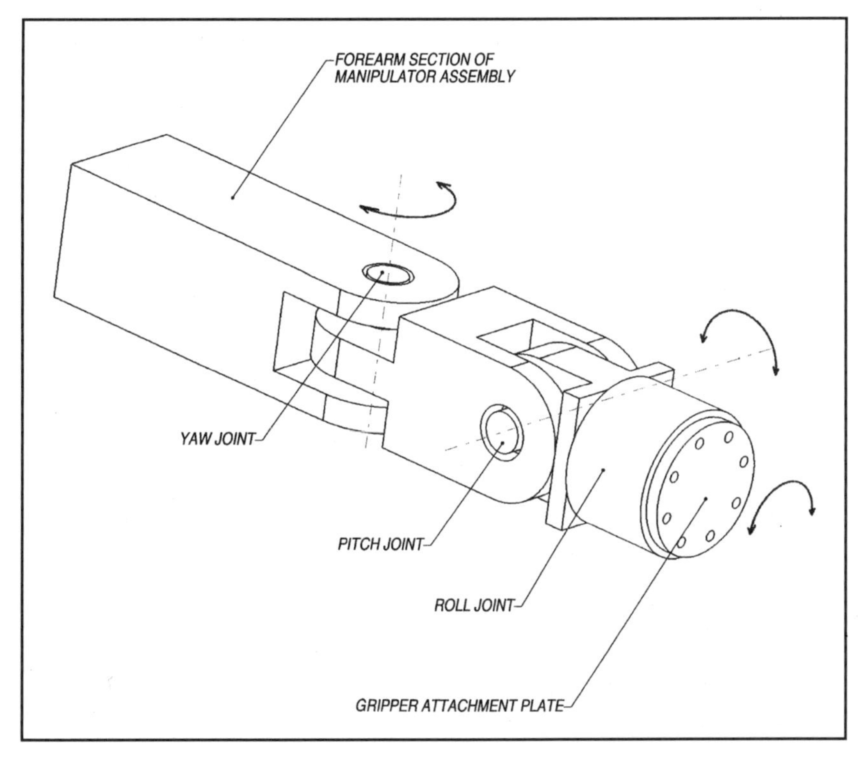

Figure 10-15 Three-DOF wrist (yaw, roll, and pitch)

GRIPPERS

The end of the manipulator is the part the user or robot uses to affect something in the environment. For this reason it is commonly called an end-effector, but it is also called a gripper since that is a very common task for it to perform when mounted on a robot. It is often used to pick up dangerous or suspicious items for the robot to carry, some can turn doorknobs, and others are designed to carry only very specific things like beer cans. Closing too tightly on an object and crushing it is a major problem with autonomous grippers. There must be some way to tell how hard is enough to hold the object without dropping it or crushing it. Even for semi-autonomous robots where a human controls the manipulator, using the gripper effectively is often difficult. For these reasons, gripper design requires as much knowledge as possible of the range of items the gripper will be expected to handle. Their mass, size, shape, and strength, etc. all must be taken into account. Some objects require grippers that have many jaws, but in most cases, grippers have only two jaws and those will be shown here.

There are several basic types of gripper geometries. The most basic type has two simple jaws geared together so that turning the base of one

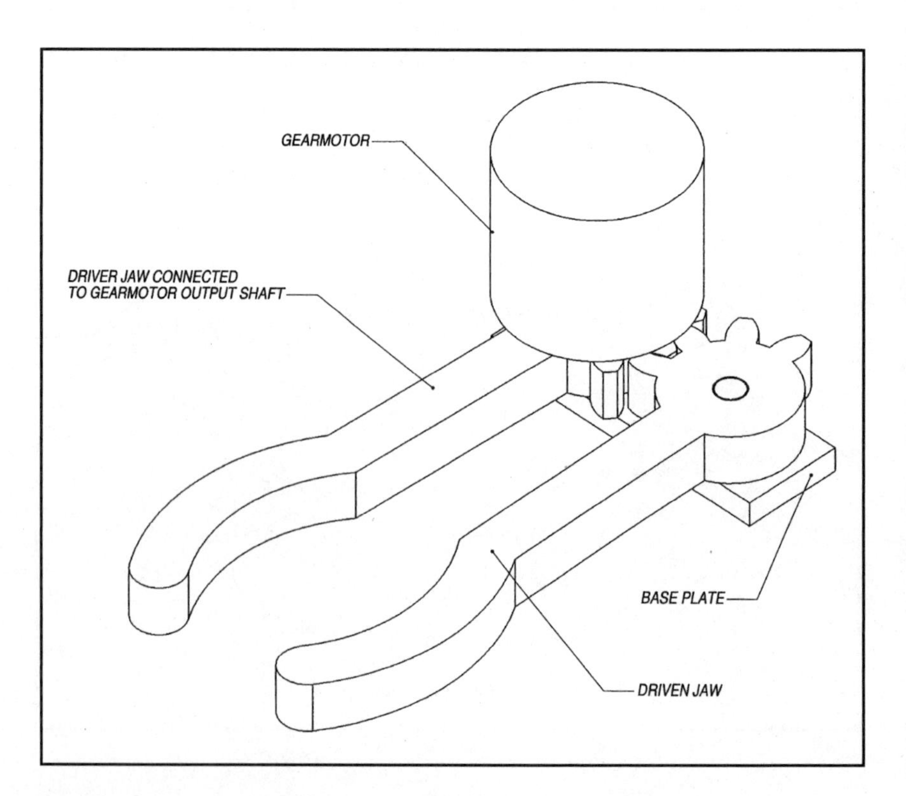

Figure 10-16 Simple direct drive swinging jaw

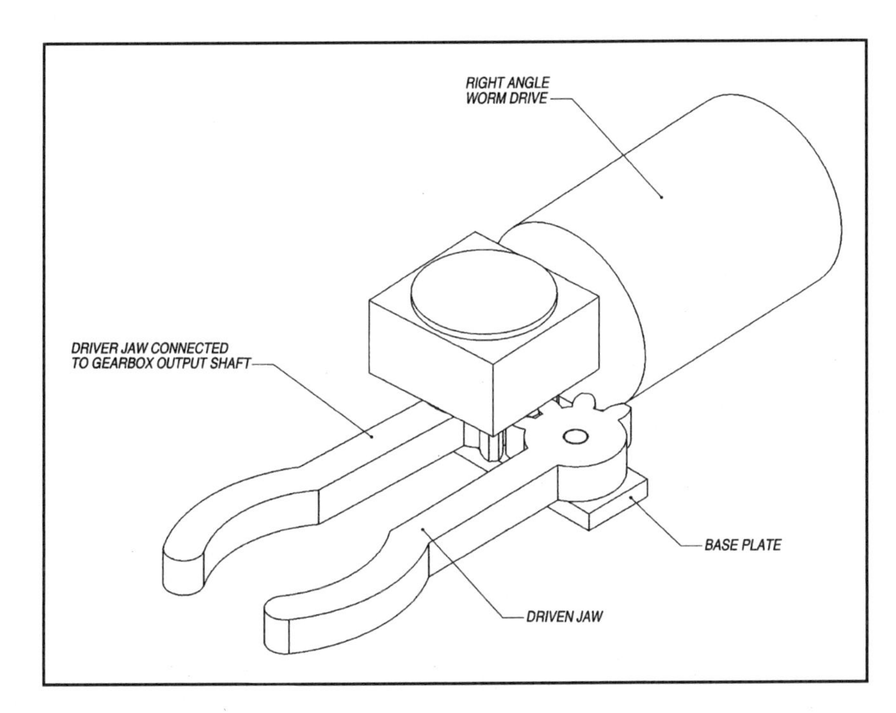

Figure 10-17 Simple direct drive through right angle worm drive gearmotor

turns the other. This pulls the two jaws together. The jaws can be moved through a linear actuator or can be directly mounted on a motor gearbox's output shaft (Figure 10-16), or driven through a right angle drive (Figure 10-17) which places the drive motor further out of the way of the gripper. This and similar designs have the drawback that the jaws are always at an angle to each other which tends to push the thing being grabbed out of the jaws.

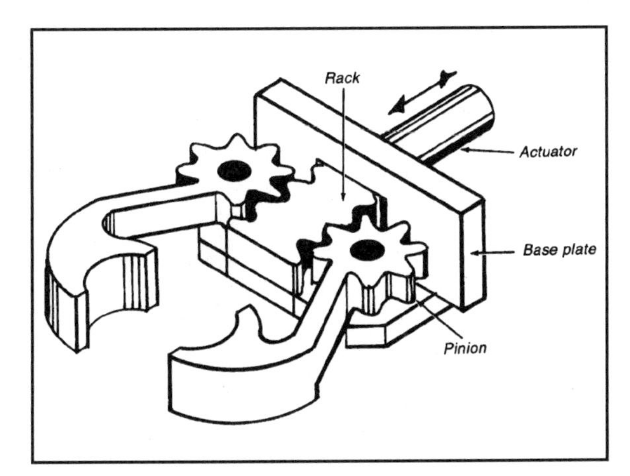

Figure 10-18 Rack and pinion drive gripper

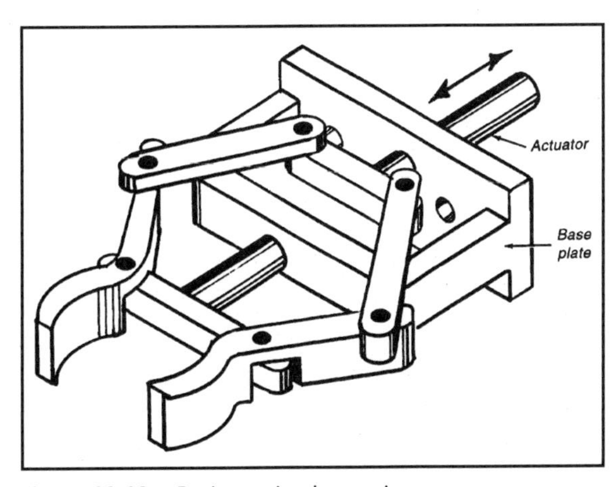

Figure 10-19 Reciprocating lever gripper

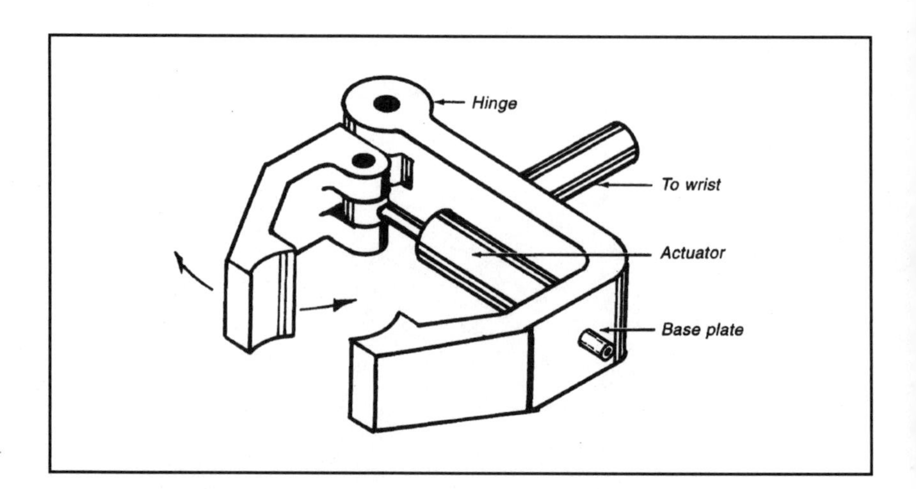

Figure 10-20 Linear actuator direct drive gripper

A more effective jaw layout is the parallel jaw gripper. One possible layout adds a few more links to the basic two fingers to form a four-bar linkage which holds the jaws parallel to each other easing the sometimes very difficult task of keeping the thing being grabbed in the gripper until it closes. Another way to get parallel motion is to use a linear actuator to move one or both jaws directly towards and away from each other. These layouts are shown in Figures 10-21, 10-22, and 10-23.

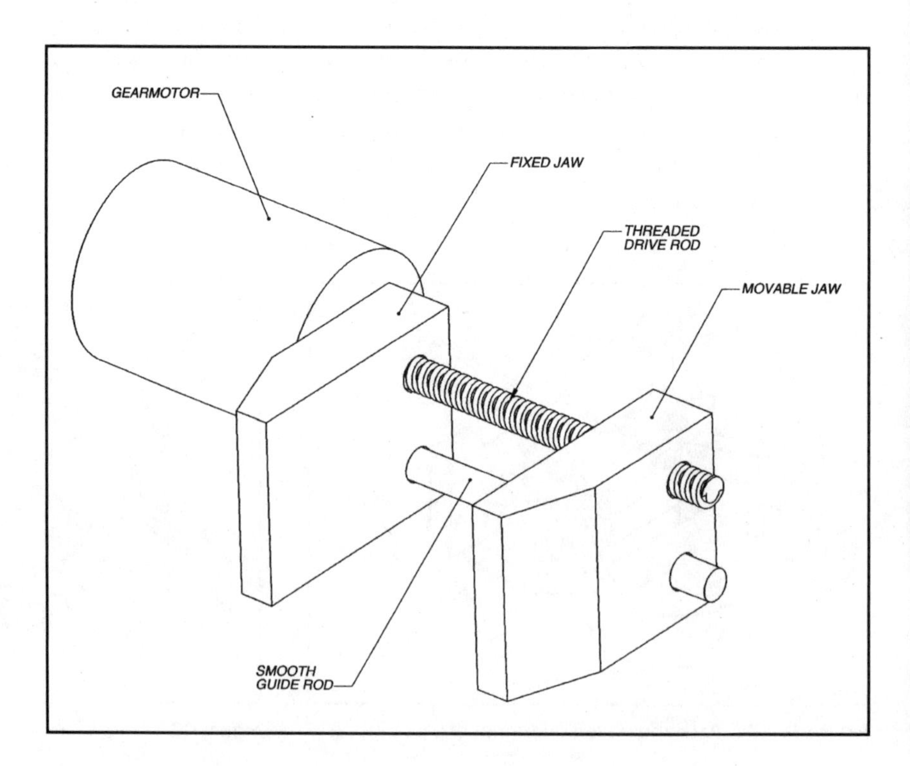

Figure 10-21 Parallel jaw on linear slides

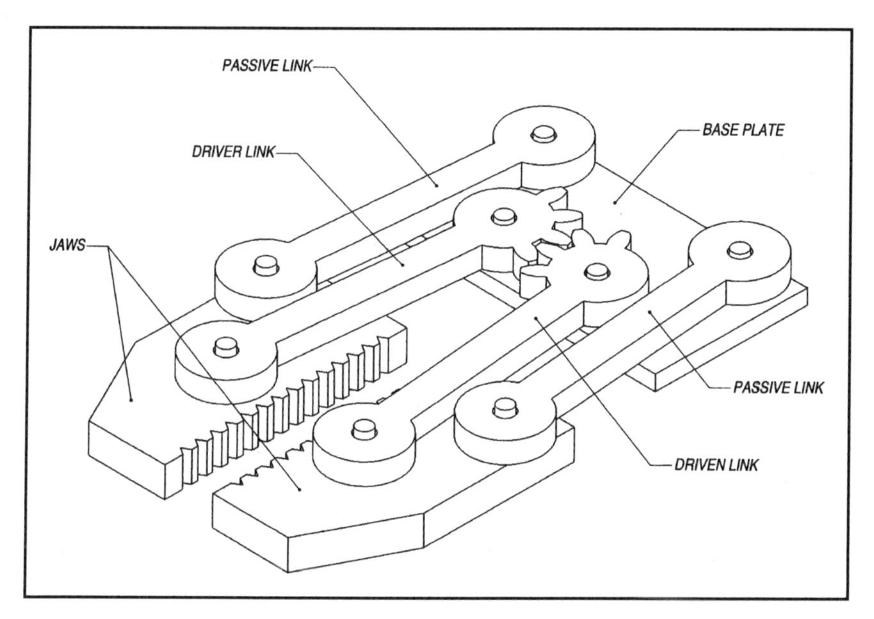

Figure 10-22 Parallel jaw using four-bar linkage

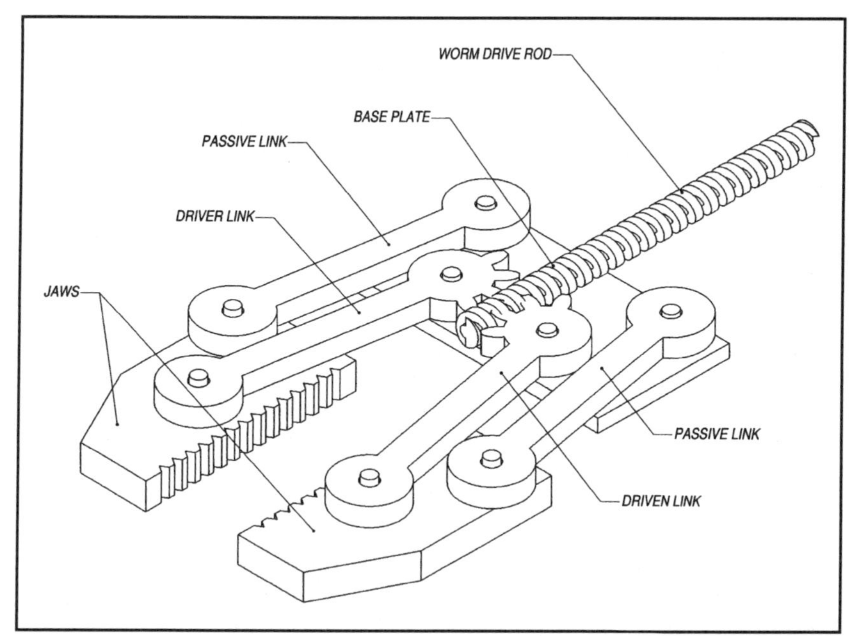

Figure 10-23 Parallel jaw using four-bar linkage and linear actuator

PASSIVE PARALLEL JAW USING CROSS TIE

Twin four-bar linkages are the key components in this long mechanism that can grip with a constant weight-to-grip force ratio any object that fits

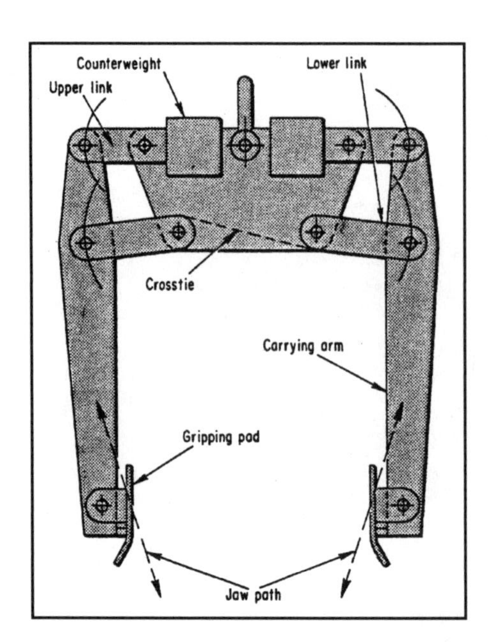

Figure 10-24 Passive parallel jaw using cross tie

within its grip range. The long mechanism relies on a cross-tie between the two sets of linkages to produce equal and opposite linkage movement. The vertical links have extensions with grip pads mounted at their ends, while the horizontal links are so proportioned that their pads move in an inclined straight-line path. The weight of the load being lifted, therefore, wedges the pads against the load with a force that is proportional to the object's weight and independent of its size.

Some robots are designed to do one specific task, to carry one specific object, or even to latch onto some specific thing. Installing a dedicated knob or ball end on the object simplifies the gripping task using this mating one-way connector. In many cases, a joint like this can be used independently of any manipulator.

PASSIVE CAPTURE JOINT WITH THREE DEGREES OF FREEDOM

New joint allows quick connection between any two structural elements where rotation in all three axes is desired.

Marshall Space Flight Center, Alabama

A new joint, proposed for use on an attachable debris shield for the International Space Station Service Module, has potential for commer-

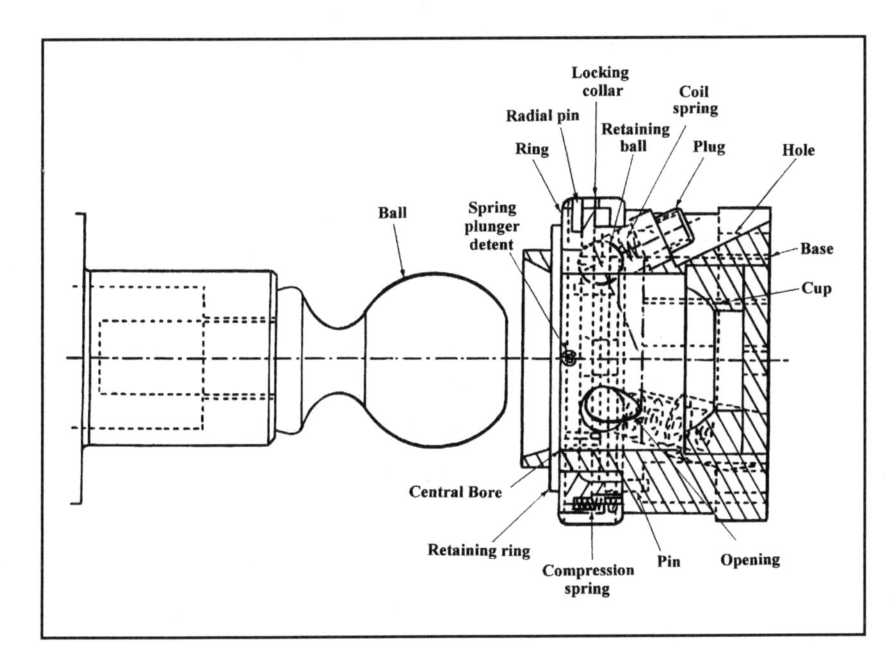

Figure 10-25 The three-degrees-of-freedom capability of the passive capture joint provides for quick connect and disconnect operations.

cial use in situations where hardware must be assembled and disassembled on a regular basis.

This joint can be useful in a variety of applications, including replacing the joints commonly used on trailer-hitch tongues and temporary structures, such as crane booms and rigging. Other uses for this joint include assembly of structures where simple rapid deployment is essential, such as in space, undersea, and in military structures.

This new joint allows for quick connection between any two structural elements where it is desirable to have rotation in all three axes. The joint can be fastened by moving the two halves into position. The joint is then connected by inserting the ball into the bore of the base. When the joint ball is fully inserted, the joint will lock with full strength. Release of this joint involves only a simple movement and rotation of one part. The joint can then be easily separated.

Most passive capture devices allow only axial rotation when fastened—if any movement is allowed at all. Manually- or power-actuated active joints require an additional action, or power and control signal, as well as a more complex mechanism.

The design for this new joint is relatively simple. It consists of two halves, a ball mounted on a stem (such as those on a common trailer-hitch ball) and a socket. The socket contains all the moving parts and is the important part of this invention. The socket also has a base, which contains a large central cylindrical bore ending in a spherical cup.

This work was done by Bruce Weddendorf and Richard A. Cloyd of the **Marshall Space Flight Center**.

INDUSTRIAL ROBOTS

The programmability of the industrial industrial robot using computer software makes it both flexible in the way it works and versatile in the range of tasks it can accomplish. The most generally accepted definition of an industrial robot is a reprogrammable, multi-function manipulator designed to move material, parts, tools, or specialized devices through variable programmed motions to perform a variety of tasks. Industrial robots can be floor-standing, benchtop, or mobile.

Industrial robots are classified in ways that relate to the characteristics of their control systems, manipulator or arm geometry, and modes of operation. There is no common agreement on or standardizations of these designations in the literature or among industrial robot specialists around the world.

A basic industrial robot classification relates to overall performance and distinguishes between limited and unlimited sequence control. Four classes are generally recognized: limited sequence and three forms of unlimited sequence—point-to-point, continuous path, and controlled path. These designations refer to the path taken by the end effector, or tool, at the end of the industrial robot arm as it moves between operations.

Another classification related to control is *nonservoed* versus *servoed*. Nonservoed implies open-loop control, or no closed-loop feedback, in the system. By contrast, servoed means that some form of closed-loop feedback is used in the system, typically based on sensing velocity, position, or both. Limited sequence also implies nonservoed control while unlimited sequence can be achieved with point-to-point, continuous-path, or controlled-path modes of operation.

Industrial robots are powered by electric, hydraulic, or pneumatic motors or actuators. Electric motor power is most popular for the major axes of floor-standing industrial industrial robots today. Hydraulic-drive industrial robots are generally assigned to heavy-duty lifting applications. Some electric and hydraulic industrial robots are equipped with pneumatic-controlled tools or end effectors.

The number of degrees of freedom is equal to the number of axes of an industrial robot, and is an important indicator of its capability. Limited-sequence industrial robots typically have only two or three degrees of freedom, but point-to-point, continuous-path, and controlled-path industrial robots typically have five or six. Two or three of those may be in the wrist or end effector.

Most heavy-duty industrial robots are floor-standing. Others in the same size range are powered by hydraulic motors. The console contains a digital computer that has been programmed with an operating system and applications software so that it can perform the tasks assigned to it.

Some industrial robot systems also include training pendants—handheld pushbutton panels connected by cable to the console that permit direct control of the industrial robot .

The operator or programmer can control the movements of the industrial robot arm or manipulator with pushbuttons or other data input devices so that it is run manually through its complete task sequence to program it. At this time adjustments can be made to prevent any part of the industrial robot from colliding with nearby objects.

There are also many different kinds of light-duty assembly or pick-and-place industrial robots that can be located on a bench. Some of these are programmed with electromechanical relays, and others are programmed by setting mechanical stops on pneumatic motors.

Industrial Robot Advantages

The industrial robot can be programmed to perform a wider range of tasks than dedicated automatic machines, even those that can accept a wide selection of different tools. However, the full benefits of an industrial robot can be realized only if it is properly integrated with the other machines human operators, and processes. It must be evaluated in terms of cost-effectiveness of the performance or arduous, repetitious, or dangerous tasks, particularly in hostile environments. These might include high temperatures, high humidity, the presence of noxious or toxic fumes, and proximity to molten metals, welding arcs, flames, or high-voltage sources.

The modern industrial robot is the product of developments made in many different engineering and scientific disciplines, with an emphasis on mechanical, electrical, and electronic technology as well as computer science. Other technical specialties that have contributed to industrial robot development include servomechanisms, hydraulics, and machine design. The latest and most advanced industrial robots include dedicated digital computers.

The largest number of industrial robots in the world are limited-sequence machines, but the trend has been toward the electric-motor powered, servo-controlled industrial robots that typically are floor-standing machines. Those industrial robots have proved to be the most cost-effective because they are the most versatile.

Trends in Industrial Robots

There is evidence that the worldwide demand for industrial robots has yet to reach the numbers predicted by industrial experts and visionaries some twenty years ago. The early industrial robots were expensive and temperamental, and they required a lot of maintenance. Moreover, the software was frequently inadequate for the assigned tasks, and many industrial robots were ill-suited to the tasks assigned them.

Many early industrial customers in the 1970s and 1980s were disappointed because their expectations had been unrealistic; they had underestimated the costs involved in operator training, the preparation of applications software, and the integration of the industrial robots with other machines and processes in the workplace.

By the late 1980s, the decline in orders for industrial robots drove most American companies producing them to go out of business, leaving only a few small, generally unrecognized manufacturers. Such industrial giants as General Motors, Cincinnati Milacron, General Electric, International Business Machines, and Westinghouse entered and left the field. However, the Japanese electrical equipment manufacturer Fanuc Robotics North America and the Swedish-Swiss corporation Asea Brown Boveri (ABB) remain active in the U.S. robotics market today.

However, sales are now booming for less expensive industrial robots that are stronger, faster, and smarter than their predecessors. Industrial robots are now spot-welding car bodies, installing windshields, and doing spray painting on automobile assembly lines. They also place and remove parts from annealing furnaces and punch presses, and they assemble and test electrical and mechanical products. Benchtop industrial robots pick and place electronic components on circuit boards in electronics plants, while mobile industrial robots on tracks store and retrieve merchandise in warehouses.

The dire predictions that industrial robots would replace workers in record numbers have never been realized. It turns out that the most cost-effective industrial robots are those that have replaced human beings in dangerous, monotonous, or strenuous tasks that humans do not want to do. These activities frequently take place in spaces that are poorly ventilated, poorly lighted, or filled with noxious or toxic fumes. They might also take place in areas with high relative humidity or temperatures that are either excessively hot or cold. Such places would include mines, foundries, chemical processing plants, or paint-spray facilities.

Management in factories where industrial robots were purchased and installed for the first time gave many reasons why they did this despite the disappointments of the past twenty years. The most frequent reasons were the decreasing cost of powerful computers as well as the simplification of both the controls and methods for programming the computers. This has been due, in large measure, to the declining costs of more powerful microprocessors, solid-state and disk memory, and applications software.

However, overall system costs have not declined, and there have been no significant changes in the mechanical design of industrial robots during the industrial robot's twenty-year "learning curve" and maturation period.

The shakeout of American industrial robot manufacturers has led to the near domination of the world market for industrial robots by the Japanese manufacturers who have been in the market for most of the past twenty years. However, this has led to de facto standardization in industrial robot geometry and philosophy along the lines established by the Japanese manufacturers. Nevertheless, industrial robots are still available in the same configurations that were available fifteen to twenty years ago, and there have been few changes in the design of the end-use tools that mount on the industrial robot's "hand" for the performance of specific tasks (e.g., parts handling, welding, painting).

Industrial Robot Characteristics

Load-handling capability is one of the most important factors in an industrial robot purchasing decision. Some can now handle payloads of as much as 200 pounds. However, most applications do not require the handling of parts that are as heavy as 200 pounds. High on the list of other requirements are "stiffness"—the ability of the industrial robot to perform the task without flexing or shifting; accuracy—the ability to perform repetitive tasks without deviating from the programmed dimensional tolerances; and high rates of acceleration and deceleration.

The size of the manipulator or arm influences accessibility to the assigned floor space. Movement is a key consideration in choosing an industrial robot. The industrial robot must be able to reach all the parts or tools needed for its application. Thus the industrial robot's working range or envelope is a critical factor in determining industrial robot size.

Most versatile industrial robots are capable of moving in at least five degrees of freedom, which means they have five axes. Although most tasks suitable for industrial robots today can be performed by industrial robots with at least five axes, industrial robots with six axes (or degrees of freedom) are quite common. Rotary base movement and both radial and vertical arm movement are universal. Rotary wrist movement and wrist bend are also widely available. These movements have been designated as roll and pitch by some industrial robot manufacturers. Wrist yaw is another available degree of freedom.

More degrees of freedom or axes can be added externally by installing parts-handling equipment or mounting the industrial robot on tracks or rails so that it can move from place to place. To be most effective, all axes should be servo-driven and controlled by the industrial robot's computer system.

Chapter 11 Proprioceptive and Environmental Sensing Mechanisms and Devices

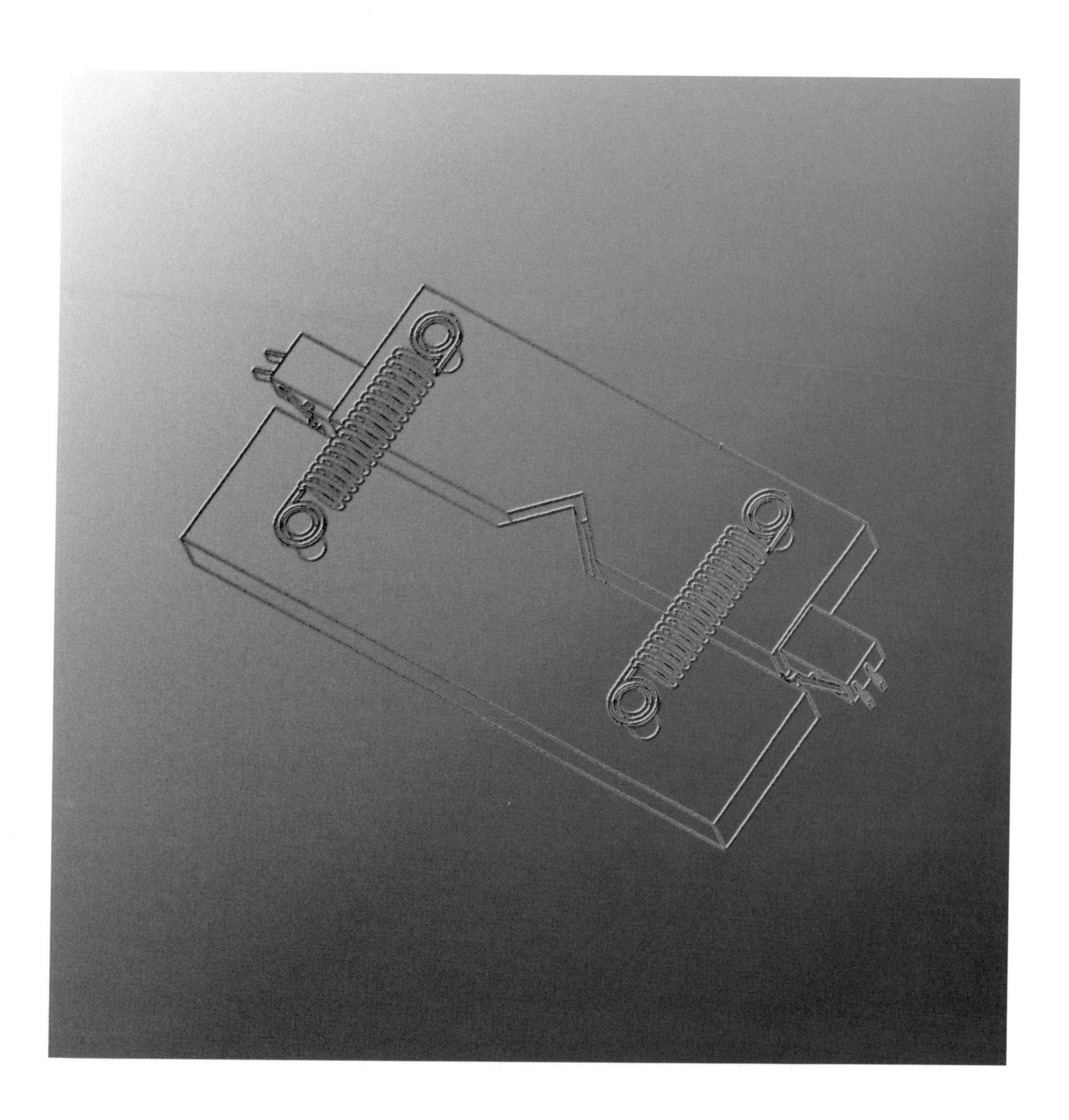

Mechanical limit switches are devices that sense objects by being either directly or indirectly touched by the object. Most use a button, lever, whisker, or slide as their local sensor. Two other types that warrant their own categories are the magnetic reed switch and the membrane switch, which is much like a long button actuated switch. On a robot, the switch alone can be the whole sensor, but in most cases the switch makes up only a part of a sensor package.

The limit switch can be thought of as a device that has at least one input and one output. The input is the button, lever, whisker, or slide (or for the magnetic type, anything ferrous nearby). The output is almost always closing or opening an electric circuit. There are several other types of limit switches whose inputs and outputs are different than those discussed above, but only those that sense by direct contact or use magnets will be included here. Other types are not strictly mechanical and are more complex and beyond the scope of this book.

In a robot, there are two general categories of things that the robot's microprocessor needs to know about, many of which can be sensed by mechanical limit switches. The categories are proprioceptive and environmental. Proprioceptive things are part of the robot itself like the position of the various segments of its manipulator, the temperature of its motors or transistors, the current going to its motors, the position of its wheels, etc. Environmental things are generally outside the robot like nearby objects, ambient temperature, the slope of the surface the robot is driving on, bumps, or drop-offs, etc. This is an over-simplified explanation because in several cases, the two categories overlap in one way or another. For instance, when the bumper bumps up against an object, the object is in the environment (environmental sensing) but the bumper's motion and location, relative to the robot, is detected by a limit switch mounted inside the robot's body (proprioceptive sensing). In this book, anything that is detected by motion of the robot's parts is considered proprioceptive, whether the thing being sensed is part of the robot or not.

These two categories subject the switch to very different problems. Proprioceptive sensors usually live in a fairly controlled environment. The things around them and the things they sense are all contained inside the robot, making their shape unchanging, moving generally in the same direction, and with the same forces. This makes them easier to implement than environmental sensors that must detect a whole range of objects coming from unpredictable directions with a wide range of forces. Environmental sensing switches, especially the mechanical type, are often very difficult to make effective and care must be taken in their design and layout.

Mechanical limit switches come in an almost infinite variety of shapes, sizes, functions, current carrying capacity, and robustness. This chapter will focus on layouts and tripping mechanisms in addition to the switches themselves. Some switch layouts have the lever, button, whisker, or slide directly moved by the thing being sensed. Others consist of several components which include one or more switches and some device to trip them. In fact, several of the tripping devices shown in this chapter can also be used effectively with non-mechanical switches, like break-beam light sensors. The following figures show several basic layouts. These can be varied in many ways to produce what is needed for a specific application.

The simplest form of mechanical limit switch is the button switch (Figure 11-1) It has a button protruding from one side that moves in and out. This opens and closes the electrical contacts inside the switch. The button switch is slightly less robust than the other switch designs because the button must be treated with care or else it might be pushed too hard, breaking the internal components, or not quite inline with its intended travel direction, breaking the button off. It is, theoretically, the most sensitive, since the button directly moves the contacts without any other mechanism in the loop. Some very precise button limit switches can detect motions as small as 1mm.

The lever switch is actually a derivative of the button switch and is the most common form of limit switch. The lever comes in an almost limitless variety of shapes and sizes. Long throw, short throw, with a roller on the end, with a high friction bumper on the end, single direction, and bidirection are several of the common types. Figure 11-2 shows the basic layout. Install whatever lever is needed for the application.

The whisker or wobble switch is shown separately in Figure 11-3 even though it is really just another form of lever switch. The whisker looks and functions very much like the whiskers on a cat and, like a cat, the whisker directly senses things in the environment. This makes it

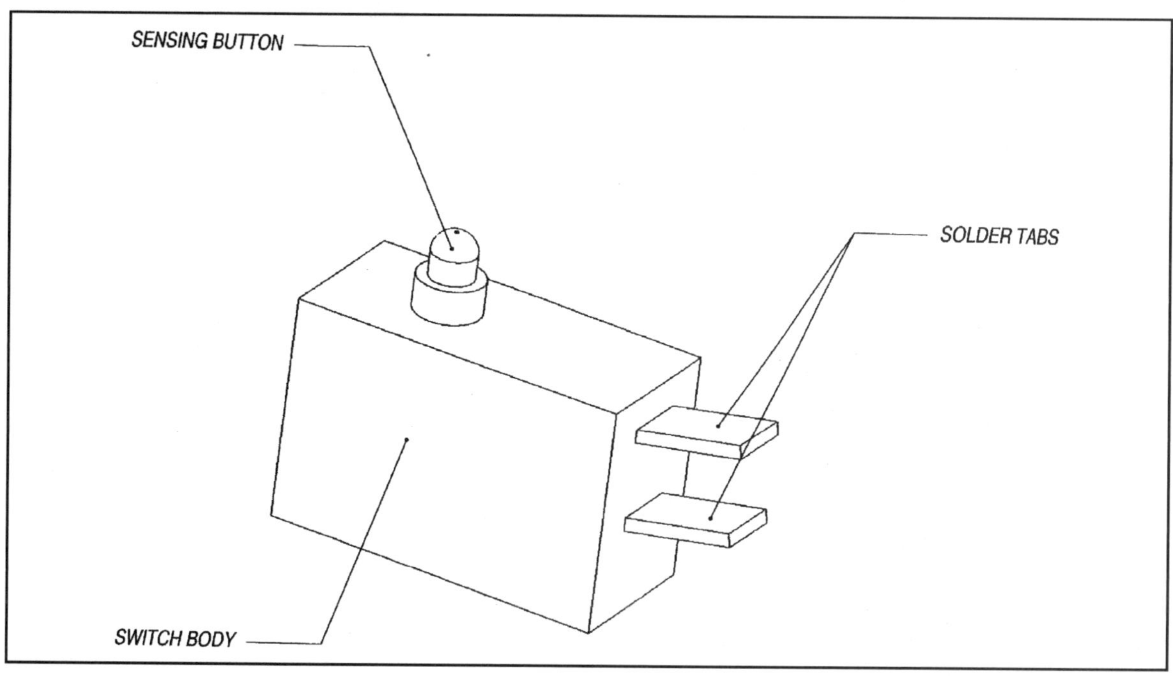

Figure 11-1 Button Switch

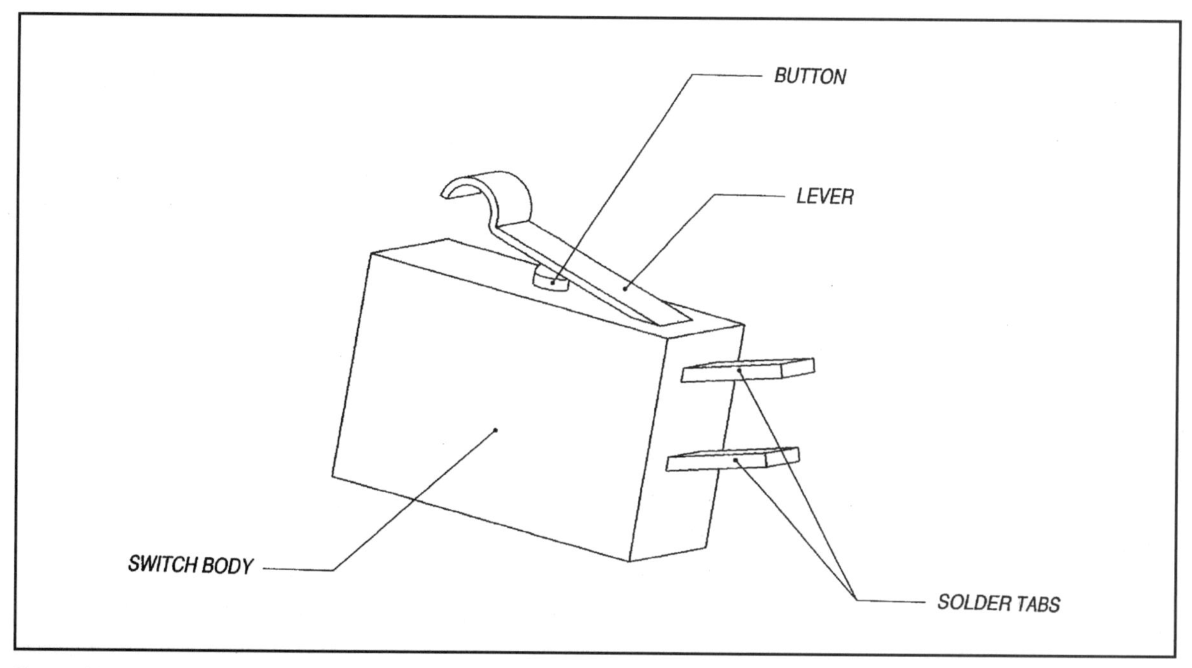

Figure 11-2 Lever Switch

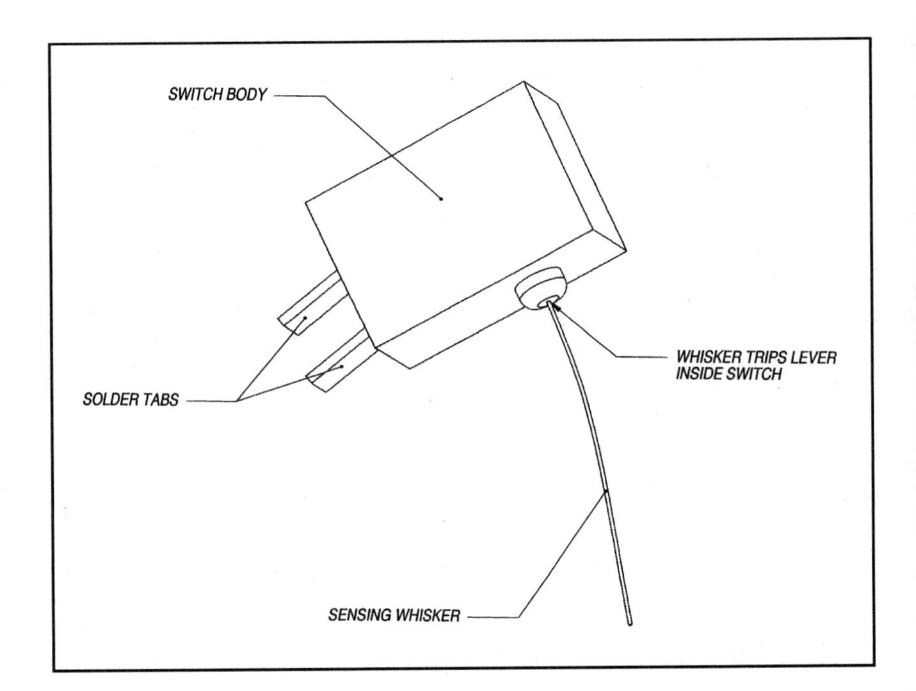

Figure 11-3 Whisker Switch

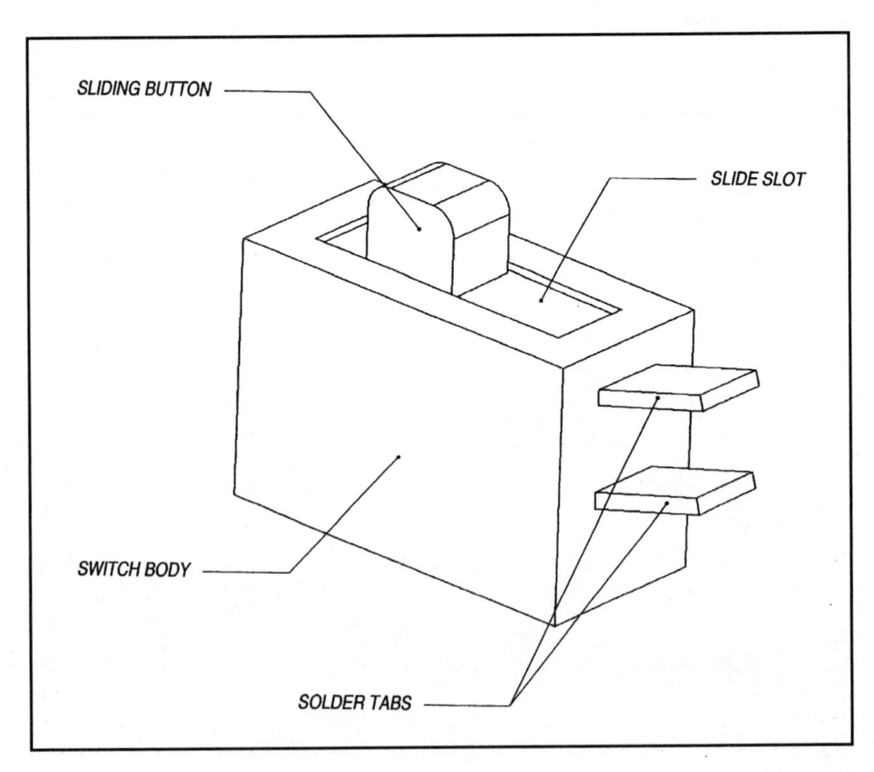

Figure 11-4 Slide Switch

more robust and easier to incorporate, but it is also much less precise since the sensing arm is necessarily flexible.

The whisker has the special property of detecting an object from any direction, making it distinctly different from lever switches. Since it bends out of the way of the sensed object, neither the object nor the switch is damaged by impact. This trick can also be done with a roller-ended lever arm, but more care is needed when using a rigid arm than with the flexible whisker. Figure 11-3 shows a basic whisker switch.

The last basic type of limit switch is the slide switch. This switch has a different internal mechanism than the button switch and its variations, and is considered less reliable. It is also difficult to implement in a robot and is rarely seen. Figure 11-4 shows a slide switch.

Magnetic limit switches come in several varieties and have the advantage of being sealed from contamination by dirt or water. The most common design has a sensitive magnet attached to a hinged contact so that when a piece of ferrous metal (iron) is nearby on the correct side of the switch, the magnet is drawn towards a mating contact, closing the electric circuit. All of the mechanical limit switches discussed in the following sections can incorporate a magnetic limit switch with some simple modification of the layouts. Just be sure that the thing being sensed is ferrous metal and passes close enough to the switch to trip it. Besides being environmentally sealed, these switches can also be designed to have no direct contact, reducing wear.

There are several ways to increase the area that is sensed by a mechanical limit switch. Figures 11-5 and 11-6 show basic layouts that can be expanded on to add a large surface that moves, which the switch then senses. There is also a form of mechanical switch whose area is inherently large. This type is called a membrane switch. These switches usually are in the shape of a long rectangle, since the internal components lend themselves to a strip shape. Membrane switches come with many different contact surfaces, pressure ratings (how hard the surface has to be pushed before the switch is tripped), and some are even flexible. For some situations, they are very effective.

The huge variety of limit switches and the many ways they can be used to sense different things are shown on the following pages in Figures 11-5 and 11-6. Hopefully these pictures will spur the imagination to come up with even more clever ways mechanical limit switches can be used in mobile robots.

INDUSTRIAL LIMIT SWITCHES

Figure 11-5a Mechanical, Geared, and Cam Limit Switches

Figure 11-5b Mechanical, Geared, and Cam Limit Switches

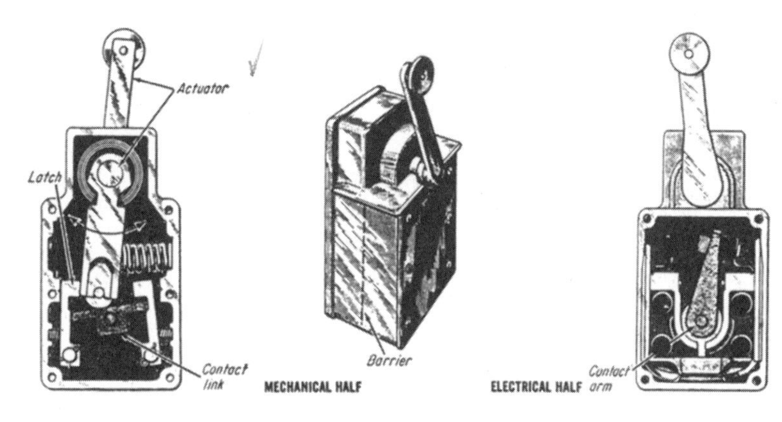

Latching Switch with Contact Chamber

Geared Rotary Limit Switches

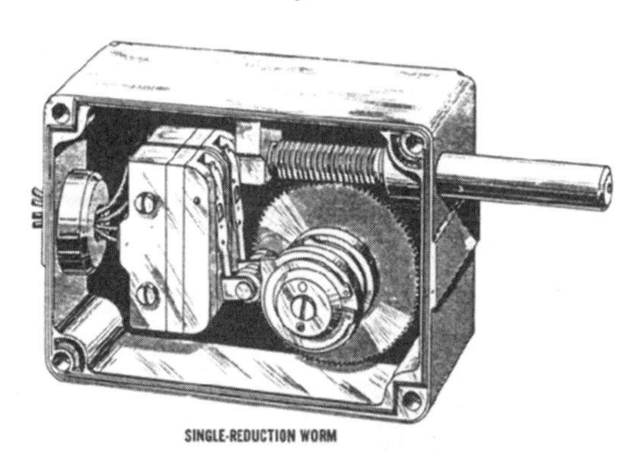

Rotary-Cam Limit Switches

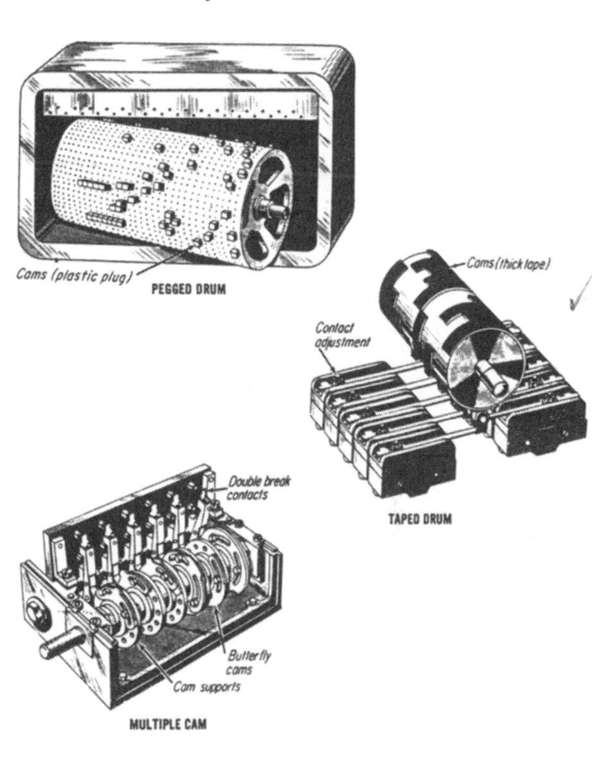

Cam adjustment pinion

Spring

Clamp plate

Cams

Cam followers

Clutch disk for taking up backlash

Worm gear

Planet gears

Housing

Input shaft

Sun gear

Cam sleeve

Worm gear

WORM AND PLANETARY

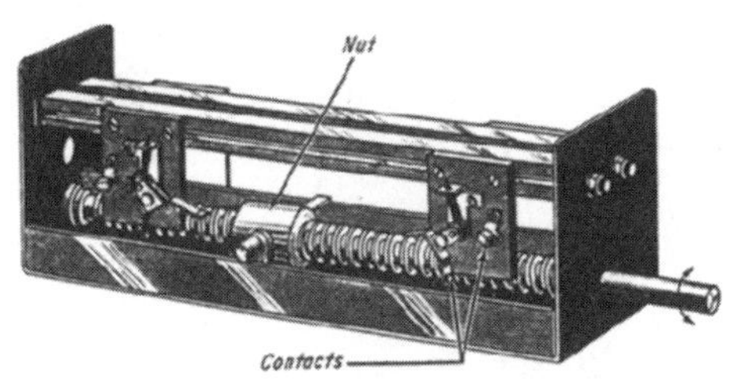

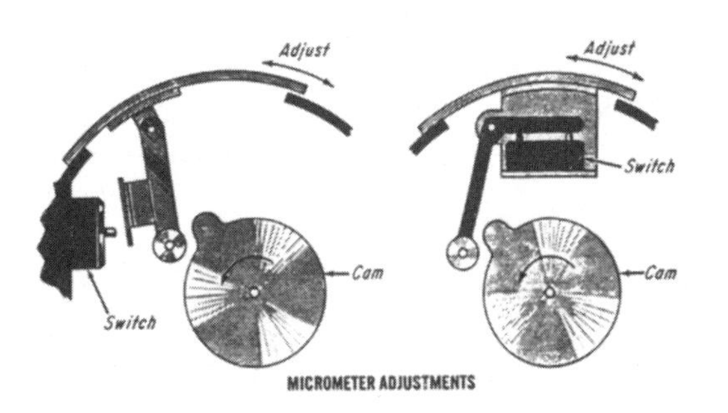

Figure 11-6 Limit Switches in Machinery

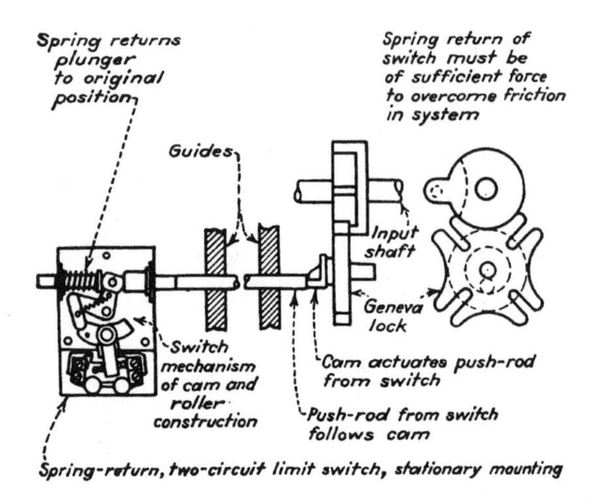

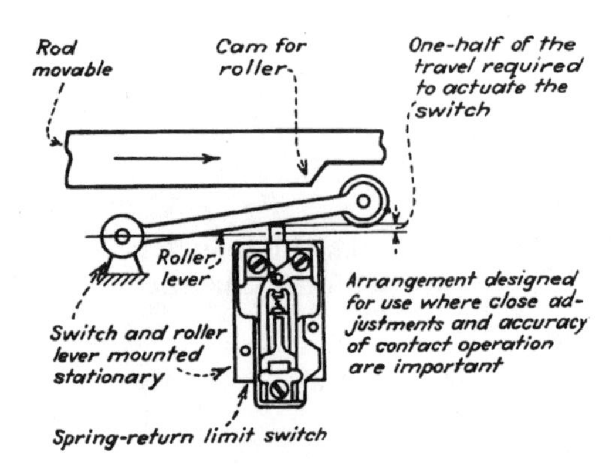

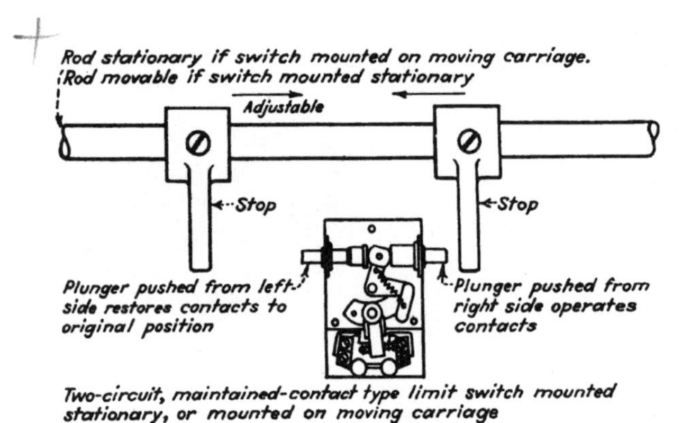

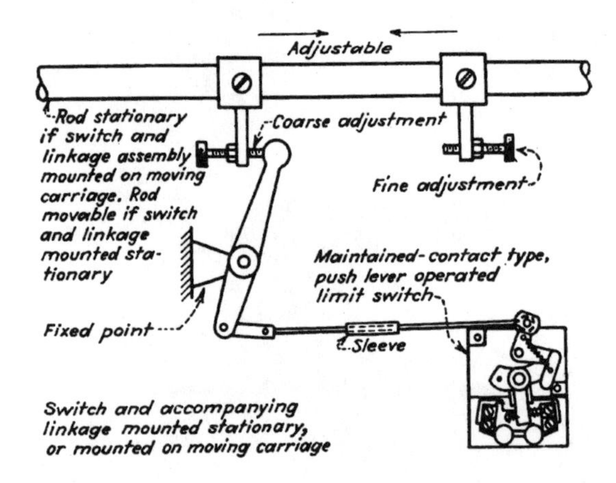

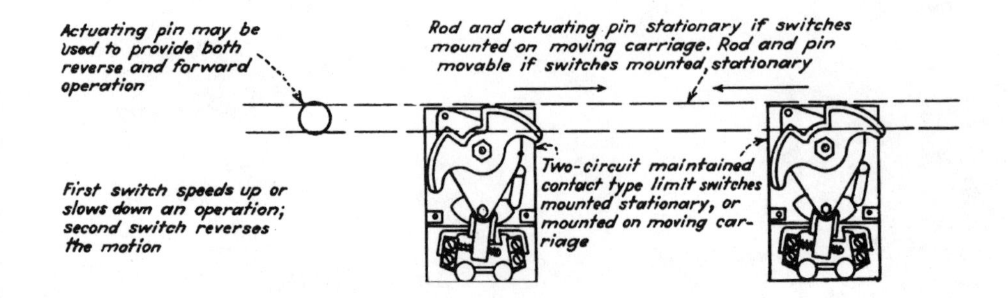

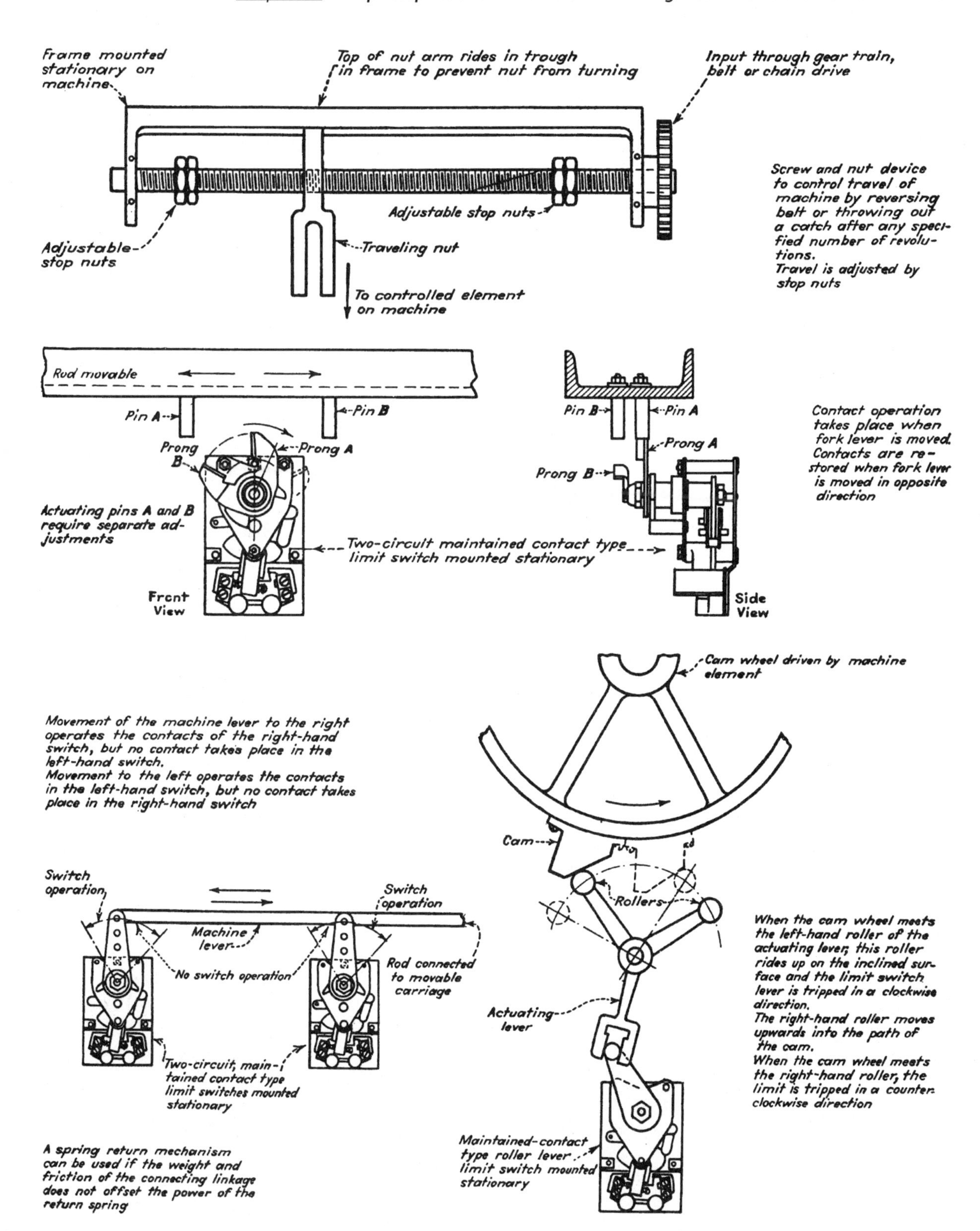
Frame mounted stationary on machine
Top of nut arm rides in trough in frame to prevent nut from turning
Input through gear train, belt or chain drive
Screw and nut device to control travel of machine by reversing belt or throwing out a catch after any specified number of revolutions.
Travel is adjusted by stop nuts
Adjustable stop nuts
Adjustable stop nuts
Traveling nut
To controlled element on machine
Rod movable
Pin A
Pin B
Prong B
Prong A
Actuating pins A and B require separate adjustments
Two-circuit maintained contact type limit switch mounted stationary
Front View
Pin B
Pin A
Prong A
Prong B
Side View
Contact operation takes place when fork lever is moved. Contacts are restored when fork lever is moved in opposite direction
Cam wheel driven by machine element
Movement of the machine lever to the right operates the contacts of the right-hand switch, but no contact takes place in the left-hand switch.
Movement to the left operates the contacts in the left-hand switch, but no contact takes place in the right-hand switch
Cam
Rollers
Switch operation
Switch operation
Machine lever
No switch operation
Rod connected to movable carriage
Two-circuit, maintained contact type limit switches mounted stationary
When the cam wheel meets the left-hand roller of the actuating lever, this roller rides up on the inclined surface and the limit switch lever is tripped in a clockwise direction.
The right-hand roller moves upwards into the path of the cam.
When the cam wheel meets the right-hand roller, the limit is tripped in a counterclockwise direction
Actuating lever
Maintained-contact type roller lever limit switch mounted stationary
A spring return mechanism can be used if the weight and friction of the connecting linkage does not offset the power of the return spring

Electrical contact arrangements

All contracts in normal position with limit switch unactuated

SINGLE POLE

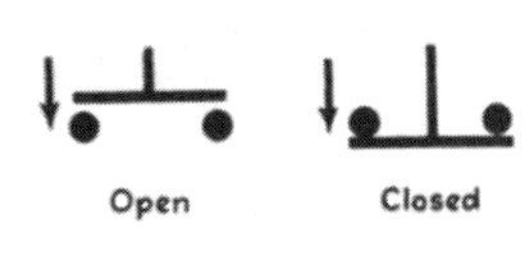

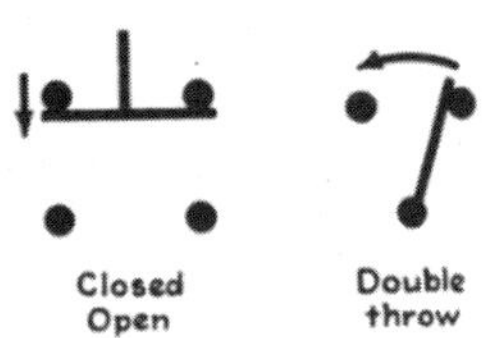

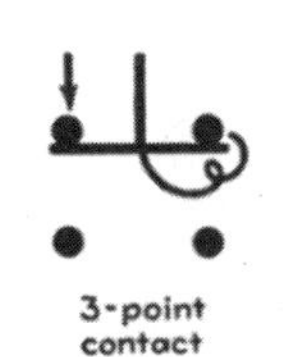

TWO POLE

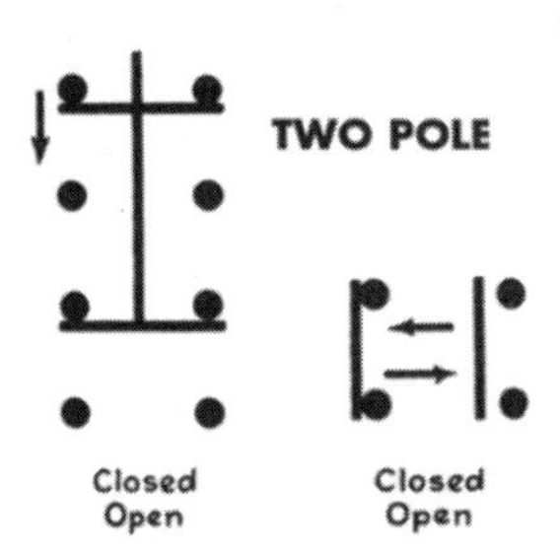

MULTI-CONTACT

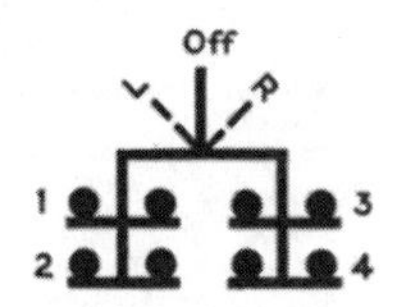

POS.	1	2	3	4
R	C	C	O	O
Off	C	C	C	C
L	O	O	C	C

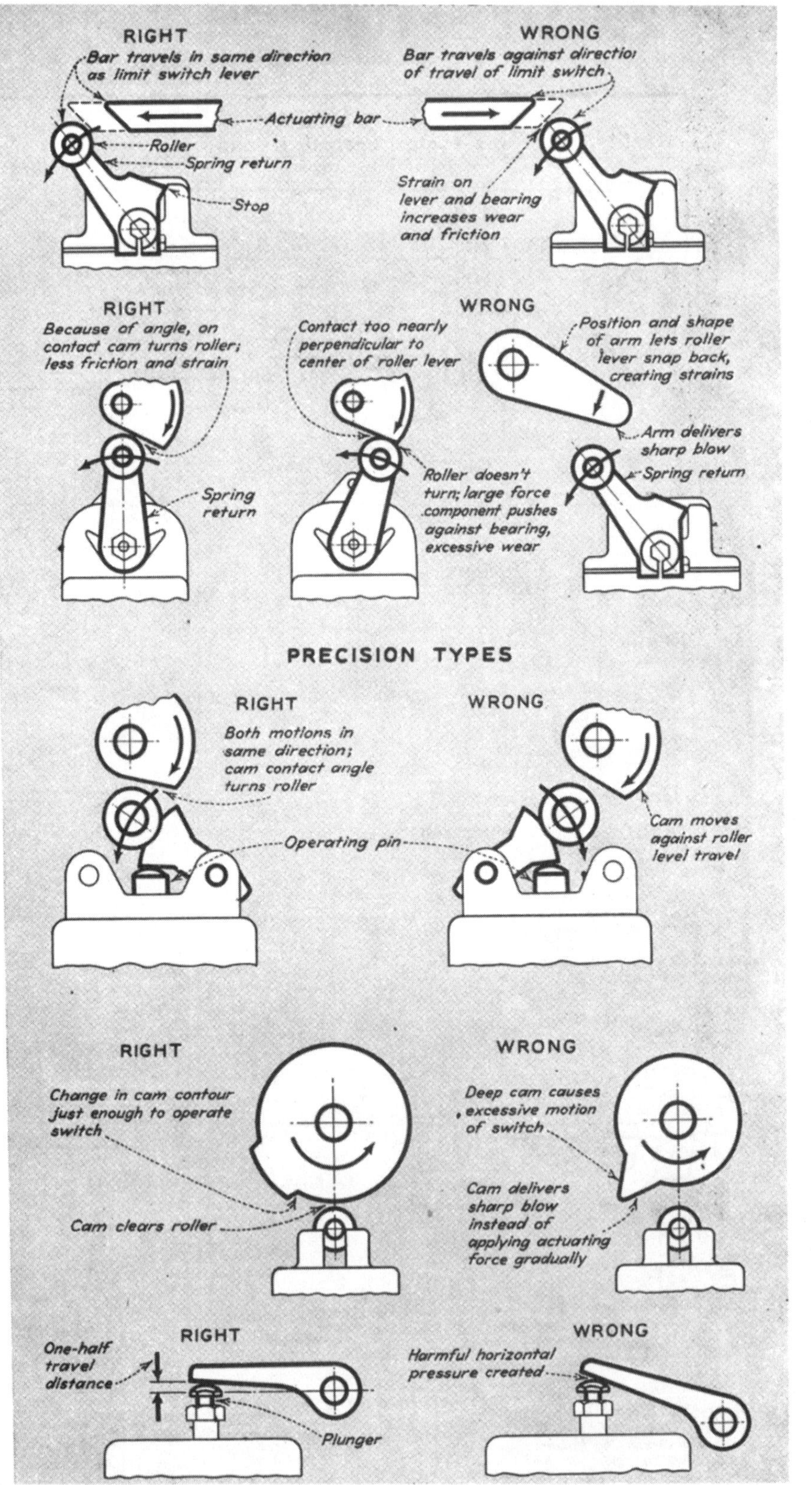

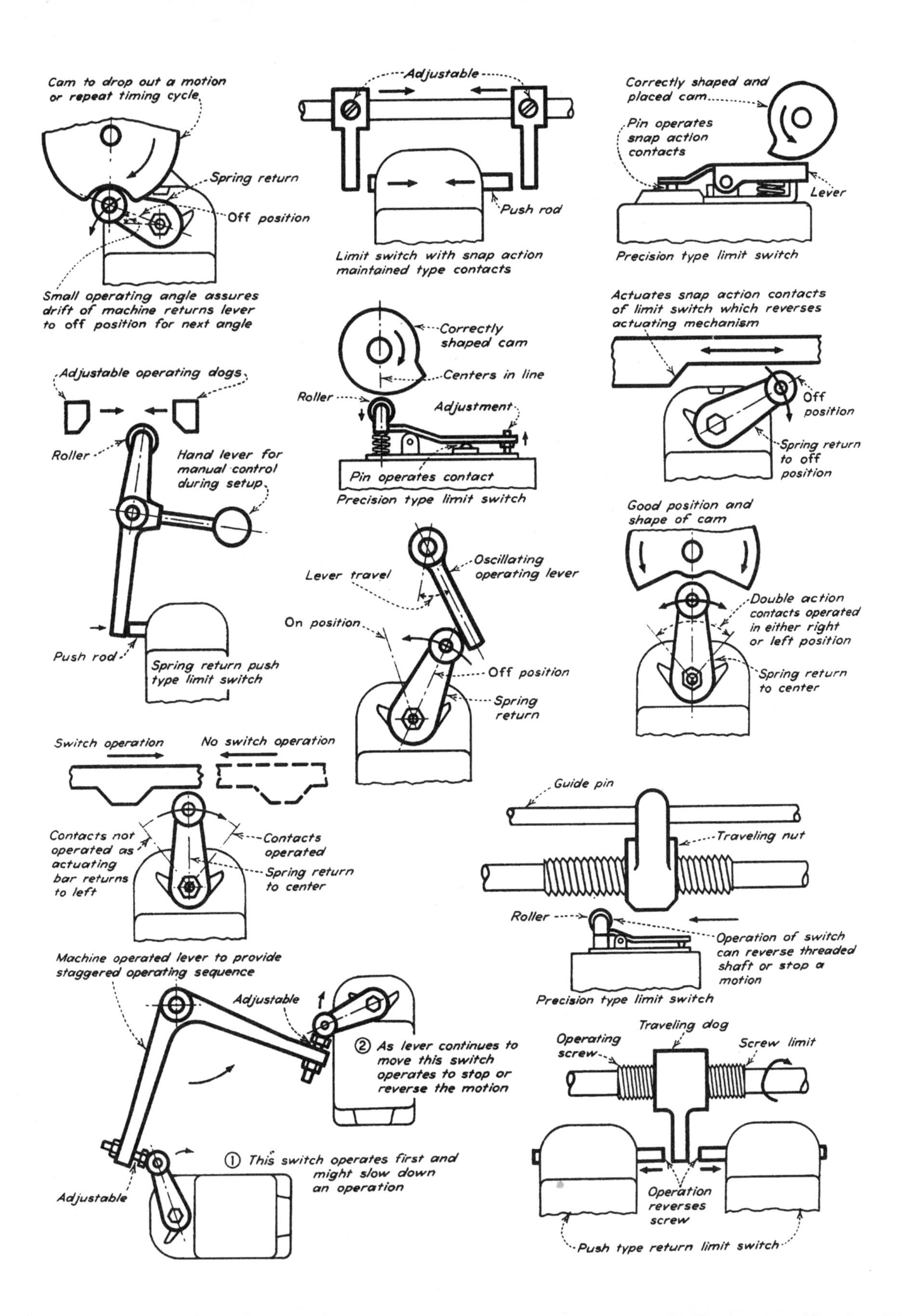

Cam to drop out a motion or repeat timing cycle
Spring return
Off position
Small operating angle assures drift of machine returns lever to off position for next angle
Adjustable
Push rod
Limit switch with snap action maintained type contacts
Correctly shaped and placed cam
Pin operates snap action contacts
Lever
Precision type limit switch
Actuates snap action contacts of limit switch which reverses actuating mechanism
Off position
Spring return to off position
Correctly shaped cam
Centers in line
Roller
Adjustment
Pin operates contact
Precision type limit switch
Adjustable operating dogs
Roller
Hand lever for manual control during setup
Push rod
Spring return push type limit switch
Lever travel
Oscillating operating lever
On position
Off position
Spring return
Good position and shape of cam
Double action contacts operated in either right or left position
Spring return to center
Switch operation
No switch operation
Contacts not operated as actuating bar returns to left
Contacts operated
Spring return to center
Guide pin
Traveling nut
Roller
Operation of switch can reverse threaded shaft or stop a motion
Precision type limit switch
Machine operated lever to provide staggered operating sequence
Adjustable
② As lever continues to move this switch operates to stop or reverse the motion
① This switch operates first and might slow down an operation
Adjustable
Traveling dog
Operating screw
Screw limit
Operation reverses screw
Push type return limit switch

LAYOUTS

With the possible exception of the whisker switch, the limit switch types discussed above almost always require some method of extending their reach and/or protecting them and the object being sensed from damaging each other. There are many ways to do this. The next several figures show various basic layouts that have their own benefits and problems.

In every sensor/actuator system, there is a time lag between when the switch is tripped and when the actuator reacts. This time lag must be taken into account, especially if the switch or object could be damaged. Object, in this case, can mean something in the environment, or something attached to the robot that is designed to detect things in the environment. If the time lag between contact and reaction cannot be made short enough, the layout must provide some other means of preventing disaster. This is done by using one of three methods.

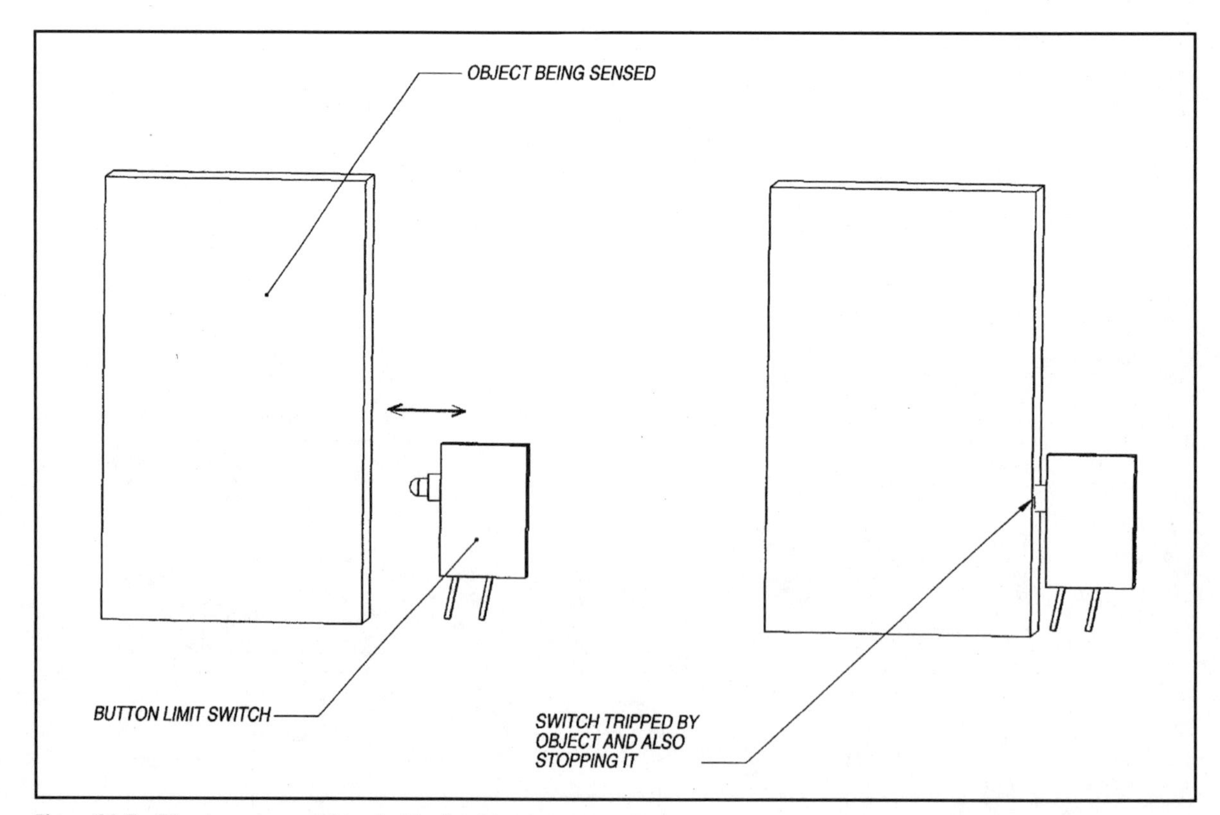

Figure11-7 Direct sensing combined with direct hard stop

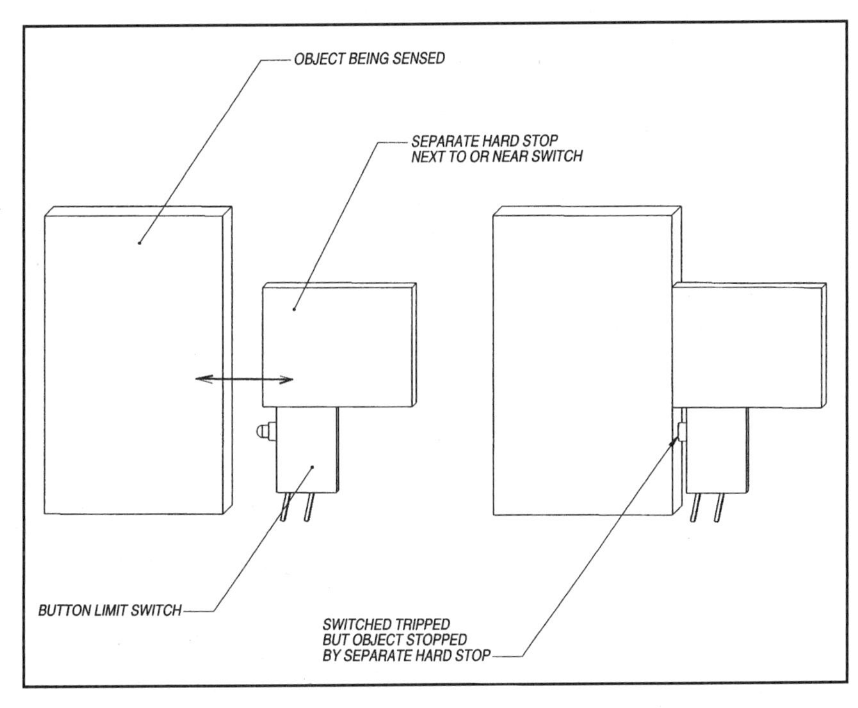

Figure 11-8 Direct sensing with separate hard stop

- A hard stop that is strong enough to withstand the stopping force (and yet not damage the object) can be placed just after the trip point of the switch.
- The layout can allow the object to pass by the switch, tripping it but not being physically stopped by anything. The robot's stopping mechanism is then the main means of preventing harm.
- The travel of the sensor's lever or button, after the sensor has been tripped, can be made long enough to allow sufficient time for the robot to stop.

Let's take a look at each layout.

Combination Trip (Sense) and Hard Stop

This is probably the simplest layout to implement. The switch directly stops the sensed object (Figure 11-7), which means the switch must be strong enough to withstand repeated impacts from the thing being sensed. Alternatively, there is a separate hard stop that is in line with the switch that absorbs the force of the impact after it has been tripped (Figure 11-8). Using a switch with a long throw eases implementation, and nearly any mechanical limit switch can be made to work with this layout, though the button and lever designs are usually best.

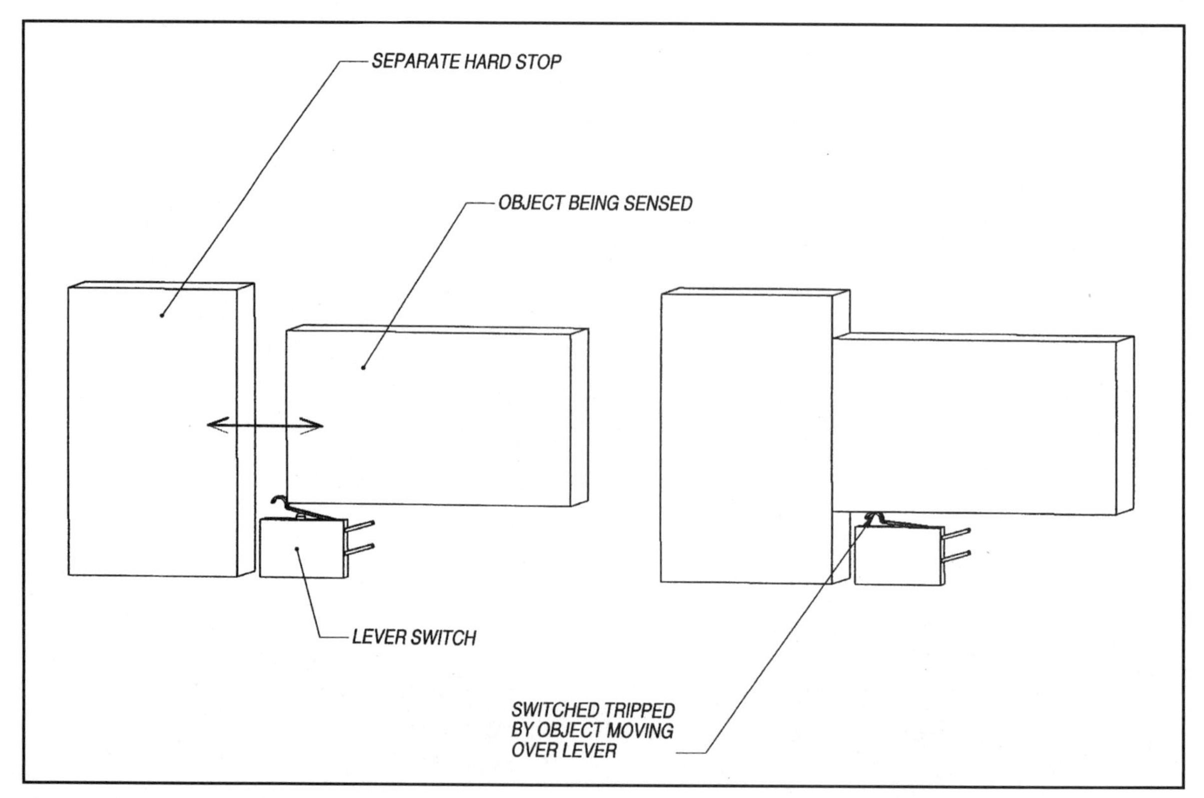

Figure 11-9 By-pass linear

By-Pass Layouts

The by-pass layout shown in Figures 11-9 and 11-10 relieves the switch of taking any force, but, more importantly, is less sensitive to slight variations in the positions of the switch and the sensed object, especially if a switch with a long throw is used. Removing the hazard of impact and reducing sensitivity make this layout both more robust and less precise. With careful design, however, this layout is usually a better choice than the previous layout because it requires less precision in the relationship between the hard stop and the switch's lever or button. Remember that the object being sensed can be anything that is close to the robot, including the ground.

This layout and its derivatives are the basis of virtually all mechanical timers. They are still found in dishwashers, washing machines, and any device where turning the knob results in an audible clicking sound as the arm or button on the switch jumps off the lobe of the cam. They can be stacked, as they are in appliances, to control many functions with a sin-

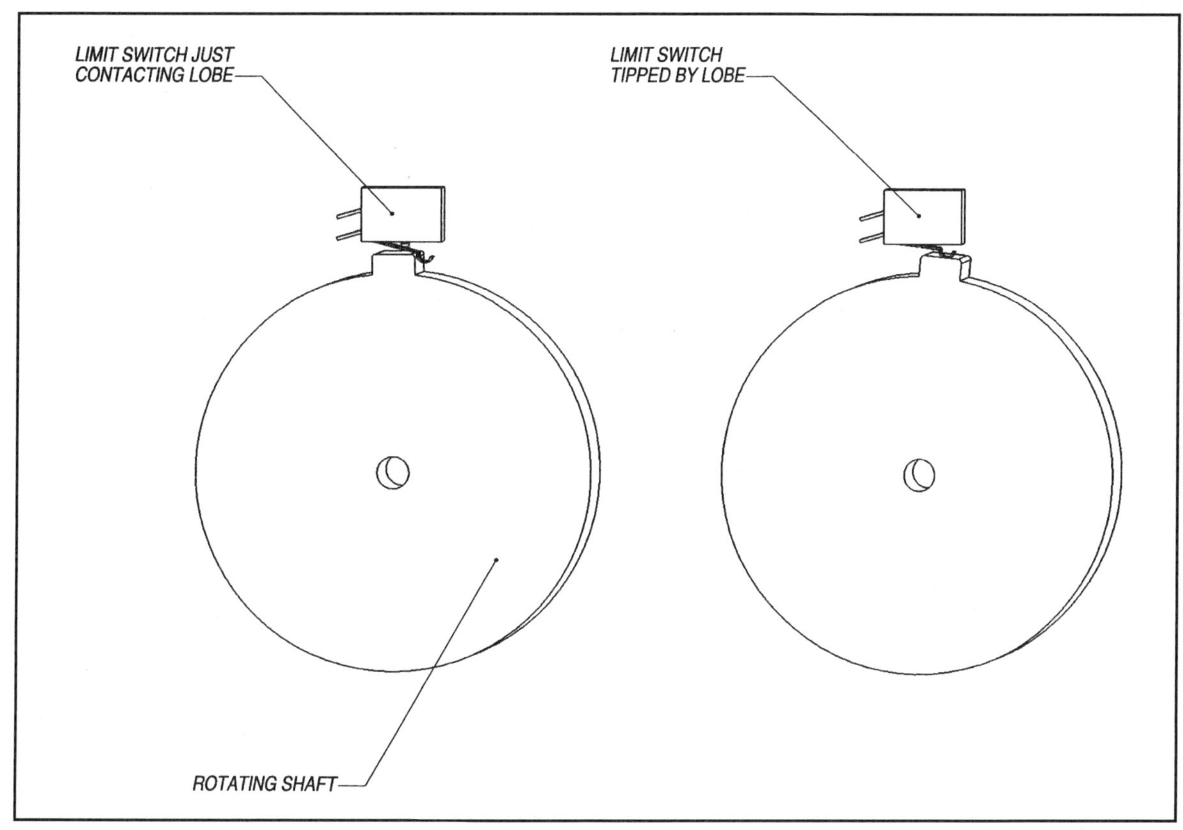

Figure11-10 Rotating cam

gle revolution of the timer. They can also be used as a very course encoder to keep track of the revolutions or position of the shaft of a motor or the angle of a joint in a manipulator.

Reversed Bump

The reversed bump layout shown in Figure 11-11 is a sensitive and robust layout. The switch is held closed by the same springs that hold the bumper or sense lever in the correct position relative to the robot. When an object touches the bumper, it moves the sense arm away from the switch, releasing and tripping it. A high quality switch is tripped very early in the travel of the sensing arm, and as far as the switch is concerned, there is no theoretical limitation on how far the bumper travels after the switch has been tripped. For this reason, it is an effective layout for sensing bumps.

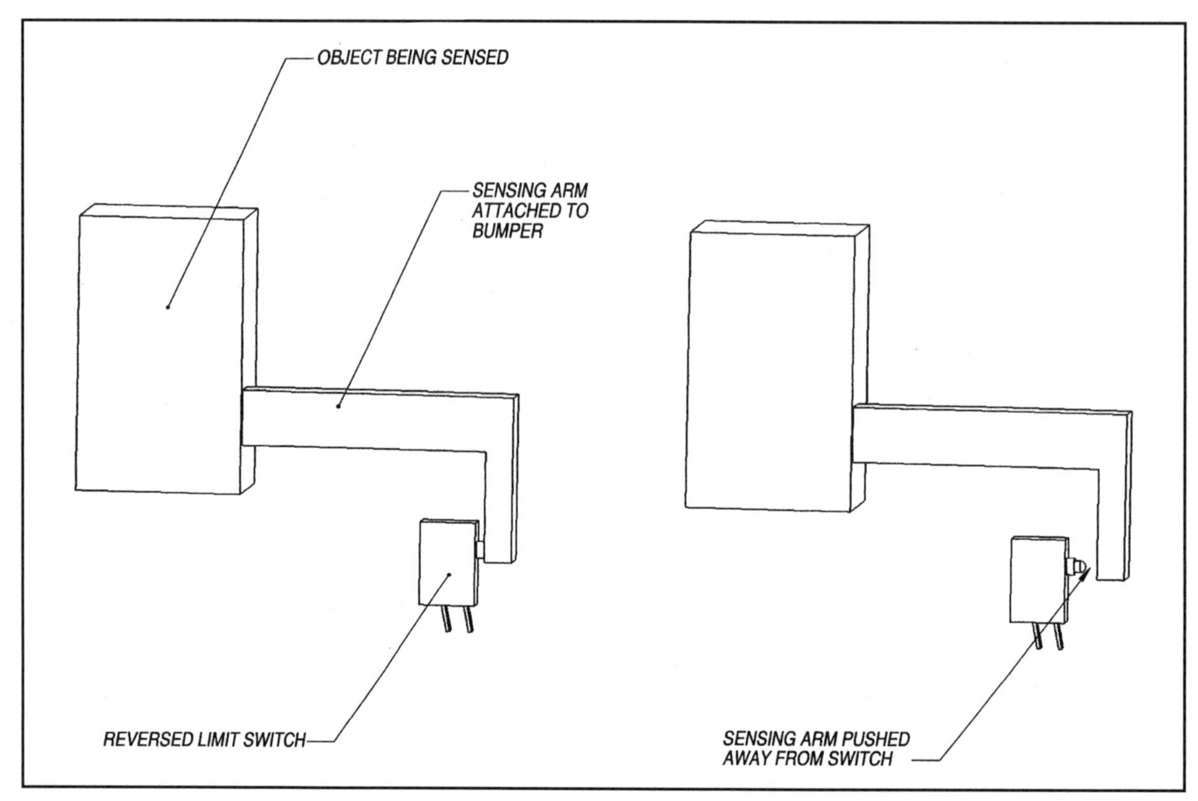

Figure 11-11 Reversed bump

BUMPER GEOMETRIES AND SUSPENSIONS

The robot designer will find that no matter how many long and short range noncontact sensors are placed on the robot, at some point, those sensors will fail and the robot will bump into something. The robot must have a sensor to detect collisions. This sensor may be considered redundant, but it is very important. It is a last line of defense against crashing into things.

The sensor must be designed to trip quickly upon contacting something so that the robot's braking mechanism can have the maximum time to react to prevent or reduce damage. To be perfectly safe, this sensor must be able to detect contact with an object at any point on the outer surface of the robot that might bump into something. This can be done with a bumper around the front and sides of the robot, if the robot only goes forward. Robots that travel in both directions must have sensors around the entire outer surface. It is important that the bumper be large enough so that it contacts the object before any other part of the robot does, otherwise the robot may not know it has hit something. Some robot designs attempt to get around this by using a measure of the current going to the drive wheels to judge if an object has been hit, but this method is not as reliable.

A bumper, though seemingly simple, is a difficult sensor to implement effectively on almost any robot. It is another case in which the shape of the robot is important as it directly affects the sensor's design and location. The bumper is so tricky to make effective as to be nearly impossible on some larger robots. Unfortunately, the larger the robot, the more important it is to be able to detect contact with things in the environment, since the large robot is more likely to cause damage to itself or the things it collides with. In spite of this, most large teleoperated robots have no collision detection system at all and rely on the driver to keep from hitting things. Even large autonomous robots (robots around the size of R2D2) are often built with no, or, at most, very small bumpers.

Simplifying any part of the robot's shape, or its behaviors, that can simplify the design of the bumper is well worth the effort. Making the shape simple, like a rectangle or, better yet, a circle, makes the bumper simpler. Having the robot designed so that it never has to back up means the bumper only has to protect the front and possibly the sides of the robot. Having the robot travel slowly, or slowing down when other sensors indicate many obstacles nearby means the bumper doesn't have to respond as fast or absorb as much energy when an object is hit. All these things can be vital to the successful design of an effective bumper.

There are several basic bumper designs that can be used as starting points in the design process. The goal of detecting contact on all outer surfaces of the robot can be achieved with either a single large bumper, or several smaller ones, each of which with its own sensor. These smaller pieces have the added benefit that the robot's brain can get some idea of where the body is hit, which can then be useful in determining the best direction to take to get away from the object. This can be done with a single piece bumper, but with less sensitivity.

A clever design that absolutely guarantees the bumper will completely cover the entire outer surface of the robot is to float the entire shell of the robot and make it the bumper, mounted using one of the techniques described later. Place limit switches under it to detect motion in any direction of this all-in-one bumper/shell. This concept works well for small robots whose shells are light enough not to cause damage to themselves but may be difficult to implement on larger robots.

Not only is it helpful to know the location of the bump, it is even better to be able to detect bumps from any direction, including from above and below. This is due to the possibility that the robot might try to drive under something that is not quite high enough, or try to drive up onto something and get the bottom edge of the bumper stuck, before it trips the sensor. Both of these cases are potential showstoppers if the robot has no idea it has hit something. This is where a bumper compliant and sensitive to bumps coming from any direction is very helpful. If there is a chance the robot will be operating in an environment where this problem will arise, this additional degree of freedom, with sensing, makes the bumper's suspension system more complex but vital. Let's start by looking at the simplest case, the one-dimension sensing bumper.

SIMPLE BUMPER SUSPENSION DEVICES

The one-dimension (1D) bumper only detects bumps that hit the bumper relatively straight on, from one direction. Although this may seem too limiting, it can be made to work well if there are several smaller bumpers, each with their own 1D sensor. Together they can sense a large area of bumps from many directions. There are also layouts that are basically 1D in design, but, by being compliant, can be made to sense bumps from arbitrary directions.

Since straight-on or nearly straight-on bumps are the most common and produce the largest forces, it is better to use a design that allows the bumper to have the longest travel in that direction. Bumps can be detected around the sides of the robot without as much motion from the bumper. This is why a compliant 1D bumper suspension can be used for 2D detection. There are many ways to attach bumpers that are basically 1D bumpers, but that can also function as 2D bumpers.

Some of these methods, or variations of them, can be used as is, with no additional devices required. Usually, though, a secondary device must be incorporated into the layout to positively locate the nominal position of the bumper. This facilitates repeatable sensing by the limit switch. The spring-centered plate layout is shown in Figure 11-14. The moving plate is so loosely positioned it requires a vibration damper or it will wobble constantly.

The V-groove centering block shown in Figure 11-12, is a basic method of realigning the bumper after encountering a bump, but there are several others that work nearly as well. The V-groove layout is essentially two reversed bump limit switch layouts at 90° to each other. It is therefore effective for bumpers designed to detect bumps from straight or nearly straight on.

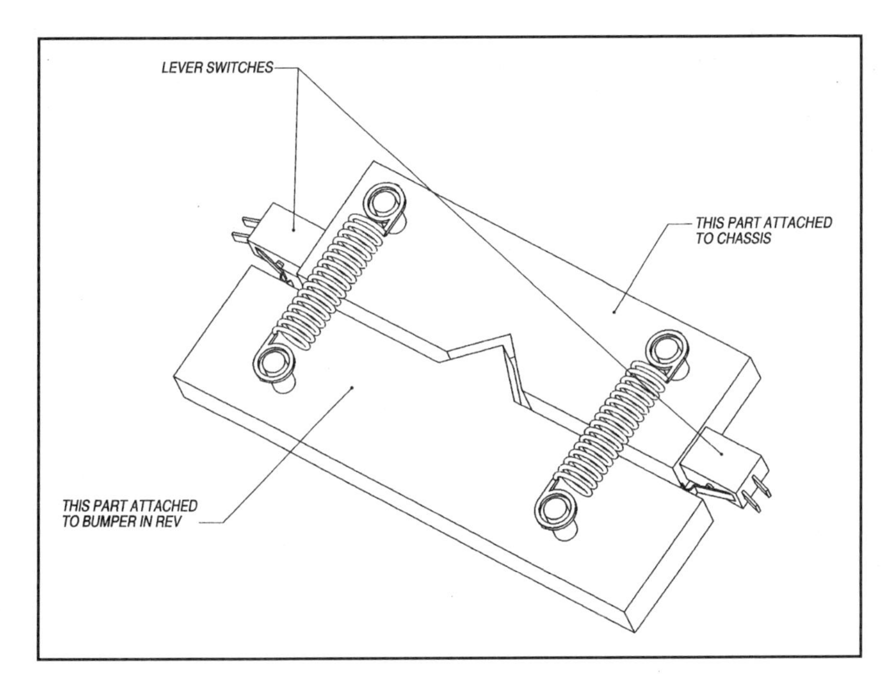

Figure 11-12 Tension spring V-groove layout

Three Link Planar

A very useful and multipurpose mechanical linkage is something called a four-bar link. It consists of four links attached in the shape of a quadrilateral. By varying the lengths of the links, many motions can be generated between the links. A 3D version of this can be built by attaching two planes (plates) together with three links so that the plates are held parallel, but can move relative to each other.

This could be called a five-bar link, since there are now essentially five bars, but the term five-bar link refers to a different mechanism entirely. A better name might be 3D four-bar, or perhaps three link planar. Figure 11-13 shows the basic idea. If the base is attached to the chassis, and the top plate is attached to the bumper or bumper/shell, a robust layout results. This system is under-constrained, though, and requires some

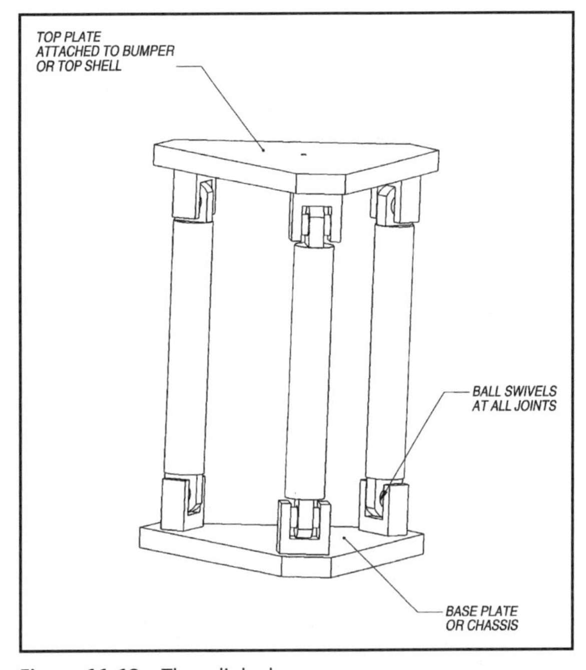

Figure 11-13 Three link planar

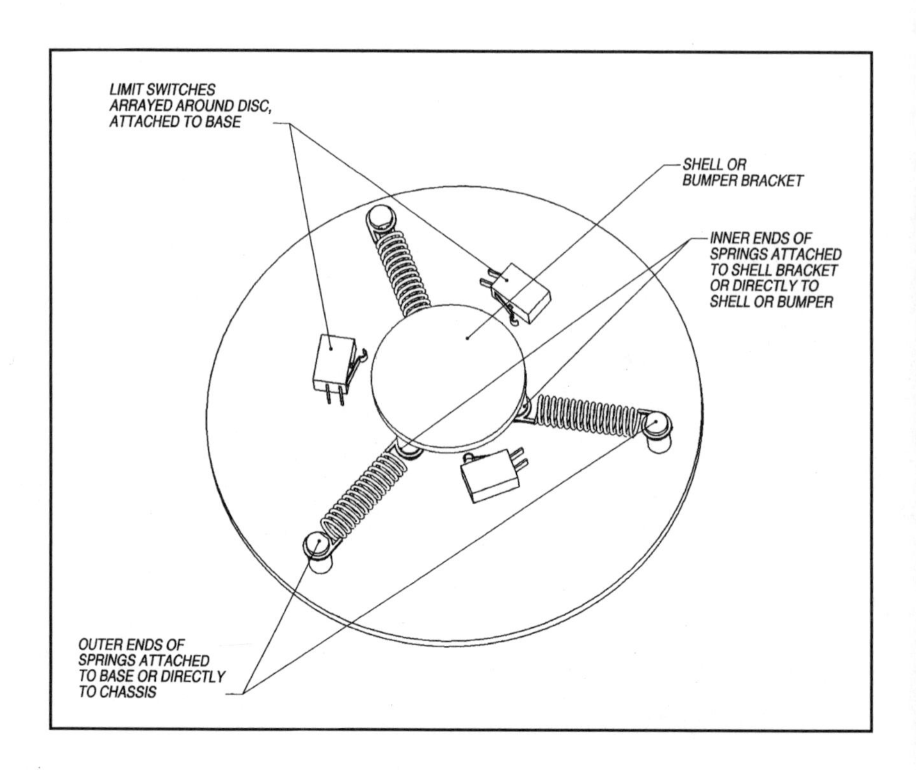

Figure 11-14 Tension spring star layout

other components to keep it centered, like the V-groove device discussed previously, and some sort of spring to hold the top plate in the groove.

Tension Spring Star

A simple to understand spring-centering layout uses three tension springs in a star layout (Figure 11-14). The outer ends of the springs are attached to the chassis and the inner three ends all attach to a plate or other point on the frame that supports the bumper. This layout is easy to adjust and very robust. It can be used for robot bumpers that must detect bumps from all directions, provided there is an array of sensors around the inner edge of the bumper, setup as a switch-as-hard-stop layout. This layout requires a damper between the chassis and plate to reduce wobbling.

Torsion Swing Arm

The torsion or trailing arm car suspension system (Figure 11-15) first appeared in the early 1930s and was used for more than 25 years on the

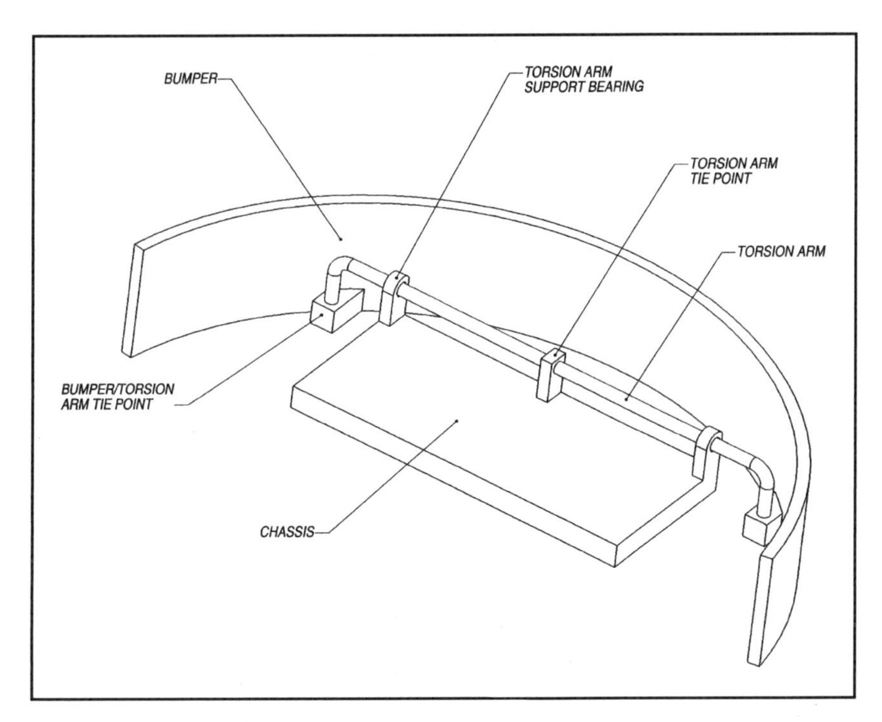

Figure 11-15 Torsion swing arm

VW Beetle. It is similar in complexity to the sideways leaf spring shown in the next section, but is somewhat more difficult to understand because it uses the less common property of twisting a rod to produce a spring. The mechanism consists of a simple bar with trailing links at each end. The center of the beam is attached to the chassis, and each end of the trailing links supports the bumper. If the beam is properly sized and sufficiently flexible, it can act as both support and spring with proper passive suspension points.

Horizontal Loose Footed Leaf Spring

Another suspension system, used since the days of horse drawn buggies, that can be applied to robot bumper suspensions is a leaf spring turned on it side. This design has great simplicity and reliability. In a car, the leaf spring performs the task of springs, but it also holds the axles in place, with very few moving parts. The usual layout on a car has one end attached to the frame through a simple pivot joint and the other end attached through either a pivoting link, or a robust slot to allow for that end to move back and forth in addition to rotating. The center of the spring is attached to the axle, allowing it to move up and down but not in any other direction. Two springs are required to hold the axle horizontal.

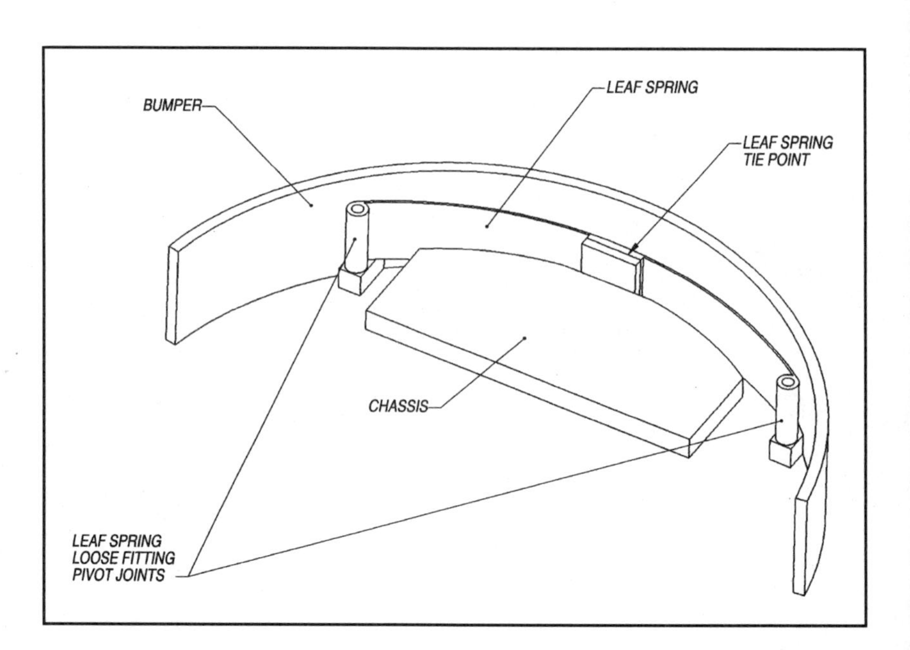

Figure 11-16 Horizontal oose footed leaf spring

The leaf spring can also be used to suspend a robot's bumper quite effectively by turning the spring on its side and attaching the center to the chassis and each end to points on the bumper. One end, or both, still needs to be attached through a slot or pivoting link, but the result is still very simple and robust. This layout can be used on larger robots also, since the leaf spring is an efficient suspension element even in larger sizes.

For robots that must detect bumps from the rear, it may be possible to use a single spring to support an entire wrap around bumper. If this would produce a cumbersum or overly large spring, the sideways-leaf spring layout can be enhanced by adding a second spring to further support the rear of a one-piece wrap around bumper. Figure 11-16 shows a single slot sideways leaf spring layout.

Sliding Front Pivot

Designing a bumper suspension system based on the fact that the bumper needs primarily to absorb and detect bumps from the front produces a system which moves easily and farthest in the fore-and-aft directions, but pivots around some point in the front to allow the sides to move some. The system could be called a sliding front-pivot bumper suspension system (Figure 11-17). Sliding joints are more difficult to engineer than pivoting or rotating joints, but this concept does allow large motions

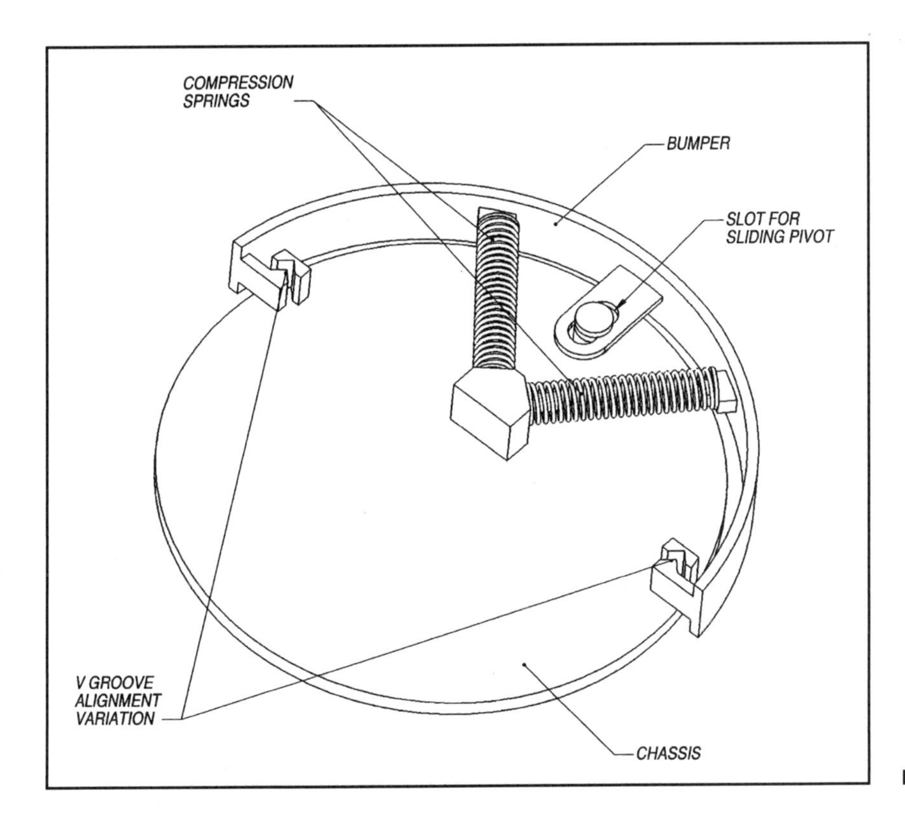

Figure 11-17 Sliding front pivot

in the most important direction. Springing it back to its relaxed position can be tricky.

Suspension Devices to Detect Motions in All Three Planes

The V-groove device can be applied to 3D layouts as well, simply by making the V-block angled on top and bottom, like a sideways pyramid. A mechanical limit switch can be placed so that any motion of the V-block out of its default position trips the switch. For even more sensitivity, the V-block can be made of rollers or have small wheels on its mating surfaces to reduce friction.

The simplest suspension system that allows motion in three directions relies on flexible rubber arms or compliant mounts to hold the bumper loosely in place. These flexible members can be replaced with springs and linkages, but the geometries required for 3D motion using mechanical linkages can be complex. Figure 11-18 shows a layout for an elastomer or spring-based system. A well-sprung bumper or bumper/shell

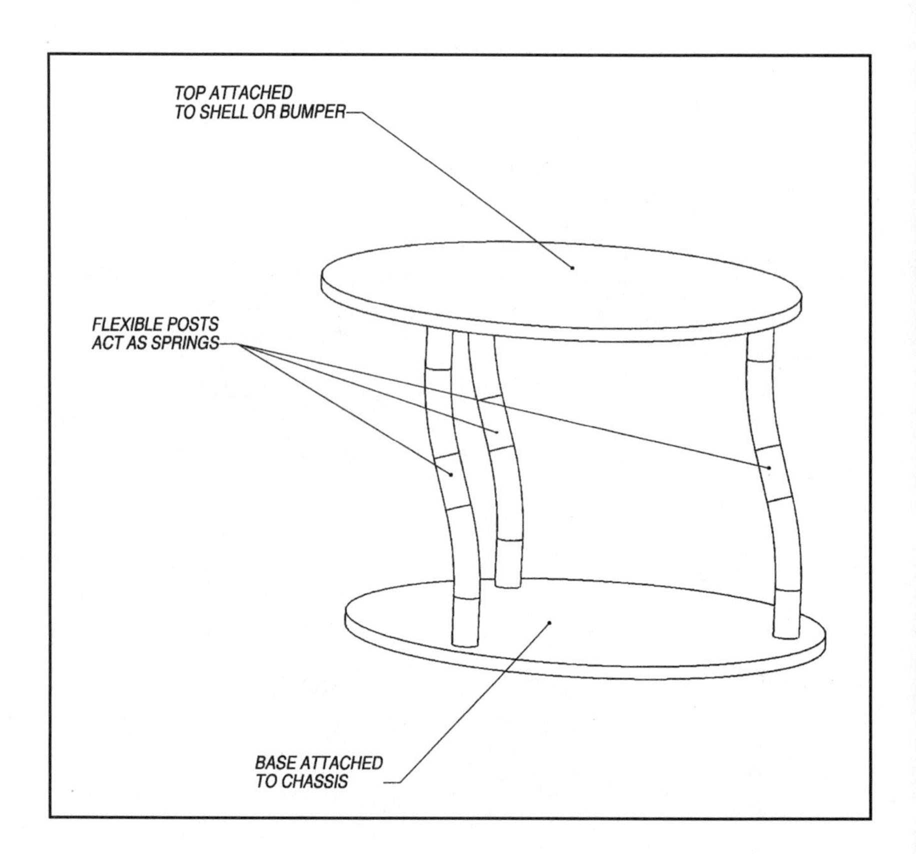

Figure 11-18 Vertical flexible post bumper suspension

that uses one of these layouts can be used with no hard centering system by using the "limit switch as hard stop" concept discussed previously. The bumper is sprung so that its relaxed position is just off the contacts of three or more switches. This system is simple and effective for smaller robots and a very similar layout is used on the popular Rug Warrior robot.

For larger designs, where the flexible post would need to be too big, compression springs can be used instead. A clever designer may even be able to size a single-compression spring layout that would be simple indeed. The system can be designed to use the springs in their relaxed state as springy posts, or, for larger forces, the springs can be slightly compressed, held by internal cables, to increase their centering force and make their default position more repeatable.

CONCLUSION

The information you've just read in this book is intended for those interested in the mehanical aspects of mobile robots. There are, of course, many details and varieties of the mobile layouts, manipulators, and sensors that are not covered—there are simply too many. It is my sincere hope that the information that is presented will provide a starting point from which to design your unique mobile robot.

Mobile robots are fascinating, intriguing, and challenging. They are also complicated. Starting as simply as possible, with a few actuators, sensors, and moving parts will go a long way towards the successful completion of your very own mobile robot that does real work.

Index

Note: Figures and tables are indicated by an italic *f* and *t*, respectively.

About the Author

Paul E. Sandin is a robotocist with iRobot Corporation, where he designs and builds systems for the Consumer Robotics Division. Previously, he worked for RedZone Robotics, where he designed suspension components for large-scale toxic waste cleanup robots. He has an intimate knowledge of robots, both large and small. He lives with his family in a suburb of Boston, Massachusetts.